Morphology, Shape and Phylogeny

The Systematics Association Special Volume Series

Series Editor
Alan Warren
*Department of Zoology, The Natural History Museum,
Cromwell Road, London SW7 5BD, UK.*

The Systematics Association promotes all aspects of systematic biology by organizing conferences and workshops on key themes in systematics, publishing books and awarding modest grants in support of systematics research. Membership of the Association is open to internationally based professionals and amateurs with an interest in any branch of biology including palaeobiology. Members are entitled to attend conferences at discounted rates, to apply for grants and to receive the newsletters and mailed information; they also receive a generous discount on the purchase of all volumes produced by the Association.

The first of the Systematics Association's publications *The New Systematics* (1940) was a classic work edited by its then-president Sir Julian Huxley, that set out the problems facing general biologists in deciding which kinds of data would most effectively progress systematics. Since then, more than 70 volumes have been published, often in rapidly expanding areas of science where a modern synthesis is required.

The *modus operandi* of the Association is to encourage leading researchers to organize symposia that result in a multi-authored volume. In 1997 the Association organized the first of its international Biennial Conferences. This and subsequent Biennial Conferences, which are designed to provide for systematists of all kinds, included themed symposia that resulted in further publications. The Association also publishes volumes that are not specifically linked to meetings and encourages new publications in a broad range of systematics topics.

Anyone wishing to learn more about the Systematics Association and its publications should refer to our website at www.systass.org.

Other Systematics Association publications are listed after the index for this volume.

Systematics Association Special Volume Series 64

Morphology, Shape and Phylogeny

Edited by
Norman MacLeod
Peter L. Forey
Department of Palaeontology
The Natural History Museum
London
UK

CRC PRESS

Boca Raton   London   New York   Washington, D.C.

Library of Congress Cataloging-in-Publication Data

Morphology, shape and phylogeny / edited by Norman MacLeod, Peter L. Forey.
    p. cm. — (Systematics Association special volume series; 64)
This book arises from a symposium held at the University of Glasgow in August 1999.
    Includes bibliographical references (p.).
    ISBN 0-415-24074-3
    1.Morphology—Congresses. 2. Phylogeny—Congresses. I. MacLeod, Norman.
  II. Forey, Peter L. III. Systematics Association special volume; no. 64.
QH351 .M617 2002
571.3—dc21   2001052293

This book contains information obtained from authentic and highly regarded sources. Reprinted material is quoted with permission, and sources are indicated. A wide variety of references are listed. Reasonable efforts have been made to publish reliable data and information, but the authors and the publisher cannot assume responsibility for the validity of all materials or for the consequences of their use.

Neither this book nor any part may be reproduced or transmitted in any form or by any means, electronic or mechanical, including photocopying, microfilming, and recording, or by any information storage or retrieval system, without prior permission in writing from the publisher.

The consent of CRC Press does not extend to copying for general distribution, for promotion, for creating new works, or for resale. Specific permission must be obtained in writing from CRC Press for such copying.

Direct all inquiries to CRC Press, 2000 N.W. Corporate Blvd., Boca Raton, Florida 33431.

Trademark Notice: Product or corporate names may be trademarks or registered trademarks, and are used only for identification and explanation, without intent to infringe.

Visit the CRC Press Web site at www.crcpress.com

© 2002 Systematics Association with the exception of chapters 1, 7, and 14,
© the Department of Palaeontology, The Natural History Museum, London.

No claim to original U.S. Government works
International Standard Book Number 1-415-24074-3
Library of Congress Card Number 2001052293
Printed in the United States of America   2 3 4 5 6 7 8 9 0
Printed on acid-free paper

Contents

	List of contributors	vii
	Preface	ix
1	Introduction: morphology, shape, and phylogenetics NORMAN MACLEOD AND PETER L. FOREY	1
2	Homology, characters and continuous variables CHRISTOPHER JOHN HUMPHRIES	8
3	Quantitative characters, phylogenies, and morphometrics JOSEPH FELSENSTEIN	27
4	Scaling, polymorphism and cladistic analysis TODD C. RAE	45
5	Overlapping variables in botanical systematics GERALDINE REID AND KAREN SIDWELL	53
6	Comparability, morphometrics and phylogenetic systematics DONALD L. SWIDERSKI, MIRIAM L. ZELDITCH AND WILLIAM L. FINK	67
7	Phylogenetic signals in morphometric data NORMAN MACLEOD	100
8	Creases as morphometric characters FRED L. BOOKSTEIN	139
9	Geometric morphometrics and phylogeny F. JAMES ROHLF	175
10	A parametric bootstrap approach to the detection of phylogenetic signals in landmark data THEODORE M. COLE III, SUBHASH LELE AND JOAN T. RICHTSMEIER	194
11	Phylogenetic tests for differences in shape and the importance of divergence times: Eldredge's enigma explored P. DAVID POLLY	220

12	Ancestral states and evolutionary rates of continuous characters ANDREA J. WEBSTER AND ANDY PURVIS	247
13	Modelling the evolution of continuously varying characters on phylogenetic trees: the case of Hominid cranial capacity MARK PAGEL	269
14	Summary PETER L. FOREY	287
	Index	295
	Systematics Association Publications	305

Contributors

Fred L. Bookstein, Center for Human Growth and Development, University of Michigan, Ann Arbor, Michigan 48109, USA.

Theodore M. Cole III, Department of Basic Medical Science, School of Medicine, Univeristy of Missouri, Kansas City, MO 64108, USA.

Joseph Felsenstein, Department of Genetics, University of Washington, Box 357360, Seattle, WA 98195-7360, USA.

William L. Fink, Department of Biology and Museum of Zoology, University of Michigan, Ann Arbor, Michigan 48109, USA.

Peter L. Forey, Department of Palaeontology, The Natural History Museum, London SW7 5BD, UK.

Christopher John Humphries, Department of Botany, The Natural History Museum, London SW7 5BD, UK.

Subhash Lele, Department of Mathematical Sciences, University of Alberta, Edmonton, Alberta T6G 2G1, Canada.

Norman MacLeod, Department of Palaeontology, The Natural History Museum, London SW7 5BD, UK.

Mark Pagel, School of Animal and Microbial Sciences, University of Reading, Whiteknights, Reading, RG6 6AH, UK.

P. David Polly, Molecular and Cellular Biology Section, Division of Biomedical Sciences, Queen Mary and Westfield College, London E1 4NS, UK and The Department of Palaeontology, The Natural History Museum, London SW7 5BD, UK.

Andy Purvis, Department of Biology, Imperial College, Silwood Park, Ascot, Berkshire SL5 7PY, UK.

Todd C. Rae, Evolutionary Anthropology Research Group, Department of Anthropology, University of Durham, 43 Old Elvet, Durham, DH1 3HN, UK, and Department of Mammalogy, American Museum of Natural History, Central Park West at 79th St., New York, NY 10024, USA.

Geraldine Reid, Department of Botany, The Natural History Museum, London SW7 5BD, UK.

Joan T. Richtsmeier, Department of Cell Biology and Anatomy, School of Medicine, The Johns Hopkins University, Baltimore, MD 21205, USA and Department of Anthropology, The Pennsylvania State University, University Park, PA 16802, USA.

F. James Rohlf, Department of Ecology and Evolution, State University of New York, Stony Brook, NY 11794-5245, USA.

Karen Sidwell, Department of Botany, The Natural History Museum, London SW7 5BD, UK.

Donald L. Swiderski, the Museum of Zoology, University of Michigan, Ann Arbor, Michigan 48109, USA.

Andrea J. Webster, School of Animal and Microbial Sciences, The University of Reading, Whiteknights, P.O. Box 228, Reading RG6 6AJ, UK.

Miriam L. Zelditch, Museum of Paleontology, University of Michigan, Ann Arbor, Michigan 48109, USA.

Preface

This book arises from a symposium 'Morphology, shape and phylogenetics' which formed part of the Second Biennial International Conference of the Systematics Association held at the University of Glasgow in August 1999. The aim of the Biennial conferences is to encourage discussion between many systematists who, although they specialise in specific taxonomic groups, confront common methodological and theoretical problems. The aim is to encourage younger scientists both to present their own work but also to become involved in broader systematic issues. As a catalyst these symposia bring together scientists with contrasting views.

This particular symposium, dedicated to the relationship between morphometrics and systematics, was stimulated by recent publications which have suggested ways in which morphometric data may be used in phylogenetic systematics as well as a synthesis of morphometric methods. The time seemed right for a compilation of ideas.

We would like to thank the Systematics Association, Dr Gordon Curry, University of Glasgow and his team of organisers who made the symposium possible.

Norman MacLeod and Peter L. Forey

The Natural History Museum, London, 2001

Speakers participating in the symposium 'Morphology, shape and phylogenetics' at the Second Biennial International Conference of the Systematics Association held the University of Glasgow in August 1999.

Chapter 1

Introduction: morphology, shape, and phylogenetics

Norman MacLeod and Peter L. Forey

This book is about the ways in which the study of morphology can be used in phylogenetic analysis and how the results of phylogenetic analysis can provide meaning for the study of morphological variation. Through the text phylogeny is understood to be the organizing principle for all biological data. Even though variations in organismal form can be studied from non-phylogenetic points-of-view, it is generally acknowledged that such variations are not different in principle, from any other type of biological data and cannot be fully understood in the absence of the historical perspective provided by phylogeny. Similarly, phylogenetic analysis is impossible in the absence of some way of describing the morphological variation between individuals, populations, species, and higher taxa. Even molecular phylogenetic studies are dependent on morphological data in that the molecular samples are typically collected from specimens that have been identified as belonging to morphologically defined species on the basis of shared morphological attributes. Additionally, the molecules themselves have form manifested as secondary and tertiary molecular structure.

Morphological correspondences form the basis of phylogenetic reconstruction. However, modern methods of phylogenetic analysis treat morphological data in an inconsistent manner. For example, the coding conventions demanded by most parsimony-analysis algorithms require that morphological data be described as discrete characters and/or character states (e.g., spine: present, absent). While this descriptive convention works well for some discrete morphological attributes (e.g., tail red or tail blue), many of the morphological descriptors used routinely by phylogeneticists represent variables that, at least in principle, can adopt a range of values (e.g., height of tooth cusps, location of eyes, see Thiele 1993). Despite the fact that character and character state descriptions such as 'spine: short, long' imply precise metrical definitions of 'short' and 'long', such terms are often used in an ambiguous manner. Even more subtle are the descriptors of shape such as 'leaf shape: oval, round'. Just where in the context of any particular systematic comparison does 'round' stop and 'oval' begin? Most scientific papers using coding of morphological variation for phylogenetic analysis are vague or completely silent on this issue. Additionally, many meristic observations (e.g., counts of vertebrae or numbers of leaves per whorl) are variable within taxonomic groups. All of these data raise the question of coding observations in discrete ways for what are inherently variable observations. Some authors (e.g., Pimentel and Riggins 1987; Felsenstein 1988) have rejected outright the use of continuous variation in phylogenetic analysis. Other authors have suggested that state delimitation must necessarily be arbitrary and therefore such data are inappropriate for phylogenetic analysis. However, a number of delimitation methods have been devised (Mickevich and Johnson 1976; Colless 1980; Almedia and Bisby 1984; Thorpe 1984; Archie 1985; Chappill 1989; Thiele 1993), and although each may have advantages and disadvantages, a simple rejection of such data, at the very least, deserves more discussion.

In another direction the techniques of geometric morphometrics are able to describe shapes accurately by using the language of mathematical geometry and can potentially have a significant input to the theory, as well as to the practice or morphological characterization. If we are able to describe shape more accurately there may be more potential sources of morphological variation available for analysis. However, there is a general issue of the suitability of using such variables which takes us to the heart of evolutionary and phylogenetic theories – the concept of homology.

Phylogenetic homology is the most important unifying principle in biology (Bock 1973) and is probably the subject most written about within the biological literature. In phylogenetic systematics homology is equated with synapomorphy (Patterson 1982). This means that propositions of homology are theories to be tested rather than self-evident truths to be acknowledged or assumed. In order to propose a theory of homology an initial postulate of identity is followed by testing (the tests applied are conjunction – meaning that no two presumed homolgues can occur simultaneously in the same organism – and congruence with other postulated homologies). The initial postulate of morphological identity is sometimes called primary homology (de Pinna 1991) and consists of two activities: identifying that structures are similar in composition and topological relationships and coding the similarity for phylogenetic analysis (see Hawkins 2000 for discussion). As an example consider the wing of a bird and a bat. These structures share a similar composition (bone) and topological relationships with, for instance, the shoulder girdle. The presence of a wing would be congruent with other characters such as warm-bloodedness and the presence of amniotic membranes. However, the wing of a bird and a wing of a bat would be non-homologous in a phylogenetic sense because, despite being compositionally and topographically the same, the wing of a bat is not congruent with the many other character-state distributions (e.g., fur, mammary glands, three-ear ossicles) suggesting that bats are more closely related to animals without wings (e.g., tree shrews and primates). The important aspect of this concept of homology relevant to the subject of this book is that there is an initial proposition of structural identity.

The concept of homology in geometry (and by extension, in morphometrics) is somewhat different since, there is no initial estimation of primary homology and no tests of conjunction and congruence are applied. Standard morphometrics would recognize potential homology between the dorsal fin of a salmon, an ichthyosaur, and a killer whale as homology of shape (triangular) and position (centrally located along the dorsal surface) while ignoring the fact that these structures had arisen quite independently in the phylogenetic history of these lineages as evidenced by the state distributions of other characters. Therefore, the extension of structure-level concept of homology in phylogenetic analysis to the geometrical point-to-point correspondences typical of many morphometric data sets raises several difficult – and therefore interesting – problems. In many morphometric shape studies it is descriptions of curvature, angularity, ratios, etc. that are being assessed and compared such that it is difficult to see how a particular angle formed by or ratio of parts (e.g., 1.543) can be regarded as being homologous. As pointed out by Zelditch *et al.* (2000: 80) "morphometric variables are not, in and of themselves, equivalent to characters". There would seem to be a preliminary hypothesis of analysis necessary so that shapes of homologous structures are being compared. In some morphometric analysis form is described with reference to landmarks without any justification that the landmarks are homologous in the sense used above. However, if those landmarks are chosen respecting the concept of homology used in phylogenetic systematics there may be much more that we can learn from the morphometric study of form.

Many of the problems systematists have encountered in the efforts to use morphological data in phylogenetic contexts derive from the nature and descriptive complexity of those data. Organisms exhibit a bewildering array of structures that are often very difficult to abstract meaningfully into the scalar values (e.g., lengths, widths, breadths, and depths) of traditional systematic measurement systems. In addition, geometric concepts such as size and shape (not to mention shape translations) have proven to be more complex – and to require more complex descriptive-analysis tools – than had been widely appreciated. But, recent advances in morphometrics have, at least partially, addressed this descriptive problem. To some extent these advances have been possible because of recent dialogues within the morphometrics community (e.g., Marcus *et al.* 1993, 1996).

Geometric morphometrics represents a quantitative synthesis of two themes that have dominated the study of form for well over a century (Bookstein 1993). The older of these can be traced from the Renaissance studies of form by Leonardo da Vinci, Albrecht Dürer, Michelangelo, and others through its introduction into the modern scientific literature by D'Arcy Thompson (1917). This theme visualizes morphological change as a smooth mapping transformation between the starting and ending forms of implicit form classes. Thompson was intrigued particularly by the manner in which geometrically simple deformation patterns could combine with existing organic geometries to produce seemingly complex results. In order to give graphical expression to the underlying simplicity of the deformation pattern Thompson employed a Cartesian grid system in which the intersections of the grid lines were taken to represent corresponding or 'landmark' points on both the starting and final forms. Although Thompson's 'transformation-grid' approach to shape characterization intrigued generations of morphologists and geometers, his invention proved stubbornly resistant to precise quantitative formulation. Biological acceptance of such transformationist notions were also not helped by Thompson's own goal of using transformation grids to demonstrate that interspecific variation obeyed law-like rules reflecting the predominance of physical forces in the creation of morphological novelty.

The second theme united by the morphometric synthesis grew out of Francis Galton's biometric 'regression analysis' by way of the distinction between the truly linear aspect of patterning between a pair of morphometric variables (quantified in terms of their covariance or correlation) and the non-linear aspect of their patterning (quantified in terms of the residual scatter about a linear regression line). Galton's original insight was expanded into what has now come to be known as the generalized linear model which includes bivariate/multiple regression analysis, component/factor analysis, discriminant/canonical variates analysis, canonical correlation analysis and path analysis). From this beginning, attention came to focus on the abstraction of synthetic linear components from covariance or correlation matrices that can be thought of as vectors existing within a multidimensional space defined by the original variables. In the more extreme forms of this research program these synthetic vectors – and not the original variables – came to be regarded by some as being closer to 'true' observations. In terms of morphological analyses though, the problem with this approach was that these methods in their original formulation failed to preserve the inherently geometric nature of the data and failed to support techniques, whereby the analytic results could be expressed in terms of the original geometries.

The geometric morphometric synthesis combines these two established themes in quantitative morphological analysis by, (1) focusing on the representation of landmark configurations (= geometries) as variables, (2) registering geometric data collected from actual specimens to remove size and orientation differences. This operation effectively projects these data onto the surface of a k-3 dimensional hypersphere (where k = number of landmarks) with the inter-specimen distance representing the great circle distances between all pairs of taxa, and (3) formalizing Thompson's transformationist approach through the use of an algorithmic-graphical device known as the 'thin-plate spline'. Discussions detailing various methods included within the geometric synthesis and examples of applications can be found in various publications (e.g., Rohlf and Bookstein 1990; Bookstein 1991; Reyment 1991; Marcus *et al.* 1993; Marcus *et al.* 1996) as well as herein. At present, these methods represent a very large and somewhat abstract body of largely theoretical work whose practical application to the understanding of morphological variation, the creation of morphological novelty, and covariances between form and a variety of non-geometric covariates has yet to be explored in detail. What is clear, however, is that these tools can detect, represent, and describe morphologies in ways that are analytically superior to all previous methods. Their existence, at the very least, provides systematists with an opportunity to revisit a number of long-standing issues regarding the employment and interpretation of morphological data in systematic contexts and the relation of these to phylogenetic studies.

During the past 15 years the morphometrics community has been as slow to embrace, explore, and exploit the phylogenetic aspects of their data (e.g., through various 'comparative method' strategies, see Harvey and Pagel 1991), as mainstream phylogeneticists have been slow to embrace, explore, and exploit the new geometric approaches to morphological analysis. Fortunately, however, there are signs of a rapprochement between phylogenetic systematics and morphometrics. Some systematists have begun to re-evaluate their traditional phylogenetic-systematic taboos regarding the use of continuously-distributed variables (e.g., Zelditch *et al.* 1995; Rae 1998) while others have begun to explore methods whereby phylogenetic information can be included in morphometric studies (MacLeod 2001). This book represents an attempt to further this dialogue by undertaking a comprehensive exploration of the relationship between continuously-distributed morphological (morphometric) variables and phylogenetic. In particular the essays contained herein focus on four fundamental questions.

1. Can continuously-distributed variables (of any type) be used in phylogenetic inference?
2. Can morphometric variables be used to constrain and/or test phylogenetic hypotheses?

3. What strategies are available for taking advantage of morphometric information within the context of a phylogenetic analysis?
4. What strategies are available for taking advantage of phylogenetic information within the context of a morphometric analysis?

The authors represent a cross-section of phylogenetic systematists, morphometricians, and comparative-method specialists with a collective expertise than encompasses a wide spread of biological subdisciplines. Above all, these are biologists who have thought deeply and creatively about the relation between morphology and phylogeny. While they would not be expected to agree entirely with one another's positions on a variety of controversial practical and methodological issues, they are united in their belief that the phylogenetic treatment of morphological data represents a frontier of systematic research whose time has come and that promises to yield important new insights into the questions of the origin, patterning, and maintenance of organic morphological diversity that have always stood at the heart of biological systematics.

Christopher J. Humphries leads off with a re-evaluation of the homology problem in the context of morphological and especially morphometric characters. The concept of a character in phylogenetic analysis is intimately tied to a concept of homology, which in turn is a theory based on constant and repeatable observation. How this relates to the data used in morphometric analysis is discussed to set the scene for later essays.

Joseph Felsenstein continues the discussion of quantitative characters and their use in phylogenetic systematics by considering the question of whether it is necessary – or even desirable – to transform such characters into discrete states. Felsenstein asks difficult questions about the current state of our knowledge of the genetic, developmental, and functional aspects of morphology and suggests possible ways for future collaborations between morphometrics and these disciplines.

Todd Rae points out that complete harmony between morphometric analysis and phylogenetic analysis may not be possible since the aims are different. However, he suggests that there may be areas of overlap in measurement data and that metric data can be used in phylogenetic analysis as characters as long as due caution is exercised.

Karen Sidwell and Geraldine Reid deal specifically with the different ways in which continuously variable characters have been coded as discrete integers prior to phylogenetic analysis. They point out that different ways of coding lead to different descriptions and different sensitivity to reflecting the variation in the original observations. But more importantly, they demonstrate that some of the methods used to assign codes to continuous variables may be forcing us to recognize differences, and hence different codes for phylogenetic analysis, than are truly there.

Donald Swiderski, Miriam Zelditch, William Fink open the discussion of how morphometric data can be used to recognize phylogenetically meaningful characters by reviewing a variety of morphometric methods and asking the question of whether the variables produced by these methods conform to the concepts of correspondence and homology. In the context of this review these authors offer an alternative strategy for characterizing inter-landmark boundary curves in a manner consistent with the needs of phylogenetic systematics.

Norman MacLeod focuses on the use of landmark-based morphometric summaries and their specific relationship to the concepts of spatial localization, biological homology, and the character coding problem. In a series of examples MacLeod demonstrates a variety of ways landmark-based morphometric strategies can successfully contribute the analysis of morphology in phylogenetic contexts, culminating in a reconsideration of Naylor's (1996) simulated fish morphology dataset.

Fred L. Bookstein closes this subsection with the first presentation of a new method for the analysis of spatially localized shape deformations that may be of use in discovering and describing new phylogenetic characters and character states. Since the discovery of new morphological characters has long been recognized as a principle advantage of the metrical description of morphology, this new class of morphometrically defined features – called creases – holds great potential for helping to fulfil the potential of morphometrics in systematic and phylogenetic analyses.

F. James Rohlf's contribution initiates a subsection of essays dealing with the mechanics of combining phylogenies with morphometric descriptions of shapes to model the morphological aspects of evolutionary processes (e.g., ancestral character state estimation). Rohlf's method employs the squared-change parsimony estimation criterion and results in a series of deformation-based shape change models that can be used to illustrate shape change as a continuously variable parameter along any phylogenetic tree.

David Polly continues this discussion by considering the question of how best to assess the divergence times among taxa that are crucial to ancestral character state estimation. Using an example dataset drawn from fossil carnivorans Polly employs a combination of phylogenetic and stratigraphical data to determine an expected per generation rate of shape divergence to which actual shape divergence estimates can be compared. Results of these types of comparisons will allow systematists to use phylogenetically-referenced morphological data to quantitatively test a variety of functional and developmental hypotheses.

Andrea Webster and Andrew Purvis continue this theme of using given phylogenies to estimate ancestral character states for continuous characters and from these deducing rates of evolution. Their chapter emphasizes that the results of different methods which have been used to infer ancestral states for continuous characters are inherently dependent upon the assumptions of the model of evolution and do not always correspond to states observable in fossils. These discussions have implications for the way in which we infer ancestral states in general.

References

Almeida, M. T. and Bisby, F. A. (1984) 'A simple method for establishing taxonomic characters from measurement data', *Taxon*, **33**, 405–409.

Archie, J. W. (1985) 'Methods for coding variable morphological features for numeric taxonomic analysis', *Systematic Zoology*, **34**, 326–345.

Bock, W. J. (1973) 'Philosophical foundations of classical evolutionary taxonomy', *Systematic Zoology*, **22**, 375–392.

Bookstein, F. L. (1991) *Morphometric tools for landmark data: geometry and biology*, Cambridge: Cambridge University Press.

Bookstein, F. L. (1993) 'A brief history of the morphometric synthesis', in Marcus, L. F., Bello, E. and García-Valdecasas, A. (eds) *Contributions to morphometrics*, Madrid: Museo Nacional de Ciencias Naturales 8, pp. 18–40.

Chapill, J. A. (1989) 'Quantitative characters in phylogenetic analysis', *Cladistics*, **5**, 217–234.

Colless, D. H. (1980) 'Congruence between morphological and allozyme data for *Menidia*: a reappraisal', *Systematic Zoology*, 29, 288–299.

Felsenstein, J. (1988) 'Phylogenies and quantitative characters', *Annual Review of Ecology and Systematics*, 19, 445–471.

Harvey, P. H. and Pagel, M. D. (1991) *The comparative method in evolutionary biology*, Oxford: Oxford University Press.

Hawkins, J. A. (2000) 'A survey of primary homology assessment: different botanists perceive and define characters in different ways', in Scotland, R. and Pennigton, R. T. (eds) *Homology and systematics*, London: Taylor and Francis, pp. 22–53.

Marcus, L. F., Bello, E. and García-Valdecasas, A. (1993) *Contributions to morphometrics*, Madrid: Museo Nacional de Ciencias Naturales 8.

MacLeod, N. (2001) 'The role of phylogeny in quantitative paleobiological analysis', *Paleobiology*, 27, 226–241.

Marcus, L. F., Corti, M., Loy, A., Naylor, G. J. P. and Slice, D. E. (eds) (1996) *Advances in morphometrics*, NATO ASI Series, New York: Plenum Press.

Michevich, M. F. and Johnson, M. F. (1976) 'Congruence between morphological and allozyme data', *Systematic Zoology*, 25, 260–270.

Naylor, G. J. P. (1996) 'Can partial warps be used as cladistic characters?', in Marcus, L. F., Corti, M., Loy, A., Naylor, G. J. P. and Slice, D. E. (eds) *Advances in Morphometrics*, New York: Plenum Press, pp. 519–530.

Patterson, C. (1982) 'Morphological characters and homology', in Joysey, K. A. and Friday, A. E. (eds) *Problems of phylogenetic reconstruction*, Systematics Association Special Volume, No. 21, London: Academic Press, pp. 21–74.

Pimentel, R. A. and Riggins, R. (1987) 'The nature of cladistic data', *Cladistics*, 3, 201–209.

de Pinna, M. C. C. (1991) 'Concepts and tests of homology in the cladistic paradigm', *Cladistics*, 7, 367–394.

Rae, T. C. (1998) 'The logical basis for the use of continuous characters in phylogenetic systematics', *Cladistics*, 14, 221–228.

Reyment, R. A. (1991) *Multidimensional paleobiology*, Oxford: Pergamon Press.

Rohlf, F. J. and Bookstein, F. L. (1990) *Proceedings of the Michigan Morphometrics Workshop*, Ann Arbor: The University of Michigan Museum of Zoology Special Publication 2.

Swiderski, D. L., Zelditch, M. L. and Fink, W. L. (1998) 'Why morphometrics is not special: coding quantitative data for phylogenetic analysis', *Systematic Biology*, 47, 508–519.

Thiele, K. (1993) 'The holy grail of the perfect character: the cladistic treatment of morphometric data', *Cladistics*, 9, 275–304.

Thompson, D. W. (1917) *On growth and form*, Cambridge: Cambridge University Press.

Thorpe, R. S. (1984) 'Coding morphometric characters for constructing distance Wagner networks', *Evolution*, 38, 244–355.

Zelditch, M. L., Fink, W. L. and Swiderski, D. L. (1995) 'Morphometrics, homology, and phylogenetics: quantified characters as synapomorphies', *Systematic Biology*, 44, 179–189.

Zelditch, M. L., Swiderski, D. L. and Fink, W. L. (2000) 'Discovery of phylogenetic characters in morphometric data', in Wiens, J. J. (ed.) *Phylogenetic analysis of morphological data*, Washington: Smithsonian Institution Press, pp. 37–83.

Chapter 2

Homology, characters and continuous variables

Christopher John Humphries

ABSTRACT

Owen (1849) coined 'homology' to describe relationships between organisms, using corresponding morphological parts (homologues) of vertebrate skeletons. Since that time it has been recognised that homology is the central relation in comparative biology. Relationships of taxa are recognised through homologues that are discovered through analysis of characters. Characters have been described for myriad purposes – operationally as entities diagnostic of taxa, as identifying attributes of organisms, as transformation series in evolution and as taxic homologues. Characters come from many sources and the debate on what constitutes a 'good' character lies on a scale of preferences from clear-cut qualitative morphology to continuous variables (measurements, ratios, counts) that need to be manipulated with a range of special coding procedures to extract cladistic signal. It will be shown that, for measurement data, characters are described in terms of positional correspondence of parts between internal or external points. In evolution and phylogenetic systematics, homologues are described as transforming relations from unknown common ancestors. In cladistics, characters are seen as hypotheses of taxic homology subject to the tests of similarity, conjunction and congruence. It will be proposed that the recognition of primary homologues is possible for discrete variables and operationally defined states derived by gap-coding methods applied to continuous or overlapping variables. However, because continuous or overlapping variables are transformation series equivalent to manipulated range data, the lack of theory for the coding methods and the need for prior assumptions makes it difficult to find cladistic structure in measurement characters.

Introduction

The term 'homology' was first used to describe relationships between organisms, with particular reference to corresponding morphological parts (homologues) of vertebrate skeletons (Owen 1843, 1849). After 150 years of debate it is recognised today that homology is neither an empirical problem, nor a theoretical one, but the central relation in comparative biology. Acres of print have been written on the subject of homology. In the last 10 years there has been considerable discussion of how to deal with characters and particularly how one determines homologies through character analysis. A recent re-consideration of morphometric data, and the purpose of this book, explores what characters, character states, continuous variables, transformation and homology mean in systematics and morphometrics. On top of all this are the different points of view as to which phylogenetic and cladistic methods are actually appropriate in systematics. These distinctions are critical as the debate revolves around what is meant by character transformation and how this is seen in our understanding of homology.

The problems of characters and character analysis are not new but stretch back at least two centuries (Rieppel 1988). However, the use of continuous variables and morphometrics has its origins amongst the ideal morphologists at the turn of the twentieth century. Irrespective of this history, the use of morphometrics in systematic contexts has undergone a clear revival over the last 10 years or so. Indeed, it has come full circle from being considered of little or no use in cladistics (Pimentel and Riggins 1987; Cranston and Humphries 1988; Farris 1990), to being apposite for the discovery of natural groups (e.g., Fink and Zelditch 1995; Zelditch *et al.* 1995, 2000; Swiderski *et al.* 1998; MacLeod 1999, 2002, in press). My purpose is to review these differences of opinions in the light of recent studies from the perspective of an unreconstructed cladist interested in morphological characters.

Homology

Homology refers to the property of topological relations between different organisms. Owen (1849) distinguished homology from analogy, the latter interpreted as meaning something different in terms of relations, comparability of function, for example. As pointed out by David Williams (person. comm.) it is not only possible for comparable organs to be homologous and analogous, but also to be homologous and not analogous, or even analogous and not homologous. Wings in birds and forelimbs in hoofed mammals are clearly homologous in form but not analogous in function. Owen was quite clear on the subject of this distinction. This impinges on the great 1830 debate between Cuvier and Geoffroy St. Hilaire, that concluded animals were all based on the same fundamental 'ground' plan. Owen used St. Hilaire's 'principle of connections' to describe the similarities and subtle differences of form and the 'principle of composition' to describe the topographical relations of organs, the two combining as the basic properties of homology (Brady 1985; Rieppel 1988; Schuh 2000).

As Brady (1994) pointed out, Geoffroy St. Hilaire had already emphasised the importance of homology in the mid-nineteenth century when he drew connections between similar organs of animals; for example, when comparing the paddle of a porpoise, the hand of man and the foot of a horse. As Brady discusses, Darwin, when commenting on Geoffroy's insight, noted it was possible to shift from figurative into a historical explanation by connecting together those forms more closely related to each other by comparison to other organisms. Different organs have changed into many forms of varying shape and size. Yet the main body organs, although showing some differences, always remained in the same order, such as the relative positions of forelimbs in tetrapods, whether frogs, birds, bats or shrews. Such was the realisation of constancy of topological position. The concept of sameness and difference has been the centrepiece of homology debates ever since (Patterson 1982).

Explanation, according to most accounts, is the real task of science and description is just one procedure to achieve this end. Darwin (1859) gave great emphasis to explanation, and argued in the *Origin of Species* that patterns of morphology and classification await explanation, implying that both are descriptive activities. Darwin never gives account of the science that produces these patterns, but his assumption that taxonomic relations are discovered prior to the inception of explanation remains an interesting one (Brady 1994). Darwin used natural selection by successive gradual changes to explain that the different forms changed or transformed from an ancestral archetype into the visible structures that we see in modern organisms. Consequently, Darwin changed Geoffroy's philosophical relations of corresponding parts into historically literal or actual transformations. For example, limbs of crabs turned into jaws and stamens and pistils were derived from leaves. All transformations were gradual transitions from one form into another. When there were no intermediate forms between real ones, he imagined them (Brady 1994). From this it follows, as Brower (2000) points out, that modern concepts of homology are all manifestations of our conviction that there is a single natural system that explains all of biodiversity. It follows, too, that homology is considered as similar due to common ancestry because the natural system can be explained by evolution by common descent.

Consequently, the correspondence of parts between one organism and another is the establishment of a hypothesis to suggest that particular characters belong to a phylogenetic transformation series from one taxon to another or between modern organisms and their common ancestors (Hennig 1966). However, Nelson (1994) has shown that modern cladists never treat species, Recent or fossil, as ancestors and descendants of one another, but rather as taxa, characters, or character states at the terminals of a cladogram. In systematic research one determines correspondences between organisms through correspondence of their characters. Woodger (in Cain and Harrison 1958) noted that; '[i]n comparing two things we set up a one-to-one relation or correspondence between the parts of the one and those of the other and proceed to state how corresponding parts resemble or differ from one another with respect to certain sets of properties'. Remane (1952) elaborated the procedure and stated that homology is recognised only through the relative position or organs and tissues, similarity of special structures and connections by intermediate taxa (see Schuh 2000). Remane suggested, therefore, that the relationships of taxa are recognised through homologues that are discovered through analysis of characters.

According to Brady (1985) Remane's viewpoint was empirical because it distinguished between the condition to be explained, similarity of structure, from the explanation, the theory of evolution. This was a critical observation because it demonstrated that empirical work on characters and taxa must precede interpretation, and reference to unknown ancestors is interpretation of that pattern. However, Hennig (1966) had already criticised Remane's approach. Hennig (1966: 94) stated of Remane that 'the criterion of linkage of intermediate forms' and the 'criterion of special quality of the structures' are accessory criteria to the 'criterion of sameness of position in comparable fabric systems', an observation that turns out to be more or less identical to Woodger's set theory approach.

In developing his principles of phylogenetic systematics, Hennig (1966: 93) noted that 'different characters that are regarded as transformation stages of the same original character are generally called homologous'. He made it clear that that transformation 'refers to real historical processes of evolution' and not from deriving one character from another in the sense of ideal morphology. Rather cladograms were hypothesised to be artificial constructs, with hypothetical ancestors at the internal nodes, which

Patterson (1982) likened to archetypes in transformational homology and morphotypes if viewed taxically (see below). Hennig thus fleshed out the Darwinian notion of historical transformation, but was absolutely clear that, as one is never in a position to observe phylogenetic transformation, the question arises as to what criteria could convince one that transformation series are comprised of homologues. Hennig (1966: 93–94) stated that '[a]pparently it is often forgotten that the impossibility of determining directly the essential criterion of homologous characters – their phylogenetic derivation from one and the same previous condition – is meaningless for defining the concept of "homology"'. Thus, he went on to describe a range of auxiliary criteria, of geological character precedence, chronological progression and ontogenetic character precedence as independent means to establish potential synapomorphies.

Hennig's auxiliary criteria have since all been rejected (see Wiley 1981). But critically, Hennig changed the concept of homology to allow for losses as well as gains to be used in phylogenetic systematics. He realised that genealogy implied similarity, but the reciprocal was not necessarily true. Similarity could be misleading owing to parallel and convergent changes. He thus coupled particular kinds of similarity with particular kinds of groups.

As any student of systematics will know, Hennig used synapomorphy to diagnose monophyletic groups, parallel and convergent changes gave rise to polyphyletic groups and grouping on symplesiomorphies rendered paraphylies. Allowing for losses as well as originations in characters meant that transformation series could equally be the reduction of organs as to their gain. The rejection of Hennig's auxiliary criteria, and the realisation that theories of characters (synapomorphies) gave theories about groups (monophyly), Patterson (1982, 1988) equated synapomorphy with homology, a viewpoint that has been held by many since (Janvier 1984; Stevens 1984; de Pinna 1991; Panchen 1994; Nelson 1994; Brower 2000). The implication of such a viewpoint is that homology is discovered through analysis of characters and does not impinge a priori. If then, character analysis – the discovery of homologues – is an empirical procedure discovered by comparing similar organs, but synapomorphies are evidence of homology as a relation and thus monophyletic groups, it follows that homology is both part of character analysis and cladistic analysis.

Rieppel (1988) started to resolve the issue of primary homology by suggesting that one could apply tests of similarity to discover whether homologues could be erected as hypotheses with similar topological correspondence before cladistic analysis and a 'test' of congruence.[1] De Pinna (1991) further resolved Rieppel's 'relation of homologous similarity (synapomorphy)' by distinguishing between primary and

---

1 However, it must be borne in mind that it could be possible for a morphological transformation between two non-homologous characteristics (e.g., character states that were incorrectly ascribed to the same character [= transformation series]) to be consistent with a cladogram. We are not so bound by Patterson's (1982) logic to consider that any transformation between any set of morphological descriptors must be accepted as homologous so long as it is congruent with the majority of other characteristics. It is especially true that Type 1 and Type 2 statistical errors can occur and this is worrying. For example, Zelditch et al. (1995) consider that partial warps are homologous characters, it seems, based on the proposition that, so long as partial-warp ordinations are consistent with an established cladogram they must be homologues. However, independent work since then has shown that, when evaluated on their own, partial warps do not behave like the morphological characters traditional systematists use. Partial warps are prone to homoplasy, and different studies have shown that morphometric characters were not able to reconstruct a credible phylogeny if used in the absence of traditional, qualitative characters (MacLeod person. comm.).

secondary homology. Primary homology is the discovery of characters and character formulation, literally the generation of new characters through thorough sampling of taxa. Secondary homology equates with the discovery of synapomorphy, through collective cladistic analysis of primary homologues, the so-called legitimation phase of homologous similarity. Brower and Shawaroch (1996) after comparing morphological and molecular data suggested that primary homology assessment in itself is a two-step process, the determination of topographic identity followed by character state identity. Operationally, this comprises first the recognition of characters and then scoring of the characters into a matrix for further analysis (e.g., Hawkins 2000). The final outcome is that synapomorphies are distinguished from homoplasy and symplesiomorphy as interpretations on rooted trees after one or more rounds of cladistic analysis (Brower 2000). Cladograms are chosen in terms of best fit through optimisation of characters and thus the primary homology statements (similarity and topographic alignment) are 'tested' by showing the greatest congruence with other characters (Patterson 1982). Congruent characters are generally considered as homologues associated with monophyletic groups and incongruent characters, or homoplasies, are associated with paraphyletic and polyphyletic groups.

The shared presence of homologues is the basis for recognising monophyletic groups (Patterson 1982). Shared presence of a homologue between two or more taxa is thus the only evidence we have of relationship, that the taxa form a group, taxic homology. The shared presence of homologues is indicated by a qualifying phrase, such that all vertebrates share vertebrae as distinct from those organisms that lack vertebrae. For a character to become an established homology the feature in question must also occur in the same topographical position within the organisms being compared. Rieppel (1988) states that the relative positions of organs or structures in topographical correspondence are essential conceptually, to initially generate primary homology propositions. This is true for all organisms it seems. In addition to the tetrapod example given above, in flowering plants bisexual flowers invariably display the same sequence of whorls from the outside to the centre of the flower, sepals, petals, stamens and ovules, whatever myriad modifications might occur amongst them.

Homologising the features of topographical correspondence then becomes the basis for hypotheses of groups. Patterson (1982) called this taxic homology which he contrasted sharply from transformational homology. The crucial point about the transformational approach is perhaps indicated using a binary character. Given two states, 0 and 1, only one, but not both, but either at any one time can be a synapomorphy in a rooted tree (Farris et al. 1970). Brower (2000: 13) summarised succinctly the potency of such an approach: 'While evolutionary systematists (and Hennig) had no method to realise their phylogenetic theories, and pheneticists had no theory to discriminate among the many possible methods of grouping based on similarity, cladists, particularly pattern cladists who saw fit to separate and discard the metaphysical husk of common ancestry, were able to compare the advantages of both, using the method of grouping by parsimonious patterns of shared character state change.'

As Scotland (2000) noted, in cladistics the taxic approach is concerned with the monophyly of groups. The transformational approach is concerned with change. Patterson (1982) first described taxic homology to imply hierarchy of groups, when he recognised that transformational homology does not necessarily. Using the same two-state example, 0 means absence and the 1s become the only candidates for synapomorphy on a rooted tree. Transformational homology (for example as in an ordered multistate character) seems to be a complex interplay between topological correspondence and literal transformation through metamorphosis, a concept that need not imply hierarchy at all. It is this aspect that raises problems for the use of continuous variables in systematics. As described by Patterson (1982) in Owen's conception, general homologies between organisms are the result of transformations from an archetype. Patterson went on to show that such an interpretation does not lead to new insights in grouping, but goes more to provide an empty hypothesis of difference between a modern form and an ancestor. Taxic homologies on the other hand are considered to be those characters that diagnose groups relative to other organisms. Transformation from one character to another might be implied, but the important distinction is that the grouping characters form the morphotype or list of homologues of a group. Morphotypes imply definite hierarchies of relationships and can be interpreted to equate with the internal nodes of cladograms.

Characters

To be of use in systematics, characters have to be extracted from a mass of observational data and turned into matrices of consistent scores for further analysis. There are various procedures for the determination of primary homologues and the range of methods for coding has many choices (e.g., Forey and Kitching 2000; Wiens 2000). There is considerable debate about how characters and character states might be defined. As Brower (2000) noted, the process of sampling organisms for characters is still largely an intuitive process done in the same way now as hundreds of years ago. Attempts at quantifying the approach is still fraught with difficulties largely because two or more systematists rarely look at organisms and score characters in the same way (Gift and Stevens 1997). Nevertheless, it is generally agreed amongst biologists that a character is any feature or attribute that is shared among organisms that has the potential for becoming synapomorphies after cladistic analysis (e.g., Colless 1985; Fristrup 1992; Scotland and Pennington 2000). For characters or character states to be cladistic and hence be features of taxa, they ideally would be invariant in some taxa and completely absent in others. However, this is rarely the case. Characters and character states must be extracted from observations on the sampled organisms and summarised into a data matrix that (hopefully) contains some pattern reflecting relationships among the taxa that can be discovered via appropriate analysis. For most computerised cladistic analyses characters are arranged into binary and multistate columns with each integer representing a different state. For multistate characters each column represents a transformation series of dependent variables that can be ordered (in the sense of Mickevich 1982) or unordered during cladistic analysis.

There has always been a tension between the notion of defining characters in order to identify and distinguish organisms and the discovery of homologies in comparative biology to systematise the relations among organisms. Smith (1994: 37), for example, stated that '[c]haracters are observed variations which provide diagnostic features for differentiation amongst taxa'. He showed that characters must occur in two or more states (one of which may be absence[2]) and they should be defined as objectively as possible (see also Mayr et al. 1953; Cain and Harrison 1958; Stuessy 1990). Features that are relatively indistinguishable from one another are generally coded as the same character state so as to reflect the underlying notion of primary homology.

Wiley (1981: 116) stated that: 'A character is a feature of an organism which is the product of an ontogenetic or cytogenetic sequence of previously existing features, or a feature of a previously existing parental organism(s). Such features arise in evolution by the modification of previously existing ontogenetic or cytogenetic or molecular sequence.' Such a definition recognised that features of organisms are the products of evolution and hence have arisen as changes in ontogeny and transformation through time. Pimentel and Riggins (1987) stated that a character can only be a feature of an organism when it can be recognised as a distinct variable. For cladistic analysis, Farris et al. (1970) made it clear that, in order to be able to determine characters for phylogenetic reconstruction, it was necessary to recognise that they were mutually exclusive states that could be considered transformations with a fixed order of evolution. Hennig's interpretations of characters and transformation series was refined by Farris et al. (1970) so that characters have a 'fixed order of evolution', 'each state is derived from another state' and 'there is a unique state from which the every other is ultimately derived'.

Jardine (1969) considered diagnosing taxa and describing individual organisms using the same character to be a confusing process. He said that the presence of a backbone is not a property of Vertebrata, but all of the organisms within the group Vertebrata possess backbones. Jardine made the distinction between taxa and organisms on the basis of characters and character states. Taxa have characters and organisms have individual attributes or character states.

For phylogenetic systematists, characters to convey cladistic information must transform from one state into another through time. However, this does not mean that a brown eye changes into a blue eye or that ovate leaves change into obovate leaves. Similarly, invertebrate animals lacking backbones do not change into those possessing them. What actually changes is the frequency of a particular character state for a given character and the frequencies of different character states change through time. Thiele (1993) stated that cladistic character states are frequency distributions and conversely, all cladistic character states have particular frequencies of distribution. Thus, desirable cladistic characters are those with large, clear-cut changes rather than small, gradual ones and a 'good cladistic character' is, in effect, a value judgement on data.

As Stevens (2000: 82) noted: 'a character is the sum of features showing particular similarities (e.g., Patterson 1982; Stevens, 1984), topographical homologies (Jardine 1969), topographical identities (Brower and Schawaroch 1996), or relationships of primary homology (e.g., de Pinna 1991)'. Stevens (2000) elaborated on the scheme proposed by Brower and Schawaroch (1996) and suggested that the stages between the beginning of a study on a particular group through to the cladistic analysis comprised a sequence of at least three operations between choosing characters and delimiting states.

---

2 The arguments against using absence are compelling (e.g., Nixon and Carpenter 1993). Using absence as a state confuses the obvious differences between genuinely absent, not yet developed and secondary loss. Therefore, absence as a state is logically flawed, except in the case of 'not present'.

The first stage was the lining up of characters thought to be the same (similarity or topographic identity). The second stage involved the actual measurement of individuals (which he called data 1) and grouping the measurements in some way for the taxa to be analysed (data 2). Finally the third stage compared all the measurements between taxa to create the data matrix (data 3). All three stages offer opportunities for making errors largely because there are so many different ways of undertaking them. Gift and Stevens (1997) had noted that even the first stage of delimiting similarities gave as many different solutions as recorders collecting the data. Patterson and Johnson (1977) also brought attention to this problem. In a pungent criticism Patterson and Johnson (1977: 361) noted that 'the emphasis has shifted from observation, the source of the matrix, to whatever message can be extracted from the matrix ...'. In a reanalysis of the characters of osmeroid fish an 11 per cent error rate in the original observations and subsequent coding of the errors had immense consequences for the topologies obtained by cladistic analysis. The third stage is highly significant. As indicated already the ideal character states for cladistics are those that have distinct gaps. However, when it comes to coding measurement data it appears that the methods are frequently operationally less clear-cut than those for qualitative variables (see Farris 1990 for example).

Continuous variables

Considering that there are a number of ways for determining and coding qualitative characters the question arises of whether continuous and non-continuous characters and overlapping and non-overlapping distributions of observations along the variable axes can actually yield cladistic characters given the requirements of synapomorphy and secondary homology. Variables are observed variations of some attribute or characteristic feature, ergo characters = variables. The three-step procedure of Stevens becomes complicated when applied to overlapping distributions on variable axes. The first stage of determining the similarities or topographical identities involves the fixing of points for determining geometric locations on the organisms concerned. The second stage involves the actual collection of observations (= measurements) of the variables to obtain distributions on which to apply a comparative method. Comparisons between the organisms use methods for distinguishing geometric transformations. There are a variety of methods, but they all make a general scheme of size, shape or meristic values. The third stage is the different morphs (recognised on the basis of observed discontinuities in the distribution of observations along the variable axis) are then coded into a taxon × character matrix for cladistic analysis. If reproducible ways can be found to recognise discontinuities in the distribution of the observations along the variable/character axes those discontinuities can be used to delimit different states. Provided discontinuities in the distribution of observations are present, quantitative character analysis proceeds in a manner identical to qualitative character analysis. However, if no discontinuities in the distribution of observations along a quantitative variable axis are found, the third stage of this process cannot be completed and the quantitative variable must be regarded as unsuitable for inclusion in a cladistic investigation. Under these circumstances the variable/character cannot be used for diagnosing a group.

Primary homology and topographical identity

Determination of topographic identity of continuous variables in things such as the shape of jaws or length/width ratios of leaves is complicated by a number of factors: complexities of shape, allometric change during ontogeny and sheer variation that superficially appears to render unique values for every specimen under consideration. This is not saying that qualitative variables are any less problematic, just that detailed variation such as measurements become inherently less easy to divide into gaps. The problem of ontogeny is fundamental to all methods and Hennig (1966) was careful to point out that because individuals change throughout the life cycle the same stages or semaphoronts had to be compared. For different stages of an insect's life cycle semaphoronts are possibly easily identified but for subtle changes in measurement data defining semaphoronts becomes a more difficult problem. Løvtrup (1988) was at pains to point out that allometric trajectories (especially after birth in vertebrates) differ from one organism to another and show great variation within taxa. Kluge (1988) even suggested that maybe the whole allometric phase might be the level of comparison for morphometric data.

For complex shape data, Zelditch *et al.* (1995) and Swiderski *et al.* (1998) suggest two strategies, either examine the shape as a whole or subdivide the shape into individual dimensions, including aspect ratios and distances between landmarks as a measure. Both approaches have advantages and disadvantages. Viewing complex morphology as single items of comparison can lead to problems of coding as sampling increases. Determining the information content of characters with many different states shuffle in other problems of direction, order and polarity during cladistic analysis. Although one is naturally using quantitative analysis to search for discontinuities, there are instances where artificial gaps are created (Chappill 1989). Thus comparing suites of unique shapes can lead to creating columns of variables that are effectively autapomorphs in the final matrix. Atomising the components of any character as a reductionist pursuit can lead to separate columns of independent variables that are both logically and biologically correlated. Stevens (2000) notes that problems also emerge when characters are wrongly linked biologically and logically. In all characters finding the logical and biological divisions becomes increasingly difficult when overlap increases. Several characters might only be one, but theory is lacking on the precise course of action to follow for coding them (see also Fink and Zelditch 1995; Pleijel 1995; Hawkins *et al.* 1997; Hawkins 2000).

One of the complicating issues in quantitative and morphometric literature is the language surrounding units of comparison and the use of the term homology. Homology tends to bridge the formalisms of geometric shape analysis and the evidential use as character hypotheses supporting monophyletic groups in systematics. Following a tradition set by D'Arcy Wentworth Thompson (1942), shape analysts frequently apply homology to mean comparisons between discrete geometric structures, such as comparable points or curves, and, by a further extension, to the multivariate descriptors that arise as part of the subsequent multivariate analyses. Smith (1988: 335) distinguished this aspect as operational homology: 'character correspondence, among taxa, based on the optimal matching of internal and external landmarks on exemplars, samples, or developmental series of OTUS. It is usually a quantified construct within which landmarks, variables, and characters are oriented for comparison in systematic biology. In this context, morphometrics can provide very precise quantitative values for character states.'

In Smith's context, the term 'homologous' means something other than the primary and secondary homology in systematics. Rather, it is used for corresponding parts in different samples of taxa or developmental stages during life cycles. In morphometrics, then, to declare something 'homologous' is an assertion about comparison of structures in a consistent manner rather than anything to do with historical transformations gradual or otherwise. However, Zelditch et al. (1995: 180) note that 'when systematists choose particular landmarks, the choice is often defended on the grounds that they sample parts of the organism judged to be homologous at the most inclusive level being studied'. MacLeod (1999) challenges the Zelditch et al. assertion on this issue and provides several examples where biologically non-homologous features have been used as landmarks. Moreover, a homologue is a structure, part of an organism, not an infinitesimal location point. MacLeod (1999) also challenges the notion that landmarks can be homologous with one another in the absence of evidence for point-to-point correspondence. Similarly, to declare an interpolation (such as a thin-plate spline) a 'homology map' means that one intends to refer to its features as if they had something to do with valid biological explanations pertaining to the regions between the landmarks, about which there is frequently no data (MacLeod 1999, 2002, in press). This is an important point as systematists use outlines to make comparisons between taxa and it is a key source of relevant information that could be used in cladistic analysis. For detailed discussions of the landmark-outline debate see Bookstein et al. (1982, 1985), Ehrlich et al. (1983), and Bookstein (1990, 1991, 1996a,b, 1997).

Thus, homology in morphometrics is a complicated interplay between precise topological correspondence and differences amongst taxa as literal transformations through ontogenetic and phylogenetic metamorphosis. This might explain the sharp criticisms of Pimentel and Riggins (1987), Cranston and Humphries (1988) and Bookstein (1994) who felt that it was impossible to apply taxic homology to overlapping variables, but these are all comments that did not take full cognizance of primary homology assessment. To put it into the context of this paper however, determining similarity and topographic identity can easily be undertaken with both measurements and morphometric data as with any other procedure used for determining primary homologies, and which like any source of data, has potential for cladistic analysis. Stevens (2000) has already noted that there is so much confusing baggage around the word homology that the word should be replaced with pertinent and relevant replacements. Bookstein et al. (1985), and Bookstein (1991), attempt to simplify and marry two concepts of homology by arguing that the 'traditional' concept of homology should be extended to cover morphometric homology. However, they are quite clear that there is a distinction. MacLeod (1999) disagrees with the idea of extending the concept because biological homology refers to structures, not infinitesimal points on structures. Thus, landmarks simply abstract the spatial position of putatively homologous structures relative to other such structures. These points are not themselves homologous because alternative, but nearby, locations can serve equally well for morphometrics. Indeed outlines and outline segments have a firmer claim on correspondence to the biological concept of homology than landmarks (MacLeod, person. comm.). Unsurprising then that Fink and Zelditch (1995), Zelditch et al. (1995) and Swiderski et al. (1998)

have suggested that the term 'homologous' should be replaced with 'corresponding' or 'comparable' when dealing with landmarks at the character definition phase in morphometrics. Various authors are at pains to point out that morphometric and quantitative data are somehow different from qualitative data. But, all character sources can be assessed along a scale of good to bad or best to worse in terms of the chances for yielding cladistic classifications (Chappill 1989; Thiele 1993).

Character state identity

Character coding for both continuous or overlapping variables and qualitative variables has been the subject of intense scrutiny over the last few years. The question is simple: how are measurements of raw data coded into the 0s, 1s and 2s of a data matrix (see Scotland and Pennington 2000; Wiens 2000)? Stevens (2000) commenting on Brower and Schawaroch's (1996) division for primary homology assessment noted that topographic identity for morphological characters was somewhat factual and uncontroversial. The real problems for morphological characters emerge at the character state identity stage and the problem becomes even more acute for overlapping variables. Here, the problem is what to call a character, or a character state. A useful rule of thumb is to consider that characters are equivalent to variables, and discontinuities are equivalent to the character state boundaries. If there are no discontinuities the character is invariant having only one state. Pimental and Riggins (1987) recognise conventional nominal variable coding for obvious characters and states as the normal way to proceed, but given the range of character variation from obvious discontinuities to gaps made by gap-coding methods, suggest that only qualitative gaps be coded. Hawkins (2000) presents several challenges to conventional coding variously known as composite coding (Wilkinson 1995), unspecified homologue coding, ratio coding, logically related coding, unifying coding, inapplicable data coding, positional coding and mixed coding. In addition to these there are coding schemes for multistate characters, contingent coding methods and a variety of different ways for coding presence and absence (Pleijel 1995; Forey and Kitching 2000).

Given all the vagaries associated with relatively clear-cut situations the situation becomes more complicated with continuous distributions of observations or measurements. Usually opaque to assessment in raw form, one can only begin to discover grouping homologies through specific methods for converting raw data into discrete codes for subsequent cladistic analysis. Thus undertaking the second and third stages of primary homology assessment, delimiting and coding the features of organisms as characters and character states, is part of the process of recognising their systematic value. Despite these difficulties, Thiele (1993) believes that continuous variables should only be excluded if the cladistic analysis cannot handle such data or if it can be shown empirically that those characters convey no information or phylogenetic signal relative to other characters in the data matrix. It is also obvious that there are many manipulations to continuous variables that can be undertaken, principally coding features in a matrix and the question of whether manipulations such as scaling and weighting (e.g., Goldman 1988; Thiele and Ladiges 1988) are justifiable with respect to the results obtained.

In theory, continuous variables have an infinite number of potential values. However that does not mean that observations/measurements must be continuously distributed along such variable axes, and the continuity of the variable scale has nothing necessarily to do with the nature of the distribution of observations that might be made along that scale. Although there are several methods (e.g., in MacClade) that can be used to examine continuous variables without recoding and have some limited use for looking at character evolution over trees (e.g., Swofford and Berlocher 1987; Huey and Bennett 1987), there are few computer algorithms available for cladistic analysis of raw data (but see Felsenstein 1988). Most methods manipulate the raw scores. The values can be bounded within a certain range but the potential list of values can still be large. Some argue that there are few variables that can actually be considered as continuous because our ability to measure values to the nth degree are so imprecise that the potential values are in fact finite. Of the few studies available in order to compare continuous variables with qualitative variables on cladograms, all raw data are invariably filtered during coding as discrete integers (e.g., Cranston and Humphries 1988; Thiele and Ladiges 1988; Chappill 1989; Thiele 1993).

Methods include simple gap coding (Mickevich and Johnson 1976), segment coding (Colless 1980), divergence coding (Thorpe 1984; Almeida and Bisby 1984), homogenous subsets coding (Simon 1983), generalized gap coding (Archie 1985; Goldman 1988; see also Thiele and Ladiges 1988), range coding (Baum 1988) and gap weighting (Thiele 1993). Samples of taxa are ranked along a scaled attribute axis, and then simple rules are applied to create gaps, segments or subsets in an effort to produce discrete codes for the continuous values. The attribute axis is rescaled into states for cladistic analysis.

Simple gap coding divides the axis at those points where no values occur or between the means of the frequency distributions at the point where the 'gap' exceeds a particular preconceived value, such as one standard deviation about the mean. Usually, the attribute axis will be divided into fewer states than there are taxa and for most computer programs there is an upper bound to the number of states per character that can be analysed. Chappill (1989: 220) indicated that desirable attributes for any method should be that it should 'reflect the proportional differences between taxa', ... have '[T]he ability to discriminate between divergent taxa', ... 'using a particular character should be equal for all comparisons between pairs of such taxa', ... 'the number of states produced should be proportional to the variability of the character', ... 'it should not recognize insignificantly small differences between taxa', ... 'and the addition of new taxa, or improved sampling, should not reduce the discrimination possible between the original taxa.'

It turns out there are problems with all of the methods. Farris (1990) provided a characteristically robust critique indicating that each method had its drawbacks and most damning of which that these were techniques more commonly used by pheneticists (e.g., Sneath and Sokal 1973), as incorporated into the studies of Cranston and Humphries (1988), Goldman (1988), Thiele and Ladiges (1988), and Chappill (1989). More specifically, it is the assumptions for scaling multistate characters to unit range (so as to reduce their effect in comparison to binary characters and confounding weighting with scaling) that causes most problems. That there is no real justification for scaling or weighting multistate characters a priori confounds the outcomes in cladistic analysis.

Coding methods invariably consist of four stages: the terminal taxa are identified, a sample of each is measured and scored with sample means and variance, and then the means and ranges are converted to integers using a gap, segment or range-coding method. It seems that the problem with all coding methods (quantitative and qualitative) is that the rules for converting the measurement data into codes lack any justifiable theory. What is needed in a systematic morphometric analysis is agreement about the discontinuities in the distribution of observations (MacLeod, person. comm.; Zelditch *et al.* 1995). The existence of discontinuities represents the practical justification, congruence provides the operational test and an agreed definition of discontinuity. Farris (1990) showed explicitly that for generalised gap coding (Archie 1985) as used by Thiele and Ladiges (1988), for example, varying the sample size, using different standard deviations between the means, and thus varying the critical gap size had profound effects on the outcomes. This method like all others attempts to formulate '*ad hoc*' rules for subdividing a continuum. Gradual continua simply cannot be used as a basis for unambiguous grouping. Farris (1990) further demonstrated that homogenous subset coding and gap coding gave very different results on the same data set and that generalised gap coding could yield nonsensical codes. He elaborated further saying that it was of no use to rescale codes as these invariably produced meaningless character states. Farris concluded by saying that if a character can be broken up into several meaningful distinct conditions there are no rational grounds for reducing weights of the distinctions but to code the states in an appropriate manner. This would surely justify morphometric methods at least in some cases. On the other hand, if the coded states obtained by one of continuous variable techniques reflect no meaningful distinction, the remedy is to eliminate the arbitrary differences. Of the many examples I have examined I would say that many of the states are meaningless except by justification on statistical differences.

Furthermore there are those who have used morphometrics to justify using their collected data (e.g., Chappill 1989) regardless of whether those data uniquely characterise taxonomic groups or not. The problem is that those who have used morphometrics have wanted to find ways of using the data they collected, regardless of whether those data characterised groups of taxa or not. That was wrong, but some cladists have overreacted in regarding all morphometric data as being unacceptable (e.g., Pimentel and Riggins 1987). They considered (1) continuous variables imply possibility of a continuous distribution of observations (irrespective of whether this possibility is realised in nature), and (2) no theoretical justification for *ad hoc* methods of subdividing a continuum could be found. On the contrary, morphometrics is important for systematics because it (1) can yield additional variables that can be used to define groups (provided it is realised that the subject of morphometric analysis is to uncover the discontinuities that separate taxa from one another), (2) can test hypotheses of the correctness/objectivity of state definition for qualitative characters, and (3) can render the assignment of states to taxa/individuals more precisely. Sadly it seems that some cladists and traditional systematists avoid morphometrics because they are innumerate and apprehensive at the idea of having to learn new skills, (2) they question the cost-benefit of morphometric analyses, and (3) they understand that if they subject many of their personal/traditional character state definitions (which are rarely defined in precise terms) to the rigour of morphometric analysis, those definitions might be found wanting. All of those reasons are understandable at some level, but none of them have anything to do with the theory or logic of cladistic analysis. The mistakes of the past will need to be explained and acknowledged by systematists and morphometricians before progress in this area will be able to be made.

Cladistic analysis

Given the difficulties encountered by coding methods it might be considered perverse to enquire whether the performance of such characters can be applied. Nevertheless, Thiele (1993) asked the question whether morphometric data were of any use for inferring phylogenies. Of the few studies available, he noted that Cranston and Humphries (1988), Thiele and Ladiges (1988) and Chappill (1989) all used consistency indices as performance indicators to determine the differences between explicit quantitative and qualitative characters. Thiele (1993) tested the efficacy of continuous variables by suggesting that if a set of morphometric characters induces one phylogeny, the matrix should contain cladistic co-variation. Also cladograms derived from quantitative characters should be similar to those derived from other data sets. In all three analyses, morphometric data gave lower consistency indices on the cladograms in comparison to the qualitative data. He noted, however, that in his own study of *Angophora*, the morphometric variables performed well and mapped well onto cladograms produced from the qualitative data. Later Thiele (1993) applied a more elaborate test on morphometric data in studies of *Banksia*. Again, he found that morphometric characters produced lower consistency indices than qualitative data, but did perform better than results obtained from random data. In studies of partitioned data sets, representing difference sections of *Banksia*, in all but one out of four studies, the morphometric and qualitative characters were significantly similar, and both produced similar cladograms. It was significant to notice that both qualitative and quantitative data sets produced similar trees.

Conclusions

It would seem that any source of data is suitable for cladistic analysis. It is obvious that the more clear-cut observations can be, the more obvious divisions can be made in coding characters. However, unlike Pimentel and Riggins (1987) and Chappill (1989) I agree with Thiele (1993) and Stevens (2000) that all data should be scrutinised for potential analysis. It is obvious that the more quantitative observations become, the more difficult it becomes to partition that information into characters and character states. The methods for doing so become elaborate and lack obvious underlying theory to justify the methods. In at least some cases (e.g., Thiele 1993, Fink and Zelditch 1995 and subsequent papers) cladistic analysis appears to have succeeded in inferring hypotheses of relationship. Rather than considering that some data are better than others, not all data sets can be considered as one homogenous class. What might be true for one class of characters might not be true for others. As Thiele noted the best data are not necessarily different in kind from the worst.

Nevertheless, at the end of the day the most robust classifications are those with the highest information content. There is no doubt that manipulating measurement data into long transformation series reduces the information content and creates gaps where none can actually be agreed upon except by convention. For measurement data, of the kinds that compare different leaf lengths or widths, for example, transcribing the results into clear-cut integers becomes vacuous, especially when individual scores or codes are given for each taxon in the analysis. There is little doubt too that recommendations for analysing continuous variables to include such things as aligning the variation into series from the smallest to largest or *vice versa*, and insisting on ordered transformation series is a perverse use of transformation series analysis in the sense of Mickevich (1982). Homology is about relations and at the minimum refers to the fact that at least one homologue must be present in two taxa and absent from a third to be useful. In this context it appears that overlapping variables have less in the way of relational information. The use of gap-coding methods to determine discrete states appears not to have any particular theory, and like many phenetic studies, are methods devised on statistical or algorithmic ground without clear reasons for doing so (Farris 1990). In this context the gap-coding procedures have little in the way of theory as compared with some morphometric methods (see MacLeod, this volume) and it is clear that the different procedures have their drawbacks, but especially in attempting to create gaps when none are really present. Stevens (2000) noted that there are two kinds of data: 'one in which the states are taken from visual inspection of overlapping variation and one in which states are taken from largely non-overlapping variation'. The latter invariably contained stronger cladistic signal.

I believe that Patterson (1982) was right to draw attention to the distinction between transformational and taxic homology. It seems to me that much of the ambiguity that exists in cladistics today is a direct result of worrying about transformation. Taxa and characters are really the same thing. Characters are variables and thus portions or fragments of organisms. To overcome the ambiguities what is needed is to bring the activities of morphometrics and cladistics closer together to find nested hierarchies of characters and taxa. It still worries me that the arguments about transformation and character evolution have got muddled up with the business of sorting out homology and classification, and if these were teased fully apart I am sure the activities of both groups could come closer together.

Acknowledgements

I would like to thank Peter Forey and Norman MacLeod for inviting me to take part in this symposium, and for having so much patience as I procrastinated over my contribution. I also thank Norman MacLeod for critically reading the draft manuscript and helping with the literature.

References

Almeida, M. T. and Bisby, F. A. (1984) 'A simple method for establishing taxonomic characters from measurement data', *Taxon*, 33, 405–409.

Archie, J. W. (1985) 'Methods for coding variable morphological features for numerical taxonomic analysis', *Systematic Zoology*, 34, 236–345.

Baum, B. R. (1988) 'A simple procedure for establishing discrete characters from measurement data applicable to cladistics', *Taxon*, 37, 63–70.

Bookstein, F. L. (1990) 'Analytic methods: introduction and overview', in Rohlf, F. J. and Bookstein, F. L. (eds) *Proceedings of the Michigan Morphometrics Workshop*, Ann Arbor: The University of Michigan Museum of Zoology, Special Publication 2, MI, pp. 61–74.

Bookstein, F. L. (1991) *Morphometric tools for landmark data: geometry and biology*, Cambridge: Cambridge University Press.

Bookstein, F. L. (1994) 'Can biometrical shape be a homologous character?', in Hall, B. K. (ed.) *Homology: the hierarchical basis of comparative biology*, New York: Academic Press, pp. 198–227.

Bookstein, F. L. (1996a) 'Landmark methods for forms without landmarks: localizing group differences in outline shape', in Amini, A., Bookstein, F. L. and Wilson D. (eds) *Proceedings of the Workshop on Mathematical Methods in Biomedical Image Analysis*, San Francisco: IEEE Computer Society Press, pp. 279–289.

Bookstein, F. L. (1996b) 'Landmark methods for forms without landmarks: morphometrics of group differences in outline shape', *Medical Image Analysis*, 1, 1–20.

Bookstein, F. L. (1997) 'Landmark methods for forms without landmarks: localizing group differences in outline shape', *Medical Image Analysis*, 1, 225–243.

Bookstein, F. L., Strauss, R. E., Humphries, J. M., Chernoff, B., Elder, R. L. and Smith, G. R. (1982) 'A comment on the uses of Fourier methods in systematics', *Systematic Zoology*, 31, 85–92.

Bookstein, F., Chernoff, B., Elder, R., Humphries, J., Smith, G. and Strauss, R. (1985) *Morphometrics in evolutionary biology*, Philadelphia: The Academy of Natural Sciences of Philadelphia.

Brady, R. H. (1994) 'Explanation, description, and the meaning of "transformation" in taxonomic evidence', in Scotland, R. W., Siebert, D. J. and Williams, D. M. (eds) *Models in phylogeny reconstruction*, Systematics Association Special Volume No. 52, Oxford: Clarendon Press, pp. 11–29.

Brady, R. H. (1985) 'On the independence of systematics', *Cladistics*, 1, 113–126.

Brower, A. V. Z. (2000) 'Homology and the inference of systematic relationships: some historical and philosophical perspectives', in Scotland, R. and Pennington, R. T. (eds) *Homology and systematics: coding characters for phylogenetic analysis*, Systematics Association Special Volume No. 58, London: Taylor and Francis, pp. 10–21.

Brower, A. V. Z. and Shawaroch, V. (1996) 'Three steps of homology assessment', *Cladistics*, 12, 265–275.

Cain, A. J. and Harrison, G. A. (1958) 'An analysis of the taxonomist's judgement of affinity', *Proceedings of the Zoological Society of London*, 131, 85–98.

Chappill, J. A. (1989) 'Quantitative characters in phylogenetic analysis', *Cladistics*, 4, 217–234.

Colless, D. H. (1980) 'Congruence between morphometric and allozyme data for *Menidia* species: a reappraisal', *Systematic Zoology*, 29, 288–299.

Colless, D. H. (1985) 'On "character" and related terms', *Systematic Zoology*, 34, 229–233.

Cranston, P. S. and Humphries, C. J. (1988) 'Cladistics and computers: a chironomid conundrum', *Cladistics*, 4, 72–92.

Darwin, C. (1859) *On the origin of species, or the preservation of favoured races in the struggles for life*, London: John Murray.

De Pinna, M. C. C. (1991) 'Concepts and tests of homology in the cladistic paradigm', *Cladistics*, 7, 367–394.

Ehrlich, R., Pharr, Jr., R. B. and Healy-Williams, N. (1983) 'Comments on the validity of Fourier descriptors in systematics: a reply to Bookstein *et al.*', *Systematic Zoology*, 31, 85–92.

Farris, J. S. (1990) 'Phenetics in camouflage', *Cladistics*, 6, 91–100.

Farris, J. S., Kluge, A. G. and Eckhardt, M. J. (1970) 'A numerical approach to phylogenetic systematics', *Systematic Zoology*, 19, 172–189.

Felsenstein, J. (1988) 'Phylogenies and quantitative characters', *Annual Review of Ecology and Systematics*, 19, 445–471.

Fink, W. L. and Zelditch, M. L. (1995) 'Phylogenetic analysis of ontogenetic shape transformations: a reassessment of the piranha genus *Pygocentrus* (Teleosti)', *Systematic Biology*, 44, 343–360.

Forey, P. L. and Kitching, I. J. (2000) 'Experiments in coding multistate characters', in Scotland, R. and Pennington, R. T. (eds) *Homology and systematics: coding characters for phylogenetic analysis*, Systematics Association Special Volume No. 58, London: Taylor and Francis, pp. 54–80.

Fristrup, K. (1992) 'Character: current uses', in Keller, E. F. and Lloyd, E. A. (eds) *Keywords in evolutionary biology*, Cambridge, MA: Harvard University Press, pp. 45–51.

Gift, N. and Stevens, P. F. (1997) 'Vagaries in the delimitation of character states in quantitative variation – an experimental study', *Systematic Biology*, 46, 112–125.

Goldman, N. (1988) 'Methods for discrete coding of morphological characters for numerical analysis', *Cladistics*, 4, 59–71.

Hawkins, J. A. (2000) 'A survey of primary homology assessment: different botanists perceive and define characters in different ways', in Scotland, R. and Pennington, R. T. (eds) *Homology and systematics: coding characters for phylogenetic analysis*, Systematics Association Special Volume No. 58, London: Taylor and Francis, pp. 22–53.

Hawkins, J. A., Hughes, C. E. and Scotland, R. W. (1997) 'Primary homology assessment, characters and character states', *Cladistics* 13, 275–283.

Hennig, W. (1966) *Phylogenetic systematics*, Urbana: University of Illinois Press.

Huey, R. B. and Bennett, H. F. (1987) 'Phylogenetic analysis of co-adaptation: preferred temperatures versus optimal performance temperatures of lizards', *Evolution*, 41, 1098–1115.

Janvier, P. (1984) 'Cladistics: theory, purpose and evolutionary implications', in Pollard, J. W. (ed.) *Evolutionary theory: paths into the future*, Chichester: Wiley, pp. 39–75.

Jardine, N. (1969) 'A logical basis for biological classification', *Systematic Zoology*, 18, 37–52.

Kluge, A. G. (1988) 'The characterisation of ontogeny', in Humphries, C. J. (ed.) *Ontogeny and systematics*, New York: Columbia University Press, pp. 57–81.

Løvtrup, S. (1988) 'Epigenetics', in Humphries, C. J. (ed.) *Ontogeny and systematics*, New York: Columbia University Press, pp. 189–227.

MacLeod, N. (1999) 'Generalizing and extending the eigenshape method of shape visualization and analysis', *Paleobiology*, 25, 107–138.

MacLeod, N. (2002) 'Phylogenetic signals in morphometric data', in MacLeod, N. and Forey, P. (eds) *Morphology, shape, and phylogeny*, London: Taylor and Francis, pp. 100–138.

MacLeod, N. (in press) 'Landmarks, localization, and the use of morphometrics in phylogenetic analysis', in Edgecombe, G., Adrain, J. and Lieberman, B. (eds) *Fossils, phylogeny, and form: an analytical approach*, New York: Plenum Press.

Mayr, E., Linsley, E. G. and Usinger, R. L. (1953) *Methods and principles of systematic zoology*, New York: McGraw-Hill.

Mickevich, M. F. (1982) 'Transformation series analysis', *Systematic Zoology*, 31, 461–478.

Mickevich, M. F. and Farris, J. S. (1981) 'The implications of congruence', in *Menidia, Systematic Zoology*, 30, 351–370.

Mickevich, M. F. and Johnson, M. S. (1976) 'Congruence between morphological and allozyme data in evolutionary inference and character evolution', *Systematic Zoology*, 25, 260–270.

Nelson, G. (1994) 'Homology and systematics', in Hall, B. K. (ed.) *Homology: the hierarchical basis of comparative biology*, San Diego, CA: Academic Press, pp. 101–149.

Nixon, K. C. and Carpenter, J. M. (1993) 'On outgroups', *Cladistics*, 9(4), 413–426.

Owen, R. (1843) *Lectures on comparative anatomy*, London: Longman, Brown, Green and Longmans.

Owen, R. (1849) *On the nature of limbs*, London: J. van Voorst.
Panchen, A. L. (1994) 'Richard Owen and the concept of homology', in Hall, B. K. (ed.) *Homology: the hierarchical basis of comparative biology*, San Diego, CA: Academic Press, pp. 21–62.
Patterson, C. (1982) 'Morphological characters and homology', in Joysey, K. A. and Friday, A. E. (eds) *Problems of phylogenetic reconstruction*, London: Academic Press, pp. 21–74.
Patterson, C. (1988) 'Homology in classical and molecular biology', *Molecular biology and evolution*, 5, 603–625.
Patterson, C. and Johnson, G. D. (1977) 'The data, the matrix and the message: comments on Begle's relationships of the Osmeroid fishes', *Systematic Biology*, 46, 458–465.
Pimentel, R. A. and Riggins, R. (1987) 'The nature of cladistic data', *Cladistics*, 3, 201–209.
Pleijel, F. (1995) 'On character coding for phylogeny construction', *Cladistics*, 11, 309–315.
Remane, A. (1952) *Die Grundlagen des natürlichen systems, der vergleichenden anatomie und der phylogenetik*, Leipzig: Akademische Verlagsgesellschaft.
Rieppel, O. C. (1988) *Fundamentals of comparative biology*, Basel: Birkhäuser Verlag.
Schuh, R. T. (2000) *Biological systematics: principles and applications*, Ithaca and London: Cornell University Press.
Scotland (2000) 'Homology, coding and three-taxon statement analysis', Chapter 8, in Scotland, R. and Pennington, R. T. (eds) *Homology and systematics: coding characters for phylogenetic analysis*, Systematics Association Special Volume No. 58, London: Taylor and Francis, pp. 145–182.
Scotland, R. and Pennington, R. T. (eds) (2000) *Homology and systematics: coding characters for phylogenetic analysis*, Systematics Association Special Volume No. 58, London: Taylor and Francis.
Simon, C. (1983) 'A new coding procedure for morphometric data with an example from periodical cicada wing veins', in Felsenstein, J. (ed.) *Numerical taxonomy*, Berlin and Heidelberg: Springer-Verlag, pp. 378–382.
Smith, A. B. (1994) *Systematics and the fossil record: documenting evolutionary patterns*, Oxford: Blackwells Scientific Publications.
Smith, G. R. (1988) 'Homology in morphometrics and phylogenetics', in Rohlf, J. F. and Bookstein, F. L. (eds) *Proceedings of the Michigan Morphometrics Workshop*, Special Publication No. 2, Ann Arbor: Michigan Museum of Zoology, pp. 325–338.
Sneath, P. H. A. and Sokal, R. R. (1973) *Numerical taxonomy*, San Francisco: W. H. Freeman and Company.
Stevens, P. F. (2000) 'On characters and character states: do overlapping and non-overlapping variation, morphology and molecules all yield data of the same value?', in Scotland, R. and Pennington, R. T. (eds) *Homology and systematics: coding characters for phylogenetic analysis*, Systematics Association Special Volume No. 58, London: Taylor and Francis, pp. 81–105.
Stevens, P. F. (1984) 'Homology and phylogeny: morphology and systematics', *Systematic Botany*, 9, 395–405.
Stuessy, T. F. (1990) *Plant taxonomy: the systematic evaluation of comparative data*, New York: Columbia University Press.
Swiderski, D. L., Zelditch, M. L. and Fink, W. L. (1998) 'Why morphometrics isn't special: coding quantitative data for phylogenetic analysis', *Systematic Biology*, 47, 508–519.
Swofford, D. L. and Berlocher, S. H. (1987) 'Inferring evolutionary trees from gene frequency data under the principle of maximum parsimony', *Systematic Zoology*, 36, 293–325.
Thiele, K. (1993) 'The holy grail of the perfect character: the cladistic treatment of morphological data', *Cladistics*, 9, 275–304.

Thiele, K. and Ladiges, P. Y. (1988) 'A cladistic analysis of *Angophora* Cav. (Myrtaceae)', *Cladistics*, **4**, 23–42.

Thompson, D'Arcy W. (1942) *On growth and form*, Cambridge: Cambridge University Press.

Thorpe, R. S. (1984) 'Coding morphometric characters for constructing distance Wagner networks', *Evolution*, **38**, 244–355.

Wiens, J. J. (2000) (ed.) *Phylogenetic analysis of morphological data*, Washington, DC and London: Smithsonian Institution Press.

Wiens, J. J. (2000) 'Coding morphological variation within species and higher taxa for phylogenetic analysis', in Wiens, J. J. (ed.) *Phylogenetic analysis of morphological data*, Washington, DC and London: Smithsonian Institution Press, pp. 115–145.

Wiley, E. O. (1981) *Phylogenetics: the theory and practice of phylogenetic systematics*, New York: Wiley.

Wilkinson, M. (1995) 'A comparison of two methods of character construction', *Cladistics*, **11**, 297–308.

Zelditch, M. L., Fink, W. L. and Swiderski, D. L. (1995) 'Morphometrics, homology and phylogenetics: quantified characters as synapomorphies', *Systematic Biology*, **44**, 179–189.

Zelditch, M. L., Swiderski, D. L. and Fink, W. L. (2000) 'Discovery of phylogenetic characters in morphometric data', in Wiens, J. J. (ed.) *Phylogenetic analysis of morphological data*, Washington, DC and London: Smithsonian Institution Press, pp. 37–83.

Chapter 3

Quantitative characters, phylogenies, and morphometrics

Joseph Felsenstein

ABSTRACT

Morphometrics gives us a source of quantitative characters, and thus raises the question of how to use them in inferring phylogenies. I argue here that we can use statistical models for quantitative characters evolving along phylogenies, but that these require knowledge of the covariance of evolutionary change along lineages of characters. These reflect not merely genetic covariances, but selective covariances as well. In the absence of good estimates of these covariances, molecular data may bear much of the brunt of inferring phylogenies, leaving the quantitative character models to be used when the quantitative characters are themselves of interest. Models involving characters chasing selective peaks, and punctuated equilibrium models are discussed. For discrete characters, threshold models involving underlying quantitative characters are of great interest. For this 'character uncoding problem' it may be necessary to use Markov Chain Monte Carlo methods.

Introduction

In spite of its title, the main subject of this paper will be to consider the use of quantitative characters in inference of phylogenies. Morphometrics can be viewed as a set of methods for extracting measurable traits from shapes. We come to morphometrics at the end, after first reviewing the way in which the resulting traits might be used. The great merit of morphometrics is that it automates the extraction of numerical measures from shapes, and thus presents evolutionary biologists with a torrent of quantitative characters, bringing the issues of how to treat them to the fore. In this chapter, I will use the term 'character' to refer to a feature of an organism, one that may assume a variety of 'states' or numerical values. In effect, a character is a column of the species × characters data matrix. Phylogenetic systematists often use the term 'character' differently: to refer only to the derived (apomorphic) states.

There have been represented at this symposium three main positions for how, and whether, quantitative characters may be used in inferring phylogenies:

Position 1 That they cannot be used. This view was represented in this symposium, though not in this volume. It holds that if the states of the character are not inherently discrete, they are too problematic to use to infer phylogenies. References to papers taking this position will be found in the paper in this volume by Humphries (2002).

Position 2 That they can be used, but only after being coded into discrete states by an appropriate method. Swiderski *et al.* (2002) exemplifies this view. Given this position, the solution to the 'character coding problem' becomes central to any use of quantitative characters.

Position 3 That they can used, without necessarily being transformed into discrete characters first. Quantitative statistical methods should be employed. This review will take this view, with some important exceptions. As we will see, this view is not without its difficulties.

Before phylogenetic systematics became widespread, quantitative characters were often used by systematists. Frequently such characters were first reduced to discrete states such as 'long' and 'short'. Their use was not placed in any statistical context. It should be self-evident that valid information was extracted in this way, as the phylogenies of the last 100 years have held up remarkably well. What could not be done when quantitative characters were used in this way was to place any statistical interpretation on the results. One could infer that one phylogeny was better than another, but better by how much was not obvious. Twelve years ago I reviewed many of these same issues (Felsenstein 1988). My conclusions have not changed substantially since, but, as they have not been accepted by most morphological systematists, insistent and peevish repetition is in order. I ended that review doubting whether systematists will typically have the information necessary to use quantitative characters in a statistical treatment of phylogenetic inference. It thus seemed likely that molecular sequences would bear the brunt of such inference. But given an inferred phylogeny, we could then make statistical inferences about the evolution of quantitative characters.

In the years since that review, the use of quantitative comparative methods has become widespread. Statistical treatment of quantitative characters has made few inroads in the inference of phylogenies, but phylogenies have popularized statistical inferences about the evolution of the characters. Phylogenies and quantitative characters are getting together, though with the conversation going more one way than the other.

Brownian motion and character correlation

Attempts to model statistically the inference of phylogenies from quantitative characters have taken the Brownian motion model as their base. This was introduced as a model of gene frequency change by Edwards and Cavalli-Sforza (1964) in their pathbreaking paper on statistical inference of phylogenies. I applied it to quantitative characters (Felsenstein 1973). Lande (1976) also used a Brownian motion model for character change in his work on long-term evolution.

Brownian motion has an expected mean change of zero, and a variance of change that increases linearly with time. At the level of population genetics, the variability may arise from two sources: genetic drift or variable natural selection.

Brownian motion, drift, and selection

A quantitative trait that has genetic variation controlled by a single locus will change as the gene frequencies at the locus undergo genetic drift. This process may be approximated by Brownian motion model. The approximation is imperfect, as the amount of change generated by Brownian motion is constant everywhere on the scale, while the amount generated by genetic drift becomes smaller as alleles near fixation. If the trait has additive genetic variance V_A, the variance of change due to genetic drift is V_A/N_e per generation. Interestingly, this relationship for one locus can be extended to a trait that is the sum of effects from n loci with the same result. Thus Brownian motion is a reasonable approximation to change of a quantitative character by genetic drift, provided that V_A remains approximately constant. Quantitative genetic models of change in selectively neutral alleles by genetic drift have been introduced by Chakraborty and Nei (1982) and Lynch and Hill (1986). In these models the additive genetic variance is depleted by fixation, but continually replenished by new neutral mutations.

A second source of change of varying direction is natural selection. In a simple model of natural selection the change of gene frequency is

$$\Delta p \cong sp(1-p) \qquad (3.1)$$

If in different generations the selection coefficient s varies, including variation in its sign, the result can be a random walk that is difficult to distinguish from genetic drift. Cavalli-Sforza and Edwards (1967) suggested that varying selection at a single locus could be approximated by Brownian motion. I have (Felsenstein 1973, 1981) extended this to quantitative characters controlled by multiple loci, and argued that varying selection might be an important source of stochastic change in quantitative characters, particularly when neutrality is unlikely.

Response to selection

One of the central formulas of quantitative genetics gives the expected selection response as the product of the heritability (h^2) and the selection differential:

$$R = h^2 S \qquad (3.2)$$

The selection differential is the difference in mean phenotype between the selected parents and the population from which they were drawn. For natural selection, Lande (1981) has given a version of this formula in which the expected response is the product of the additive genetic variance and the slope of the gradient of log fitness:

$$R = V_A \frac{\partial \log \bar{w}}{\partial \bar{x}} \qquad (3.3)$$

The gradient term is simply the derivative of the logarithm of mean fitness, the derivative being taken with respect to the mean phenotype. The expressions above give the expected selection response. The actual selection response will also have a term from genetic drift added to this, a term whose expectation is zero.

Character correlation

These formulas are for the case of a single character. In morphological analysis we will be much concerned with character correlation, and want to know how to treat multiple characters. There are versions of these formulas for multiple characters, with matrices replacing these scalar quantities. For example, in the analogue to Lande's formulation, the vector of change in p characters $\Delta \mathbf{z}$ is the product of a $p \times p$ matrix of genetic covariances (**A**) and a p-dimensional vector **b** of the gradient of log fitness with respect to the means of all p characters (Lande 1981), plus a vector of terms for genetic drift (**e**):

$$\Delta \mathbf{z} = \mathbf{A}(\mathbf{b} + \mathbf{e}) \tag{3.4}$$

Taking expectations over generations in a lineage we can compute the covariance of changes in the different characters through time. We will assume for simplicity that the expectation of **b** is zero, and we can make use of the fact that the genetic drift changes **e** have expectation zero and are uncorrelated with each other and with the changes in selection gradient. The expectation of the covariances of changes of characters over time is

$$E[\Delta \mathbf{z}(\Delta \mathbf{z})^T] = \mathbf{A}(E[\mathbf{b}\mathbf{b}^T] + \beta \mathbf{I})\mathbf{A}^T \tag{3.5}$$

The constant can easily be shown to be the inverse of the effective population size

$$\beta = \frac{1}{N_e} \tag{3.6}$$

The term $E[\mathbf{b}\mathbf{b}^T]$ is the covariance, across time, of the gradient of log fitness. We will call it **B**. Then

$$E[\Delta \mathbf{z}(\Delta \mathbf{z})^T] = \mathbf{A}(\mathbf{B} + \beta \mathbf{I})\mathbf{A}^T \tag{3.7}$$

The covariances between characters thus come from three sources: genetic drift (β), additive genetic covariances (**A**), and the covariances of the selective pressures (**B**). This last source of covariation will be the least familiar. Nevertheless, it is not new. Stebbins (1950) discussed *selective correlation*, a term that came from Tedin (1925). Even if characters have no genetic covariance, their change along a phylogenetic lineage can covary owing to the covariance of the selection pressures on them. Imagine a set of species, some of which enter arctic habitats. Suppose that there is no genetic covariance among body size, relative limb length, and darkness of coloration in a mammal. In accordance with Bergman's, Allen's, and Glogler's rules, natural selection may favor larger body size, smaller relative limb length, and darker coloration in arctic environments (as in Figure 3.1). Thus these characters will be expected to change in a correlated manner: in the absence of genetic covariance, there would be a selective covariance in their changes. In the above equations, this is given by the covariance matrix **B**, which can create covariances even when the genetic covariance matrix **A** is a diagonal matrix.

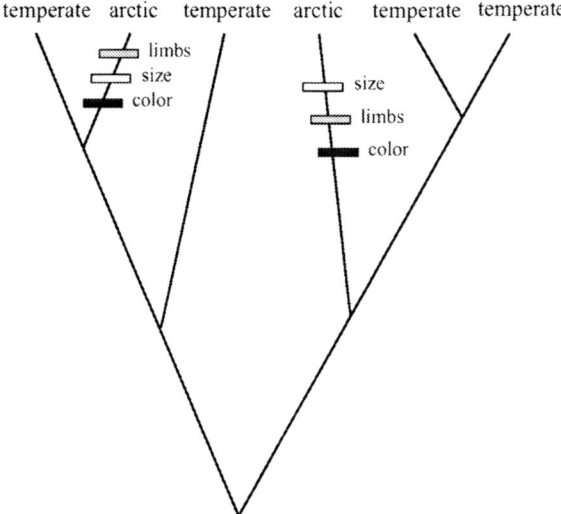

Figure 3.1 An example of selective correlation. Mammalian lineages enter arctic environments, leading to correlated changes in body size, relative limb length, and coloration.

The problem of estimation

Note that if we were to find a transformation that removed all additive genetic covariances, we would not remove all covariances between characters as long as there were also selective covariances. In order to make a statistical estimate of the phylogeny, we need to find a transformation that will remove the covariances of evolutionary change. We could then use the Brownian motion model to infer phylogenies. The difficulty lies in inferring the selective covariances. We can imagine doing, though perhaps with great effort, a quantitative genetic experiment to infer the additive genetic covariances in one or more species. We can hope that these additive genetic covariances stay roughly constant over a large enough span of time that we can use the results. But where are we to get an estimate of the selective covariances?

There are two possible sources:

- We may have paleontological data that follow a lineage through time, and enable us to infer the covariances of a set of characters through time. This does not give us a direct estimate of the selective covariances, but it does estimate the covariances of evolutionary change. If we also have an estimate of the additive genetic covariances, we can use Equation 3.7 to infer the selective covariances. Even if we do not have an estimate of the additive genetic covariances available, we at least then have an estimate of the covariances of evolutionary change, which is what we need to transform the characters to independence so that we can use the Brownian motion model.
- We can use molecular data to infer the phylogeny, and then observe the covariances of evolutionary change along that phylogeny. This is not done directly, as we cannot see the phenotypes of hypothetical ancestors. Instead we can use phylogenetic

comparative methods, which use the distribution of multiple characters on the tips of a known phylogeny to infer the covariances of evolutionary change (Felsenstein 1985; Harvey and Pagel 1991). Again, this does not give us the selective covariances directly.

Dilemmas and opportunities

Fossil and neontological data

The use of the comparative method (item 2 above) may seem beside the point: the objective is to infer the phylogeny, and we are assuming that we already have the phylogeny! But there are cases where we can make useful inferences. In particular, suppose that we have a group with both paleontological and neontological data. From the present-day species we infer a molecular phylogeny, and then use phylogenetic comparative methods to infer the covariances of evolutionary change of the quantitative characters. We then transform the characters to independence using those covariances. These new characters can be computed in both the living species and the fossils in the phylogeny. For each possible placement of the fossil species, the likelihood of the tree for the quantitative character data can be computed. The placement which maximizes this likelihood is to be preferred. This is in effect a Total Evidence approach (likelihood version), because the placement of the fossil species does not affect the likelihood of the tree on the molecular data. Taken together, the placement of all species by this method would maximize the overall likelihood, if we compute the overall likelihood as the product of the likelihoods of the molecular tree and the morphological tree.

This process is illustrated in Figure 3.2. In fact, only the part of the figure shown in bold lines is necessary, as hinted at by the double-headed arrow between the overall

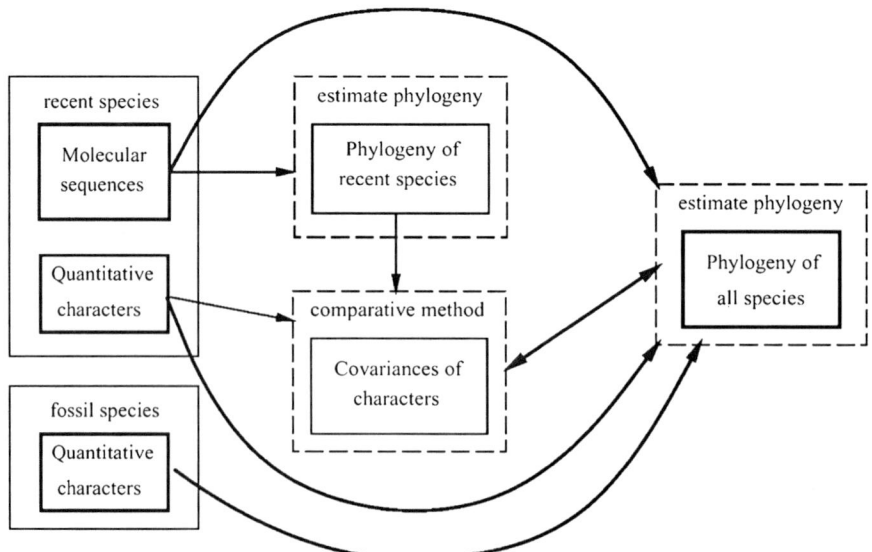

Figure 3.2 Flow chart showing the use of molecular phylogenies of present-day species to infer covariances of morphological characters, thereby allowing fossil data to be included.

phylogeny and the covariances. The two adjust to each other in light of all data. The lighter lines in the diagram show steps that may be useful to make preliminary estimates.

This approach can also be useful when we have two groups of present-day organisms, and have a molecular data set for one of them. If we are willing to assume that the morphological characters had the same covariances of evolutionary change in both groups, we could infer the phylogeny from the molecular data in one group, infer the character covariances in that group, and then use those covariances to infer the phylogeny in the other group. This too can be seen as a Total Evidence approach (likelihood version). Sometimes we may want to apply this method when there are not two distinct groups, but instead where there is only a phylogeny for some of the species in the group. If we had a phylogeny from which some species were omitted, it could be used to infer character covariances. Then the missing species could be placed from their morphological characters.

Do we need molecules?

In the preceding argument, molecular inferences provided information about part or all of the phylogeny. That information was needed to obtain the covariances needed to make use of the quantitative characters. One can have serious doubts as to whether quantitative characters could be used in the absence of molecular data. This would at first sight seem to back Position I – that quantitative characters cannot be used in the inference of phylogenies. But it does differ from that position in one important respect. Adherents of Position I typically deny that statistical inference approaches using quantitative character data are possible. I am concerned about circularity in the inference – it may not be possible to infer both the phylogeny and the character covariances. But given that independent information is available about the phylogeny, one can use comparative methods to infer the covariances. If we have both we can use them, together with the morphological characters, to infer both the phylogenies and the covariances. The morphological characters together with their statistical model will have an effect, however small, on the phylogeny.

This is a statistical analysis. As always, it is subject to worries about the correctness of the model. But if our interest is in the evolution of these particular characters (rather than in the phylogeny itself), this position is closer to Position III than to Position I. In many cases the quantitative characters are collected because they are of intrinsic interest to the biologist, rather than simply as markers for inferring the phylogeny. As molecular data become easier to obtain, they tend to displace quantitative character data from the job of inferring phylogenies, so that more and more of the use of quantitative characters will be motivated by interest in the evolution of those characters. There will be less and less use of quantitative characters as arbitrary markers for inferring phylogeny.

Allowing for uncertainty

Of course, molecular data do not provide us with a precise picture of the phylogeny. The issue arises as to how to incorporate into the analysis the uncertainty about the phylogeny. There seem to be two ways of doing this. The harder (but slightly superior) way (Felsenstein 1985) would be to combine the probabilistic model of change of the molecules with the Brownian motion model of the quantitative characters, allowing for the covariances of the latter. One could then compute a likelihood for all of the data, given both a tree and the covariances of evolutionary change of the quantitative characters. The collection of trees and covariances that were supported by the data would be those that had the highest likelihoods. If these did not have trees of different topologies, we could use asymptotic theory to choose the contour of the log-likelihood surface that defined the confidence interval – if there were n species and p characters it would be the 95 per cent value of a χ^2 distribution with $2n-3+p(p+1)/2-1$ degrees of freedom. This is the number of quantities (branch lengths and covariances less one for a scaling between them that is confounded) being estimated. The combination of tree and covariances that are acceptable can be based on the contours of the joint-likelihood curve for the covariances and the tree. For an oversimplified picture see Figure 3.3. The actual tree is not a single variable, and the character covariances are also multidimensional. This approach would seem to resolve the question of whether there is some circularity involved in using the same characters to determine the tree as are used in inferring covariances in character evolution.

We could imagine using the method to infer just the character covariances. In that case the confidence interval on the covariances would be defined by the degrees of freedom restricted by defining the covariances (in this case, $p(p+1)/2$). The set of trees and covariances that lies within the likelihood contour for one-half the significant value of a χ^2 variate with that number of degrees of freedom would be found, and then the trees ignored, leaving the set of covariances. Similarly a confidence interval on the tree could be inferred by doing this and ignoring the resulting covariances, using $2n-3$ as the degrees of freedom. More specific hypotheses about the character

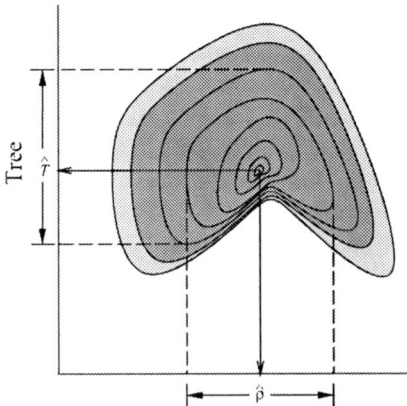

Correlation between characters 1, 2

Figure 3.3 Simultaneous inference of the tree and the character correlations when probabilistic models for both molecular and morphological characters are available. Point estimates of the tree and a correlation are shown, and approximate likelihood-based confidence intervals for the individual parameters can be based on the profile likelihoods (contours with dark shading and two-headed arrows) and joint confidence intervals based on the contours of the full likelihood curve (lighter shading).

covariances (such as that a covariance between two particular characters is zero) could be tested with even fewer degrees of freedom and consequently a tighter confidence interval. However at the moment none of this can be done, simply because present-day software is not designed for this task.

The other and simpler method is to estimate the tree solely from the molecular data. This gives us a slightly less precise estimate of the tree. However it is quite easy to allow for the uncertainty of the tree in inferring the covariances. I have pointed out (Felsenstein 1988) that for this one can use bootstrap sampling of the molecular sequences. For each bootstrap sample, one would infer the tree, and then use that tree to estimate the covariances of the quantitative characters. The resulting collection of estimates of the covariances would properly reflect the uncertainty about the tree. As the quantitative character data are derived from samples of individuals in each species, one could add another level of bootstrapping, resampling individuals within species each time. This would be unnecessary if the within-species covariances were allowed for inferring the phylogenetic covariances (Lynch 1991). Current versions of PHYLIP allow the bootstrapping of the molecular data to be carried out and the bootstrap sample estimates of the trees to be used to make multiple estimates of the covariances. Version 3.6 of PHYLIP will also allow for within-species components of variance (Lynch 1991) in inferring the covariances.

A way out?

One might wonder why we need to bother with the molecular data at all. Why not infer both the tree and the covariances from the same data set? One immediately wonders whether any such effort is totally circular. Interestingly, there is only a partial circularity, though it may be circular enough to make the whole effort mostly an academic exercise. We can get a good picture of this problem simply by counting degrees of freedom.

If there are n species and p characters, the data set has a total of np degrees of freedom. Of these, p are lost when we discard the means of the characters, leaving us with $p(n-1)$. There are $p(p+1)/2$ quantities to infer in the covariance matrix (the variances and the covariances). In the tree there are $2n-3$ branch lengths. However, we cannot use these quantities without taking into account that two of them are redundant. In particular, the total length of the tree is confounded with one of the parameters of the covariance matrix. If we double the length of the tree and halve all of the covariances, we leave the likelihood unchanged, since this leaves the covariances of the data unchanged. So we must remove one of the degrees of freedom.

This leaves us with a total of

$$p(n-1) - p(p+1)/2 - (2n-3) + 1 = np - 2n - \tfrac{1}{2}p^2 - \tfrac{3}{2}p + 4 \tag{3.8}$$

degrees of freedom. Simultaneous inference of the tree and the covariance matrix will be possible when this number is positive. We can separate the terms in n and p to get a condition for simultaneous inference (assuming $p > 2$):

$$n > \frac{\tfrac{1}{2}p^2 + \tfrac{3}{2}p - 4}{p - 2} \tag{3.9}$$

Table 3.1 shows the upper limit of the number of characters that satisfies this condition, for some values of n: Below 6 species, there is no whole number of characters that satisfies the conditions. As the number of species rises, the lower limit on characters is just above 2, and the upper limit can be shown to remain just below $2n - 5$. One might wonder whether this is worth the effort. Given this upper limit on the number of characters, the inference of the tree cannot be made precise by increasing the number of characters without limit (I am indebted to Andrew Rambaut and Michael Charleston for pointing this out to me). On the other hand, one can make the inference of the covariances more and more precise by increasing the number of species sampled. This holds out some hope for the analysis of characters, but not much for the inference of the phylogeny. Even if we are willing to concentrate on the characters instead of the phylogeny, there is a limit to how many species we can find in the relevant group – it may be far easier to find new characters than new species.

With three species, there is no possibility of inferring both the phylogeny and the character covariances. It was this case that persuaded me (Felsenstein 1988) that the two were inextricably confounded and that any attempt to infer them separately was hopeless. As we can see, this was not entirely true. They can be separated in principle, but the prospects for making practical use of this are not very encouraging.

Table 3.1 For different numbers of species, the smallest and the largest number of characters for which there are enough degrees of freedom to simultaneously infer the tree and the covariances. Values obtained by solving the quadratic condition in p in Equation 3.9 are shown

Species	Characters greater than	Characters less than
6	2.43	6.56
7	2.30	8.70
8	2.23	10.77
9	2.18	12.82
10	2.16	14.84
11	2.13	16.87
12	2.12	18.88
13	2.11	20.89
14	2.10	22.90
15	2.09	24.91
16	2.08	26.92
17	2.07	28.93
18	2.07	30.93
19	2.06	32.94
20	2.06	34.94
25	2.05	44.95
30	2.04	54.96
35	2.03	64.97
40	2.03	74.97
50	2.02	84.98
100	2.01	194.99

Genomics to the rescue?

Ahead lies the terra incognita of genomics. Though difficult and expensive now, it is clear that in a decade it will be relatively easy to do genomics on characters of interest. We could find the loci that make the largest contribution to genetic variation of the characters within populations and, if we can cross individuals from different populations, also find the quantitative trait loci (QTLs) that make the largest contributions to differences between populations, and perhaps differences between species.

To the extent that we can do this, we transform the data into QTL gene frequencies in different populations. However, we can find only the loci of largest effect, leaving behind a residuum of polygenic variation at 'background' loci. Thus, until that residuum becomes small enough to be insignificant, quantitative genetic models will be useful. The transition from polygenic models to models that have known loci will be gradual. In general, to detect a locus with half the effect, we must quadruple the sample size.

In some cases the inability to detect loci of small effect may not be a serious problem. If the divergence of the loci were due primarily to natural selection, most of that divergence would be reflected in the gene frequencies of the loci of largest effect. In simple forms of selection (e.g., directional selection), changes in gene frequencies are proportional to the sizes of the genetic effects at the loci. A locus whose genetic variants have twice the effect of those at another locus will thereby accumulate genetic differences that are twice as large. That in turn means that the phenotypic differences caused by those loci will be four times as great, since both the genetic effects and the gene frequency differences are twice as great. There is thus some prospect that the availability of genomics will rapidly illuminate cases where the differences are caused by natural selection, by detecting loci of large effect, which may be responsible for most of the differences.

No one has yet thought through how we can use QTL data, possibly in combination with a polygenic model for residual genetic variation, to infer phylogenies and to illuminate character covariation. The time for doing so is approaching. As I have suggested elsewhere (Felsenstein 2000), genomic data do promise insights on whether natural selection has acted on the characters under study or on unobserved characters correlated with them. Given the possibility of escaping some of the constraints that have plagued analysis of morphology, genomic data seem worth investigating.

Chasing peaks

We have modeled natural selection as acting in randomly varying directions in different lineages. It is not self-evident that natural selection will vary randomly in direction from moment to moment. A more convincing model would be natural selection towards an optimum (cf. Lande 1976). Some of the possible variants on this model would be:

- A single optimum stays in one place with all species attracted to it. The species wander by random genetic drift (Lande 1976; Hansen and Martins 1996).
- Different species have different optima, the optima separating at the time of speciation. Each optimum wanders independently in the space, perhaps by Brownian motion (Felsenstein 1988; Hansen and Martins 1996).

- Different species have different optima, the optima separating at speciation. The optima wander, but their positions are constrained so that they describe an Ornstein–Uhlenbeck process (random walk of an elastically bound particle) around a common point (Felsenstein 1988).
- Perhaps more realistically, each species has a different optimum, the optima separating at speciation, but optima of recently diverged species wander in a correlated fashion, the correlation declining the longer they are diverged.

A full treatment of the movement of a quantitative character under any of these models is difficult, but it is greatly simplified the longer a population remains under the influence of a peak. It is not hard to show that if a population is following a peak which is itself undergoing Brownian motion or the Ornstein–Uhlenbeck process, its distance from the peak settles down into a normal distribution with constant variance. In effect the population mean is towed along by the peak, but at the end of a somewhat flexible cable. The farther the peak wanders the more of the change of the character must be attributed to the movement of the peak and the less of it is accounted for by the cable.

If selection moves the population (say) 10 per cent of the way toward the peak each generation, then the departure of the population from the peak will represent events that have occurred in roughly the last $1/0.10 = 10$ generations. If each lineage lasts much longer than that, and if genetic drift during the 10 generations is much smaller than the net movement of the peak over its existence, then the mean of the quantitative character is basically going where the peak goes.

Figure 3.4 shows a numerical example from a computer simulation of two characters (not all details of which are described here). The two characters are negatively genetically correlated, with a correlation coefficient of -0.9. They wander by genetic drift about a peak which is itself moving. In the leftmost panel little time has elapsed; the peak has not moved much and the two characters show the negative correlation that is a consequence of their genetic covariation. The peak wanders with positive correlation between the two characters. Changes in the position of the optimum in one character have a correlation of 0.9 with the changes in the other character. Thus the genetic covariation 'wants' the characters to be negatively correlated, while selective

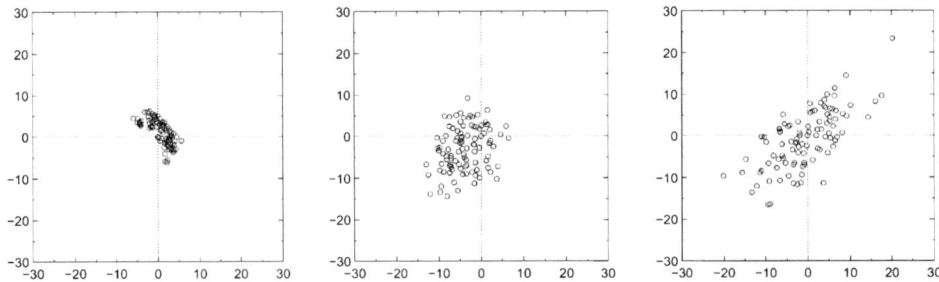

Figure 3.4 Covariation of two characters when genetic covariation between them is -0.9 but when they are attracted to optimum values that vary through time with a covariance of 0.9 in movements of the optima of the characters. For short periods of elapsed time the phenotypic covariation is negative, but as we wait 10 and 100 times longer, the movements of the optima tow the character values in positively correlated ways.

covariation wants them to be positively correlated. In the center panel of the figure, 10 times as much time has elapsed and there has been some wandering of the peaks. This smears out the distribution of character values from lower-left to upper-right, resulting in a roughly circular distribution. In the rightmost panel we see the distribution over 100 times as much time as in the first panel (10 times as much as the center panel). Now the peak movement is the dominant influence, and the characters show a strong positive correlation.

This is reason to expect that selective covariances will be important – the covariation of character change will then mostly be a matter of the covariation of peak movements with respect to different characters. For example, provided selection favors large size and also tends to favor dark coloration in the same lineages, then there will be a correlated distribution of these characters that will override any genetic correlation.

Punctuational models

In the models discussed here, it has been assumed that quantitative characters change continually along a branch of the tree. Under a punctuated equilibrium model, they would instead be expected to change mostly at the time a branch originates, and be static thereafter. If there were a burst of change (of roughly equal size) at the start of each branch, and no change thereafter, we might think that this could be accommodated by having the expected variance accumulated in each branch be equal. The tree would then consist of a series of branches, each of unit length. Hansen and Martins (1996) have made calculations along these lines (see also Felsenstein 1988). If this were all that we needed to take into account, it would be straightforward to analyze data under the assumption of punctuation (though there would be the issue of which branch at each fork was the newly-originated one).

The difficulty with this tempting model is that we do not see all branches. Even if we can collect all extant species, there should be many forks at which the new species has persisted while the parent species has died out. That would show up in our tree as a burst of change in the middle of a branch. Branches that had undergone more of these bursts of change would be longer, so that not all branches would be of unit length. In addition to species that have become extinct, we may be omitting some extant species from our data set. If there are 200 beetles in our group, but we analyze only a capriciously-chosen sample of 40, there will be many places where a fork gave rise to one of our sampled species, with the parent species being the ancestor of ones we have omitted. This will create additional uncertainties about the branch lengths on the tree.

In short, a punctuational model may be harder to distinguish from a gradualist model than first appears. There is hope for doing so if many characters are analyzed, as under the assumption of punctuated equilibrium the parent species should not change while the daughter species changes in many characters. But the analysis is complex, and needs much further examination.

The character coding problem

Many analyses of quantitative characters first reduce them to discrete characters. This is known as the 'character coding problem', and a variety of methods have been suggested for recoding the characters. Sometimes this is done under the assumption that parsimony methods require discrete states. Most parsimony programs do have such a requirement, though in the early years of the parsimony literature methods were put forward that use the original quantitative scale (Farris 1970).

We might also want to recode the quantitative characters into discrete states if we believed that the continuous scale masked regions that had widely varying properties. For example, if a character can rather easily wander between values 4 and 10, and can also wander easily between 1 and 3, but has great difficulty changing from a value of 3 to a value of 4, we might want to approximate this by having two discrete states, one consisting of all values below 3.5, the other of all values above 3.5. If the change between these two ranges is sufficiently improbable, we want to weight it heavily. We would be losing some information by not distinguishing between values of (say) 6 and 10, but we would be gaining some information by taking into account the greater difficulty of change in certain regions of the scale.

I believe that many of the character coding methods, such as gap coding (Mickevich and Johnson 1976; see also Simon 1983 and Archie 1985) are implicitly trying to take account of situations like this, using the empirical distribution of character values among species as an indication of where the regions of difficult change are located. There are complications owing to the fact that species are not drawn independently from a distribution, but arise on a phylogeny in a highly clustered fashion. Thus, a gap in the distribution along the character scale may reflect, not a region which is rarely occupied, but the distinction between two clades. There is in addition the question of why coding is taking place one character at a time, when evolution at different characters may be correlated. These issues have never been given the serious statistical examination they deserve.

Given that there are ways to analyze quantitative characters on quantitative scales, there is no compelling reason to engage in character coding. Until we have a well-thought-out method for detecting regions of scales that ought to be treated differently, perhaps the best advice about character coding is to just say no.

The character uncoding problem

In fact, one may want to do the opposite. It is possible for discrete characters to mask an underlying continuous scale. The threshold model of evolution has been around since the work of Wright (1934) on digit number in guinea pigs. It has been applied to human genetics by Falconer (1965). This model imagines an invisible underlying character (usually called 'liability') and a threshold value. The discrete trait results from a developmental system that monitors whether the liability exceeds the threshold value. The liability has the usual quantitative genetics. This class of models has some attractive features. We may compare it to a simple alternative, a simple Markov chain that alternates between two states, 0 and 1 (cf. Pagel 1994). In the threshold model, once a lineage changes from being largely of state 0 to being largely of state 1, its underlying liability is probably near the threshold. The longer the time that a lineage has remained in state 1, the farther the liability may have wandered beyond the threshold, and the less likely an immediate return to state 0. The simple Markov process model, by comparison, has the same probability of returning to state 0 however long it has been in state 1.

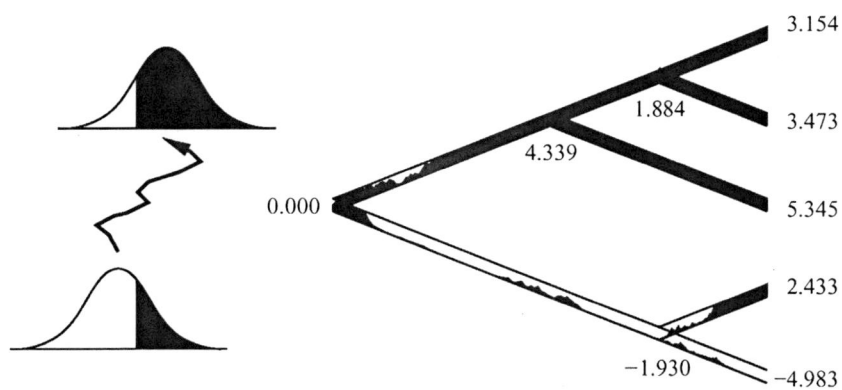

Figure 3.5 The threshold model, showing the role of the threshold and the underlying (unobserved) liability character, and the result of the simulation of the change of a threshold character along a simple phylogeny. The value of the underlying liability character is shown next to each node in the tree, and the shading in each branch shows the proportion of that population which has state 1.

Figure 3.5 shows a depiction of the threshold model and a simulation of the change of a discrete character along a tree. Note that the threshold model has one other advantage over the simple Markov chain. It does not actually envisage a lineage changing instantaneously from one state to another. At any time, the lineage has both states present in it, their proportions depending on where the threshold value lies in the distribution of the liability character. In many cases almost all of the phenotypes in the population will be the same, but as the mean of the liability crosses the threshold, there will be a period of polymorphism. This can be seen in the simulated tree.

The difficulty with the threshold model is its mathematical intractability. To compute the likelihood of a discrete character on a phylogeny, we would have to compute the probability that each individual lies above or below the threshold (depending on its observed phenotype). The probability density of the liabilities is a multivariate normal distribution, but the joint probability of the discrete phenotypes computes a corner of this distribution:

$$\text{Prob}[1,1,0,1,1,0,0]$$
$$= \text{Prob}[x_1 > c, x_2 > c, x_3 < c, x_4 > c, x_5 > c, x_6 < c, x_7 < c]$$
$$= \int_c^\infty \int_c^\infty \int_{-\infty}^c \int_c^\infty \int_c^\infty \int_{-\infty}^c \int_{-\infty}^c \text{Prob}[x_1, x_2, x_3, x_4, x_5, x_6] \, dx_1 \, dx_2 \, dx_3 \, dx_4 \, dx_5 \, dx_6.$$

(3.10)

Integrals of corners of normal distributions are hard to compute. It appears likely that they will yield only to Markov Chain Monte Carlo methods. These may make use of threshold models practical. This is in effect the 'character uncoding problem', and it seems more likely to be of interest than the character coding problem.

Morphometrics at last

At the end, let us come full circle, back to morphometrics. Given all of this, where does it leave morphometrics? Morphometrics is a source of numeric characters. Morphometricians point out that it is much more than just another source of them, that it places individuals in a morphometric space that has particular desirable properties. Other numerical methods may choose coordinates that lead to absurd results when one extrapolates, or lead to misleading covariation when there is measurement error. For the present discussion these distinctions are not important – we could as well be discussing any source of numeric characters.

Of the three positions on the use of quantitative characters in inferring phylogenies, Position I (that they cannot be used) would certainly lead to a lack of interest in using morphometrics. It might possibly be argued that this does not preclude the use of morphometrics retrospectively, using phylogenies to analyze the change of morphometric parameters. However that would require us to accept some model of change of these quantitative characters. If there were such a model possible, one could think of using it to infer phylogenies. Most practitioners of Position I do not believe that any such model is worth serious consideration.

Position II – that we can use quantitative characters only if discretely coded – leads to an interest in deriving discrete characters from morphometric parameters. Zelditch *et al.* (1995) have developed methods for doing so, and this has led to some controversy (for debate and earlier references see Rohlf 1998 and Zelditch *et al.* 1998). It becomes important to have the correct coding and the character coding problem becomes paramount.

Position III requires that we not only be able to derive numerical measurements from morphometric data, but that we ask about their genetic and selective correlations. Most of the morphometric literature has asked what parameterizations are best justified on geometric or mathematical grounds. Genetic correlations include developmental correlations. Asking about them should lead us toward a genetic and developmental morphometrics rather than a geometric morphometrics (Felsenstein 1992). As long as we do not have developmental models, we cannot construct developmental morphometrics from them. When they become available, they will lead to insights into the expected genetic correlations of morphometric parameters.

In inferring developmental models, we may be able to take the reverse route. Morphometric analyses along phylogenies may lead to insights into the genetic correlations, and thus may be a major source of insight into developmental models. The 'evo-devo' literature has yet to mine this lode. To do so will require the quantitative models of morphometrics, but also require us to relinquish a purely geometric approach.

The position taken in this essay has elements of similarity not only to Position III, but also to Position I. It argues that we typically do not have evidence as to the selective correlations, and often not for the genetic correlations either. Thus, most use of quantitative characters will be retrospective. However when this is possible, and when genetic correlations or developmental models are available, it should allow us to make interesting inferences about the selection pressures. We will then be making progress toward a functional morphometrics, even an ecological or behavioral morphometrics.

If the genetical and/or developmental models are known, and the phylogenetic distribution also, we could make inferences about how selection is acting on the characters. Alternatively, if ecological information about selection is available, and also phylogenetic distribution, we might hope to infer genetic correlations and discriminate among developmental models. We can hope that the era of geometric morphometrics will be followed by an era in which developmental morphometrics exists in dynamic interaction with functional morphometrics, the interaction being mediated by modeling change of quantitative characters across phylogenies.

Acknowledgements

This paper has been supported by NSF grants DEB-9815650 and BIR-9527687 with additional support from NIH grants R01 GM51929 and R01 HG01989. I am grateful to Michael Charleston and Andrew Rambaut for pointing out the lack of consistency of tree estimates when done jointly with inference of character covariances, and I am indebted to W. Scott Armbruster for introducing me to G. L. Stebbins' use of the term selective correlation.

References

Archie, J. W. (1985) 'Methods for coding variable morphological features for numerical taxonomic analysis', *Systematic Zoology*, 34, 326–345.

Cavalli-Sforza, L. L. and Edwards, A. W. F. (1967) 'Phylogenetic analysis: models and estimation procedures', *Evolution*, 32, 550–570.

Chakraborty, R. and Nei, M. (1982) 'Genetic differentiation of quantitative characters under optimizing selection, mutation, and drift', *Genetical Research*, 39, 303–314.

Edwards, A. W. F. and Cavalli-Sforza, L. L. (1964) 'Reconstruction of evolutionary trees', in Heywood, V. H. and McNeill, J. (eds) *Phenetic and phylogenetic classification*, London: Systematics Association Publication No. 6, pp. 67–76.

Edwards, A. W. F. (1970) 'Estimation of the branch points of a branching diffusion process', *Journal of the Royal Statistical Society, Series B*, 32, 155–174.

Falconer, D. S. (1965) 'The inheritance of liability to certain diseases, estimated from the incidence among relatives', *Annals of Human Genetics*, 29, 51–76.

Farris, J. S. (1970) 'Methods for computing Wagner trees', *Systematic Zoology*, 19, 83–92.

Felsenstein, J. (1973) 'Maximum likelihood estimation of evolutionary trees from continuous characters', *American Journal of Human Genetics*, 25, 471–492.

Felsenstein, J. (1981) 'Evolutionary trees from gene frequencies and quantitative characters: finding maximum likelihood estimates', *Evolution*, 35, 1229–1242.

Felsenstein, J. (1985) 'Phylogenies and the comparative method', *American Naturalist*, 125, 1–15.

Felsenstein, J. (1988) 'Phylogenies and quantitative characters', *Annual Review of Ecology and Systematics*, 19, 445–471.

Felsenstein, J. (1992) 'Review of Proceedings of the Michigan Morphometrics Workshop', *Quarterly Review of Biology*, 67, 418–419.

Felsenstein, J. (2000) 'From population genetics to evolutionary genetics: a view through the trees', in Singh, R. S. and Krimbas, C. B. (eds) *Evolutionary genetics: from molecules to morphology*, Cambridge: Cambridge University Press, pp. 609–627.

Hansen, T. F. and Martins, E. P. (1996) 'Translating between microevolutionary process and macroevolutionary patterns: a general model of the correlation structure of interspecific data', *Evolution*, 50, 1404–1417.

Harvey, P. H. and Pagel, M. D. (1991) *The comparative method in evolutionary biology*, Oxford: Oxford University Press.

Humphries, C. J. (2002) 'Homology, characters and continuous variables', in MacLeod, N. and Forey, F. (eds) *Morphology, shape and phylogeny*, London: Taylor and Francis, pp. 8–26.

Lande, R. (1976) 'Natural selection and random genetic drift in phenotypic evolution', *Evolution*, 30, 314–334.

Lande, R. (1981) 'Quantitative genetic analysis of multivariate evolution, applied to brain:body size allometry', *Evolution*, 33, 402–416.

Lynch, M. and Hill, W. G. (1986) 'Phenotypic evolution by neutral mutation', *Evolution*, 40, 915–935.

Lynch, M. (1991) 'Methods for the analysis of comparative data in evolutionary biology', *Evolution*, 45, 1065–1080.

Mickevich, M. F. and Johnson, M. S. (1976) 'Congruence between morphological and allozyme data in evolutionary inference', *Systematic Zoology*, 25, 260–270.

Pagel, M. (1994) 'Detecting correlated evolution on phylogenies: a general method for the comparative analysis of discrete characters', *Proceedings of the Royal Society of London Series B Biological Sciences*, 255, 37–45.

Rohlf, F. J. (1998) 'On applications of geometric morphometrics to studies of ontogeny and phylogeny', *Systematic Biology*, 47, 147–158.

Simon, C. M. (1983) 'A new coding procedure for morphometric data with an example from periodical cicada wings', in Felsenstein, J. (ed.) *Numerical taxonomy*, NATO Advanced Science Institutes Series G, No. 1, New York: Springer-Verlag, pp. 378–382.

Stebbins, G. L. (1950) *Variation and evolution in plants*, New York: Columbia University Press.

Swiderski, D. L., Zelditch, M. L. and Fink, W. L. (2002) 'Comparability, morphometrics and phylogenetic systematics', in MacLeod, N. and Forey, F. (eds) *Morphology, shape and phylogeny*, London: Taylor and Francis, pp. 67–99.

Tedin, O. (1925) 'Vererbung, variation und systematik in der gattung camelina', *Hereditas*, 6, 275–386.

Wright, S. (1934) 'An analysis of variability in number of digits in an inbred strain of guinea pigs', *Genetics*, 19, 506–536.

Zelditch, M. L., Fink, W. L. and Swiderski, D. L. (1995) 'Morphometrics, homology, and phylogenetics – quantified characters as synapomorphies', *Systematic Biology*, 44, 179–189.

Zelditch, M. L., Fink, W. L., Swiderski, D. L. and Lundrigan, B. L. (1998) 'On applications of geometric morphometrics to studies of ontogeny and phylogeny', *Systematic Biology*, 47, 159–167.

Chapter 4

Scaling, polymorphism and cladistic analysis

Todd C. Rae

ABSTRACT

The uneasy nature of the relationship between morphometrics and phylogenetic systematics is traceable to the differences in the theoretical perspectives of the two schools. Morphometric techniques can be used in a cladistic framework only if they do not violate certain assumptions. If we accept the use of metric data in cladistic analysis, some compensation must be made for scaling. Two popular solutions to this impasse, residuals and centroids, violate the fundamental cladistic data requirement of independence. Ratios derived for each individual, however, allow metric data to be scaled for size without reference to other taxa/organsims. Metric coding may also allow a more robust treatment of species polymorphisms. Modern cladistic algorithms treat polymorphic taxa as higher taxa, making such methods inappropriate for species-level analysis. The use of metric coding methods is more appropriate than assigning such taxa either missing values or a presumed primitive condition, and may allow a logically consistent method for the incorporation of molecular sequence variability in cladistic analysis.

Introduction

The purpose of the present volume is to outline the areas of agreement (or lack thereof) between practitioners of morphometric analysis and cladistic analysis and, perhaps, to suggest areas of overlap that need additional attention. Any serious hope for a rapprochement between the two disciplines relies on the explicit recognition of the practical aims of these methods of analysis and the theoretical foundations upon which they rest. One can inform and/or be incorporated into the other *if and only if* their respective underlying assumptions are not violated. Given this caveat, there are several conceivable schemata for the use of morphometric data and techniques in cladistic analysis, some of which are discussed below. In particular, two perennial data problems in phylogenetics, scaling and polymorphism, are examined; each, in its own way, embodies the ways that morphometrics and phylogeny are related.

The discussion that follows takes its philosophical basis from the perspective of a cladist anxious to introduce the rigour and repeatability of quantitative analysis into the study of phylogenetic relationships between species. Any unintended bias can be attributed to this particular viewpoint.

Methods of analysis

Morphometric relationships and cladistic relationships are not necessarily the same, or even similar. This is not surprising, given the difference in the theoretical basis of these methods. Although both schools of analysis derive relationships among organisms by the simultaneous analysis of multiple characters, at their core morphometrics and cladistics address different sets of questions; the former is concerned with distances between individuals in multidimensional morphospace, while the latter attempts to reconstruct the phylogenetic relationships of taxa. Because these are fundamentally different ways of approaching the analysis of organisms, and because the relationships they describe are not the same at the level of theory, the areas of overlap between the logical domains of these two systems are unclear at present.

At the heart of phylogenetic systematics is the question of whether species are different from one another for a given characteristic. The character state codes assigned to taxa before parsimony analysis are reflections of a series of decisions as to whether or not the taxa are deemed to be 'the same' for a particular character (Rae 1998). This type of partitioning is necessarily binary; taxa are either the same or they are different. The binary nature of cladistic coding holds true regardless of whether the original data are discrete (e.g., presence/absence) or continuous (e.g., metric), or whether they incorporate two or more character states.

As has been argued elsewhere (Rae 1998), assigning codes to taxa on the basis of metric data is not isomorphic with the arbitrary division of a continuum. The idea that the values of all individuals in an analysis form a continuum is based on a confusion of the level of analysis; *species* can be shown to differ from one another for a certain trait regardless of any overlap in the values of the *individuals* of those species (Thiele 1993). Cladistics ultimately operates at the level of genetic isolation (i.e., the species; Hennig 1966), and therefore metric data form no multi-taxic continuum. Each species can be characterised in terms of measurement data, and these data are analysed to determine if taxa are the same or different. Morphometric analysis, on the other hand, is concerned primarily with distances between individuals in morphospace. The simultaneous analysis of multiple characters is analogous to parsimony analysis of cladistic data, but operates within a theoretical multidimensional space, rather than the historically based phylogenetic system. Even given this fundamental distinction between the aims, there may still be some areas of overlap between the two disciplines. In particular, some morphometric techniques may be employed to code metric data for cladistic analysis (see Chapter 7).

The logical consequence of the observations outlined above is that morphometric techniques must fulfill two conditions to be applicable to the study of phylogenetic systematics. First, these techniques must be applied to species, not individuals. As Hennig (1966) emphasised, tokogenetic relationships (i.e., those between individuals) cannot be elucidated by cladistic analysis. As a result, traditional multivariate morphometrics that operate at the level of the individual cannot be used for coding data for cladistics. Second, any attempt to utilise morphometric techniques at the level of species must do so in such a way that researchers can determine whether taxa are the same or different; that is, it must be possible to develop hypotheses of primary homology between species. Several promising techniques have been proposed (e.g., Chapter 8), but will have to go further to demonstrate compliance with the necessary criteria of cladistics.

More commonly used univariate metric techniques are applicable to cladistics, provided that certain conditions are fulfilled. Where those prerequisites are not met, the methods must be deemed inappropriate. For example, gap coding and segment coding both rely on arbitrary distinctions, to a greater or lesser extent, which causes difficulties for their use in phylogenetic analysis (see Chapter 5). In these methods, the codes assigned to taxa are entirely dependent on a form of subjective choice, as there is no external criterion from which to derive the critical values for gaps/segments (Rae 1998). As Kluge and Farris (1969: 189–190) state, 'to achieve an accurate estimate of the real relationships of organisms, ... we must carefully select our procedures in such a way that personal bias has little chance of influencing the outcome of the analysis.' It is precisely the same lack of theoretical justification for any particular method, and the disparate results that different methods produce, that resulted in the rejection of phenetic approaches to phylogenetics (Wiley 1981).

Homogeneous subset coding, or HSC (Simon 1983), on the other hand, is one method that produces non-arbitrary, repeatable discrimination between taxa, allowing metric data to be utilised in phylogenetic systematics (Rae 1998). The technique operates by performing a statistical multiple comparisons test on the sample taxa and grouping those taxa that show identical distributions of significant differences to other taxa for the character in question (for examples, see Rae 1993). In this way, non-arbitrary codes are produced, based on standard statistical tests and the results violate none of the explicit preconditions of phylogenetic systematics. This procedure is similar to that used in divergence coding (Thorpe 1984), with the exception that HSC does not simultaneously weight character states. Another non-arbitrary coding method, one particularly suited for large data sets, is finite mixture coding (Strait *et al.* 1996), although there may be other theoretical difficulties associated with this method (Rae 1998). The remainder of the present contribution is concerned with corollaries of the conclusion that non-arbitrary coding methods allow the use of metric characteristics in cladistic analysis.

Scaling

Because '(s)ize ... and the effects of differences in size or scale appear to be inextricably linked to almost every aspect of ... biology' (Jungers 1985: x), raw metric data are often found to be significantly correlated with the size of the organisms under investigation. This correlation can introduce a confounding factor in cladistic analysis, as characters are required to be independent of one another. The problem of scaling metric data for phylogenetic analysis is illustrated by an example from cranial pneumatisation in primates.

The volume of the maxillary sinus, an air cell found lateral of the nasal cavity in most eutherian mammals (Novacek 1993), varies dramatically among anthropoid primates (Cave and Haines 1940), the group that contains monkeys and apes (including humans). Several analyses of ape phylogeny (e.g., Andrews and Martin 1987) have included maxillary sinus size as a character distinguishing various subgroups, depending on the author, each with a unique number/assignment of character states (Rae 1999a). These character state determinations have been made without reference to scaling.

More recently, the advent of computer tomography (CT) imaging and computer assisted three-dimensional virtual reconstructions have allowed accurate quantitative assessment of the volumes of internal structures for the first time (Koppe *et al.* 1996). Raw data from examinations of extant hominoids appear to support the interpretation that several character state changes have occurred in the evolution of the Hominoidea. Scaling these data, however, reveals a different story. Among extant hominoids, sinus volume is significantly correlated with cranial size (measured by the traditional univariate scalar of basicranial length and three-dimensional facial volumes) and the association is both very strong ($r = 0.89, p < 0.01$) and isometric (Rae and Koppe 2000). Thus, scaling analysis suggests that *no* change in maxillary sinus volume (independent of cranial size) has occurred in ape evolution. This example highlights the necessity of ensuring that relative measures of traits are used in cladistic analysis.

How this scaling is to be achieved, however, is another matter. Two popular methods of scaling metric data in morphometrics, residuals and centroids, have properties that may exclude them for use in cladistic analysis. The practices of scaling via residuals and centroids are identical in the sense that the denominator is derived from some experiment-wide correction factor. In the case of residuals, the element of subtraction is derived from the regression line, inferred from all individuals of all sample taxa.

Thus, the value assigned to any given *individual* is dependent on which taxa, and which individuals from those taxa, are included in the analysis. Change either the taxa or the samples within taxa and the scaled values of all individuals will be different, since the slope and/or position of the regression line, from which the scaling is derived, will not necessarily be the same. In this way, data thus obtained are not independent and therefore violate one of the fundamental assumptions of phylogenetic systematics. Centroids are similarly culpable, as these 'average' values again are often inferred from all individuals sampled, making the value of each individual a function of the measurements of other organisms.

This is not to say simply that sample-dependent methods are inappropriate for phylogenetic systematics. All morphological studies must rely on sampling taxa and individuals. What is questionable in residual/centroid scaling is the interdependence between taxa and individuals, which transgresses the necessary prerequisite for independent data in cladistics. Taxa (or individuals) cannot be considered independent if their characterisation (i.e., the actual values for a particular attribute) can be shown to change *solely* in response to the presence or absence of other taxa (or individuals) in the study.

These transgressions of the underpinning of cladistic analysis are not trivial; interdependence of data can introduce serious bias into resulting topologies. It is precisely this sort of bias that is probably to blame for the failure of cladograms produced from a non-independent morphometric data set to match those derived from molecular data[1] (Collard and Wood 2000: 5003), leading the authors to conclude that 'little confidence

---

1 The report of Collard and Wood (2000) strongly echoes a previous study (Hartman 1988), in which parsimony analysis of 102 measurements of ape molar teeth similarly failed to reproduce a molecular phylogeny. Most of the morphological characters used in the latter paper, however, were strongly influenced by one factor (enamel thickness), resulting in a topology that was, in effect, a single trait phylogeny.

can be placed in phylogenies generated solely from higher primate craniodental evidence'. This (erroneous) claim has been partly responsible for the recent call for a fundamental reassessment of the relevance of morphological data to systematics (Gura 2000).

Another form of ratio, those derived for each individual independently, avoids the problem of non-independence. For example, using some form of the geometric mean (an average value of all measurements of an individual) as the scaling value for the metric characters of a single organism (Jungers *et al.* 1995) enables the use of a value that has been 'corrected' for the size of the individual in a way that is not derived from measurements of other organisms. In this way, the scaled value for any character of a particular organism remains the same regardless of the other individuals or taxa in the analysis, which means it can be analysed cladistically. Ratios may have certain undesirable statistical properties (Atchley *et al.* 1976), but these often can be overcome by any number of transformations (Hills 1978). The practical shortcomings in no way detract from the theoretical applicability of scaled ratio data for cladistics. Techniques similar to these have been used with some success by the author (Rae 1997, 1999b) in testing various hypotheses of the phylogenetic position of fossil primate taxa. The acceptance of the use of metric characteristics in cladistics not only introduces an element of rigour to character analysis and expands the number of potential traits to be analysed, it also may allow the successful incorporation of polymorphic data in matrices subjected to parsimony analysis.

Polymorphism

Although the presence of multiple character states in a single species is, (a) a well-documented biological phenomenon, (b) one of the three necessary conditions of natural selection, and (c) in the case of peppered moth colour polymorphism, the most oft-cited textbook example of change in a species over time, it remains a difficult aspect of taxa to analyse cladistically. In fact, parsimony analysis computer programs are incapable of resolving within-species polymorphism when coded as such (see below). Excluding these taxa and/or data, however, has been shown to seriously affect resulting topologies (Wiens and Servedio 1997).

Taxonomic polymorphism, where the presence of multiple character states is attributable to the presence of multiple taxa within a single terminal (Nixon and Davis 1991), can be resolved, both in theory and practice, using current parsimony algorithms (Simmons 1993). The analysis of intra-specific polymorphism, however, is beyond the theoretical scope of parsimony implementations; for example, PAUP treats all polymorphisms coded as such (e.g., 1/2) as if they were taxonomic (Swofford and Begle 1993: 95). Thus, these programs will produce an answer, even though the data provided do not satisfy a necessary theoretical criterion. This is analogous to performing a *t*-test when the samples possess significantly heterogeneous variances; an answer can be produced, but it is meaningless because the data violate a crucial assumption of the analysis (Sokal and Rohlf 1981).

Several methods have been mooted to solve this deficiency in cladistics (Mabee and Humphries 1993). Most have concentrated on changing the way that polymorphism is coded in the data matrix. One popular method is to code a polymorphic taxon as 'unknown'. The obvious fault with this method is that the algorithm will assign the taxon with the character state that is most likely, given the most parsimonious arrangement of the other data, and that state may not match those actually present in the taxon. In addition, the resulting topologies can differ substantially from those that would have obtained otherwise (Nixon and Davis 1991).

Other methods of incorporating polymorphisms, such as utilising step matrices (Mardulyn and Pasteels 1994) or plesiomorphies (Kornet and Turner 1999), concentrate on changing the coding of polymorphisms to conform to current implementations of parsimony analysis in computer programs. This approach, in some ways, puts the cart before the horse; as with coding metric characters (see above), the emphasis should be to determine whether taxa are the same or different for each character. Again, as with measurement data, statistical tests of significance can be used to provide such a determination.

If we accept techniques (such as HSC) for coding metric data for parsimony analysis, there is no logical reason to prevent their use for coding frequency data, such as that produced in the study of polymorphisms. Provided that the statistical tools (e.g., controls for experiment-wide degrees of freedom) are available to perform multiple comparisons (cladistic analysis requires a minimum of four terminal taxa), creating codes that correspond to whether taxa are the same or different in terms of their respective frequencies of character states should follow precisely the same process used for univariate metric data. In this way, polymorphisms can be incorporated in data matrices without violating crucial data requirements, making assumptions of polarity, or needlessly complicating parsimony calculations.

One potentially important extension of this argument concerns the analysis of molecular sequence data. Common practice is to include a single representative of a taxon (e.g., Yoder 1994), which may limit the ability of the analysis to discover the most parsimonious topology (Wiens and Servedio 1997). Conversely, some choose to include multiple terminals from a single species (e.g., Ruvolo 1994), which directly contradicts the assumptions of cladistic analysis by attempting to resolve tokogenetic relationships. Tokogenetic relationships (those between individuals) are not hierarchical and cannot be resolved by phylogenetic systematics (Goldstein and DeSalle 2000). The use of HSC on polymorphism frequency data from multiple sequences of a species may provide a theoretically sound and practical method of incorporating these important data in cladistics. This, in turn, may help to eliminate some of the perceived conflict between gene trees and phylogenetic trees (Rogers 1994) and the 'discovery' of blatantly illogical 'paraphyletic species' (Melnick and Hoelzer 1993).

Conclusions

The above discussion, partly due to its philosophical nature, has left several practical points to one side (e.g., the effect of sample size and variance on multiple comparison tests). It is hoped, however, that several important points concerning the relationship between morphometrics and phylogenetic systematics have been emphasised:

1. that apparent distinctions between traditional morphometrics and cladistics are the direct result of the underlying differences in their theoretical bases and the questions with which they are concerned,

2. that any attempts to include morphometric data in cladistic analysis must include explicit reference to appropriate scaling that results in no inter-dependence of the data, and
3. that acceptance of metric data in cladistics allows the inclusion of polymorphisms (in the form of frequency data) in coded data matrices, which may also be extended to molecular sequence information.

Although these caveats/conclusions are in no way comprehensive, they outline possible ways that the two disciplines of morphometrics and cladistics can begin to interact in a cogent and robust manner.

Acknowledgements

Thanks are due to Norm MacLeod and Peter Forey for their kind invitation to participate in the Morphometrics & Phylogenetics symposium. Their encouragement was instrumental in the creation of a (hopefully) cohesive argument from somewhat amorphous origins. My thoughts on the issues discussed above were sharpened (and broadened) by many of the participants; T. Cole, D. Polly, and D. Swiderski offered insightful comments on the presentation that became this chapter. The research was supported by the Royal Society and the Dept. of Anthropology, Univ. of Durham.

References

Andrews, P. and Martin, L. (1987) 'Cladistic relationships of extant and fossil hominoids', *Journal of Human Evolution*, 16, 101–118.
Atchley, W., Gaskins, C. and Anderson, D. (1976) 'Statistical properties of ratios I: emperical results', *Systematic Zoology*, 25, 137–148.
Cave, A. and Haines, R. (1940) 'The paranasal sinuses of the anthropoid apes', *Journal of Anatomy*, 74, 493–523.
Collard, M. and Wood, B. (2000) 'How reliable are human phylogenetic hypotheses?', *Proceedings of the National Academy of Sciences USA*, 97, 5003–5006.
Goldstein, P. and DeSalle, R. (2000) 'Phylogenetic species, nested hierarchies, and character fixation', *Cladistics*, 16, 364–384.
Gura, T. (2000) 'Bones, molecules ... or both?', *Nature*, 406, 230–233.
Hartman, S. (1988) 'A cladistic analysis of hominoid molars', *Journal of Human Evolution*, 17, 489–502.
Hennig, W. (1966) *Phylogenetic systematics*, Urbana: University of Illinois Press.
Hills, M. (1978) 'On ratios – a response to Atchley, Gaskins, and Anderson', *Systematic Zoology*, 27, 61–62.
Jungers, W. (1985) *Size and scaling in primate biology*, New York: Plenum.
Jungers, W., Falsetti, A. and Wall, C. (1995) 'Shape, relative size, and size-adjustments in morphometrics', *Yearbook of Physical Anthropology*, 38, 137–161.
Kluge, A. and Farris, J. (1969) 'Quantitative phyletics and the evolution of anurans', *Systematic Zoology*, 18, 1–32.
Koppe, T., Inoue, Y., Hiraki, Y. and Nagai, H. (1996) 'The pneumatization of the facial skeleton in the Japanese macaque (*Macaca fuscata*) – a study based on computerized three-dimensional reconstructions', *Anthropological Science*, 104, 31–41.
Kornet, D. and Turner, H. (1999) 'Coding polymorphism for phylogeny reconstruction', *Systematic Biology*, 48, 365–379.

Mabee, P. and Humphries, C. J. (1993) 'Coding polymorphic data: examples from allozymes and ontogeny', *Systematic Biology*, 42, 166–181.

Mardulyn, P. and Pasteels, J. (1994) 'Coding allozyme data using step matrices: defining new original states for the ancestral taxa', *Systematic Biology*, 43, 567–572.

Melnick, D. and Hoelzer, G. (1993) 'What is mtDNA good for in the study of primate evolution?', *Evolutionary Anthropology*, 2, 2–10.

Nixon, K. and Davis, J. (1991) 'Polymorphic taxa, missing values and cladistic analysis', *Cladistics*, 7, 233–241.

Novacek, M. (1993) 'Patterns of diversity in the mammalian skull', in Hanken, J. and Hall, B. (eds) *The skull, vol. 2: patterns of structural and systematic diversity*, Chicago: University of Chicago Press, pp. 438–545.

Rae, T. (1993) 'Phylogenetic analysis of proconsulid facial morphology', Ph.D. dissertation, S.U.N.Y. at Stony Brook.

Rae, T. (1997) 'The early evolution of the hominoid face', in Begun, D., Ward, C. and Rose, M. (eds) *Function, phylogeny, and fossils: Miocene hominoid evolution and adaptations*, New York: Plenum, pp. 59–77.

Rae, T. (1998) 'The logical basis for the use of continuous characters in phylogenetic systematics', *Cladistics*, 14, 221–228.

Rae, T. (1999a) 'The maxillary sinus in primate paleontology and systematics', in Koppe, T., Nagai, H. and Alt, K. (eds) *The paranasal sinuses of higher primates: development, function and evolution*, Berlin: Quintessence, pp. 177–189.

Rae, T. (1999b) 'Mosaic evolution in the origin of the Hominoidea', *Folia Primatologica*, 70, 125–135.

Rae, T. and Koppe, T. (2000) 'Isometric scaling of maxillary sinus volume in hominoids', *Journal of Human Evolution*, 38, 411–423.

Rogers, J. (1994) 'Levels of genealogical hierarchy and the problem of hominoid phylogeny', *American Journal of Physical Anthropology*, 94, 81–88.

Ruvolo, M. (1994) 'Molecular evolutionary processes and conflicting gene trees: the hominoid case', *American Journal of Physical Anthropology*, 94, 89–113.

Simmons, N. (1993) 'The importance of methods: archontan phylogeny and cladistic analysis of morphological data', in MacPhee, R. (ed.) *Primates and their relatives in phylogenetic perspective*, New York: Plenum, pp. 1–61.

Simon, C. (1983) 'A new coding procedure for morphometric data with an example from periodical cicada wing veins,' in Felsenstein, J. (ed.) *Numerical taxonomy*, Berlin: Springer-Verlag, pp. 378–382.

Sokal, R. and Rohlf, F. (1981) *Biometry*, 2nd edn, New York: W. H. Freeman.

Strait, D., Moniz, M. and Strait, P. (1996) 'Finite mixture coding: a new approach to coding continuous characters', *Systematic Biology*, 45, 67–78.

Swofford, D. and Begle, D. (1993) *PAUP: phylogenetic analysis using parsimony. Program documentation*, Washington, DC: Laboratory of Molecular Systematics, Smithsonian Institution.

Thiele, K. (1993) 'The Holy Grail of the perfect character: the cladistic treatment of morphometric data', *Cladistics*, 9, 275–304.

Thorpe, R. (1984) 'Coding morphometric characters for constructing distance Wagner networks', *Evolution*, 38, 244–255.

Wiens, J. and Servedio, M. (1997) 'Accuracy of phylogenetic analysis including and excluding polymorphic characters', *Systematic Biology*, 46, 332–345.

Wiley, E. (1981) *Phylogenetics: the theory and practice of phylogenetic systematics*, New York: Wiley.

Yoder, A. (1994) 'Relative position of the Cheirogaleidae in strepsirhine phylogeny: a comparison of morphological and molecular methods and results', *American Journal of Physical Anthropology*, 94, 25–46.

Chapter 5

Overlapping variables in botanical systematics

Geraldine Reid and Karen Sidwell

ABSTRACT

Character coding has received occasional attention from botanists and problems of discovering characters suitable for phylogenetic analysis from overlapping variables have been addressed sporadically at best. A review of overlapping variables in plant phylogenetics over the last 10 years is presented. We show that plant systematists have tended to score data matrices in order to produce cladograms rather than focus attention on the individual characters as hypotheses. The unspoken rule for coding data matrices appears to be to process as much observational data as possible into discrete integers and by using filters, delete recalcitrant variables. Lack of methodological clarity, fear of statistical techniques, and a desire for rapid results have all contributed to the illogical exclusion of continuous data from cladistic analyses. To illustrate the theoretical issues, practical case studies from diatoms (*Pleurosigmataceae*) and higher plants (*Oxalis* section *Ionoxalis*) are presented. Different methodological approaches most commonly mentioned in botanical literature (simple gap coding, generalized gap coding, segment coding) are implemented and the consequences of including or excluding continuous characters are explored. The utility of overlapping characters for phylogenetic reconstruction is investigated with parsimony analysis using the programmes PAUP, Hennig86 and Pee-Wee. We clarify the issues and methods for coding overlapping variables for cladistic analysis. The effect of each of the coding methods on tree topology is discussed. It is shown that the four methods used here are all inappropriate. Greater time must be spent researching homologues in data sets containing overlapping characters to prevent potentially large amounts of information from being ignored.

Introduction

This chapter addresses issues relating to overlapping characters in botanical systematics analysed using standard parsimony techniques.

The aims are three fold:

- To briefly review the use of quantitative continuous data in plant systematics over the last 10 years.
- To investigate overlapping variables in data sets from diatoms and higher plants.

- To compare and contrast the four coding methods most commonly used in the literature.
- To investigate the effect of coding overlapping data on phylogenies derived using three parsimony programmes.

For clarity, a short definition of our use of 'morphometric' is necessary. It refers to overlapping linear measurements as measured from plants. The terms metric, morphometric, continuous and quantitative are all used here synonymously.

Review of the last 10 years in plant phylogenetics

This review takes Chappill (1989) as a start point. Stevens (1991), Pleijel (1995), Wilkinson (1995) and others have recognised that character coding has received little attention in the literature. Botanical publications that include phylogenetic trees produced from published morphological data matrices using parsimony methods were investigated and theoretical papers across the biological spectrum were taken into account.

Theoretical papers that cover continuous variables over the past 10 years can be summarised as arguments for and against coding continuous characters in cladistic analysis.

Arguments against including overlapping variables tend to prevail and stem from Pimentel and Riggins (1987) who suggested that:

- continuous data are 'non-cladistic', only mutually exclusive states should be recognised,
- coding such data introduces artificial distinctions, and
- homology cannot be assessed by the similarity test.

Pimentel and Riggins (1987: 275) state that 'continuously varying quantitative data are not suitable for cladistic analysis because there is no justifiable basis for recognising discrete states among them'. Although these arguments appear to have been largely followed by botanists, they are rarely alluded to in practical phylogenetic studies. Arguments for including overlapping variables are becoming increasingly noticeable in the literature:

- biological data are inherently continuous,
- the data are precise and replicable,
- qualitative data should be quantified whenever possible,
- to exclude overlapping variables is illogical, and
- a large amount of potentially useful information is being ignored.

Thiele (1993) and Stevens (1991) both stated that continuous data should not be rejected a priori. Stevens (1991: 562) discussed the fact that qualitative data is drawn from quantitative variables: 'if such qualitative variation is examined carefully, it will be found that much describes an underlying continuum that has been transformed by the terms we use; discontinuities are only semantic'. Thiele (1993: 283) agreed,

Table 5.1 Summary of literature survey of American Journal of Botany; The Botanical Journal of the Linnean Society; Cladistics and Systematic Biology from 1989–1999 showing botanical systematic studies

Journal	Including morphological cladistic analyses	Discussing continuous characters	Including continuous characters
Am. J. Bot.	30	3	0
Bot. J. Linn. Soc.	20	1	1
Cladistics	7	2	2
Syst. Biol.	7	1	0

recognising that 'all character states (as used in cladistic analyses) are frequency distributions of attribute values over a sample of individuals of a taxon', concluding that 'every effort should be made to include overlapping morphometric data in analyses, followed by detailed testing of the resulting trees, so that empirical assessments of such data can be made' (Thiele 1993: 296). Swiderski et al. (1998) also considered it illogical to exclude these kinds of data. Rae (1998: 226) concluded that continuous variables 'fulfill the necessary criterion for use in phylogenetic analysis (homologous character states), and since they can be coded in a non-arbitrary, biologically appropriate manner, there can be no theoretically justifiable means for dismissing them from phylogenetic systematics.' Thiele (1993: 275) found that continuous morphometric data '... map phylogeny almost as accurately as more conventional qualitative morphological data'.

A review of four botanical journals highlights the use of different kinds of morphological data in phylogenies. Table 5.1 shows the number of morphological cladistic analyses published in the last 10 years, and of those, the number that discussed continuous characters and the number of those that included them in analysis. It can be seen that, out of a total of 64 papers, 7 discussed overlapping variables and only 3 (Ladiges et al. 1989; 1992; Linder and Mann, 1998) utilised these data in phylogenetic studies.

Most papers did not explicitly describe character choice. Morphological characters appear to be used because 'They lent themselves to cladistic analysis of the ... taxa' (Williams et al. 1994: 1028) and also because they had a 'prominence in the traditional classification of taxa involved' (Williams et al. 1994: 1028). In several papers botanists coded seemingly continuous variables with no discussion of the rationale behind this. For example, Boufford et al. (1990) had 2 out of 22 characters that were numerical and seemingly continuously variable, and nowhere provided any discussion of how the characters were selected, measured or coded.

Four papers explicitly discussed and rejected the use of continuous characters. Luckow and Hopkins (1995) and Kelly (1997) felt unable to code continuous characters. Crisp and Weston (1993) concluded in their study of *Teleopea* that morphometric characters were useful for investigations of species delimitation, but not for looking at the relations between taxa. Freudenstein (1994) thought such characters were not useful to his work. One other paper worth mentioning here is that of Morton and Kincaid (1995) who devised a model for coding pollen morphometric data 'using conventional statistical procedures coupled with data visualisation and Monte Carlo simulation' (Morton and Kincaid 1995: 1173). The conventional statistical procedures employed were inspection of prediction and confidence ellipses (e.g., 99 per cent) and use of ANOVA. This approach enabled them to code three character states for pollen grains. They did not however attempt to use the data in a cladistic analysis.

Three papers that both discussed and used continuously variable characters in their analyses differ in the methods used to code those characters. Linder and Mann (1998) used Almeida and Bisby's (1984) ranging method, which is a non-statistical method where the ranges are plotted out graphically and gaps identified visually. The rationale for using morphometric data was that it was 'very useful for distinguishing between the species of *Thamnochortus*' (Linder and Mann 1998: 322), therefore it might be informative of relationships in that group. They concluded that, in this case, the 'contribution of the morphometric characters to the resolution of the cladogram is weak' (Linder and Mann 1998: 339). However, the continuous characters did contribute to the resolution of the terminal clades without contradicting the topology produced by qualitative characters alone. Ladiges *et al.* (1989: 346; 1992: 106) gap coded their characters following the earlier work of Ladiges *et al.* (1987). Ladiges *et al.* (1992: 106) coded the characters where 'clear disjunctions in measures were used to define states. Where three or more character states were recognised the plesiomorphic state was chosen on the basis of outgroup comparison and transformations were based on an assumed developmental trend from the plesiomorphic state'.

From this brief summary it is clear that there is no consensus regarding the use of morphometric data in phylogenetic analyses. The argument for including such data appears to be gaining strength and there are no justifiable reasons to exclude such information a priori from any analysis even though the majority of botanical papers reviewed above have done so implicitly.

Methods

Two different data sets, from diatoms (Pleurosigmataceae) and higher plants (*Oxalis* section *Ionoxalis*) were used. All characters were assumed to be independent. The diatom data set contained 13 characters: 9 qualitative and 4 quantitative, listed in Table 5.2. The characters were selected from all parts of the diatom frustule and included all parts measured in previous taxonomic studies. They were measured using a micrometer eye piece graticule. The *Oxalis* data set contained 20 characters: 8 qualitative and 12 quantitative, listed in Table 5.3. Those characters which were selected were relatively easy to observe and measure given the time available for the study and were taken from all parts of the plants. They were measured using a mm ruler or a micrometer eye piece graticule. Four coding methods: simple gap coding, segment coding, and two forms of generalized gap coding were applied to all morphometric characters. Three parsimony programmes: Hennig86, PAUP v.3.1 and Pee-Wee were employed to look at the effect of including such data in phylogenetic analysis.

All coding methods follow the same initial procedure requiring calculation of the *pooled within group standard deviation* (S_p). The value of S_p is obtained by finding the standard deviation of all the measurements for one character for all taxa.

Table 5.2 Characters and codes for Pleurosigmataceae

Character	Description
1	Valves: arcuate 0; sigmoid 1
2	Striae: transverse and oblique 0; longitudinal and transverse 1
3	Raphe: single curvature 0; double curvature 1
4	Central external raphe fissures: curved in same direction 0; in opposite directions 1
5	Central bars: smooth, slender 0; wide and thick 1; smooth and flattened 2; smooth with undulating outer edge 3
6	Central area: even 0; transapically offset 1
7	Crescent at apex absent 0; present 1
8	Valve outline: smooth 0; undulating 1
9	Hyaline area at apex: absent 0; present 1
10	Diatom length (μm)
11	Diatom breadth (μm)
12	Number of longitudinal striae per 10 μm
13	Number of transverse striae per 10 μm

Table 5.3 Characters and codes for *Oxalis* section *Ionoxalis*

Character	Description
1	Heterostyly: homostylous 0; distylous 1; tristylous 2
2	Number of leaflets: three 0; more than three 1
3	Nerves on bulb scale: three 0; more than three 1
4	Flanges on petioles: not extended above bulb 0; extended 1
5	Hair type: nonseptate 0; septate 1
6	Seed surface: lacking transverse ridges 0; transverse ridges present 1
7	Leaflet shape: entire 0; lobed 1
8	Inflorescence type: umbelliform cyme 0; branched cyme 1
9	Petiole length (mm)
10	Leaflet length (mm)
11	Leaflet width (mm)
12	Leaflet lobe length (mm)
13	Scape length (mm)
14	Bract length (μm)
15	Pedicel length (mm)
16	Sepal length (μm)
17	Sepal width (μm)
18	Petal length (μm)
19	Fruit length (μm)
20	Fruit width (μm)

The equation is:

$$S_p^2 = \sum_{j=1}^{k} (n_{ij} - 1)s_{ij}^2 \Big/ \sum_{j=1}^{k} (n_{ij} - 1)$$

Then

$$S_p = \sqrt{S_p^2}$$

where n = sample size for character i in taxon j; S_p^2 = the pooled within group variance for character i; k is the number of taxa.

For each method the mean (Y) of each character for each taxon is calculated and taxa ranked in order of mean size (see Table 5.4). The coding methods then proceed as follows:

Simple gap coding – Mickevich and Johnson 1976

The first taxon is given a code (m) of 0. The difference between adjacent means is calculated ($Y_i - Y_j$). This value is compared to the pooled within group standard deviation. If the 'the gap' difference is greater than the pooled within group standard deviation multiplied by an arbitrary constant (c) then the taxon is given a new code ($m + 1$).

$$\text{If } (Y_i - Y_j) > cS_p \quad \text{then } Y'_i = m \text{ and } Y'_j = m + 1$$

If the difference between adjacent means is not greater than the pooled within group standard deviation then the second taxon retains the same character code (m) as the prior taxon.

$$\text{If } (Y_i - Y_j) \leq cS_p \quad \text{then } Y'_i = Y'_j = m$$

This procedure is repeated throughout the study group until all taxa are coded.

Working through an example using character 11 of the diatom data set (see Table 5.4) The $S_p = 7.26$ and c is set to 1. The taxa are ranked in order with the taxon with the lowest mean first (see Table 5.4, *G. gibbii*) which is given a code of 0. The means of the first two adjacent taxa are compared ($13.4 - 10.4 = 3$) the resulting value is compared to the cS_p, it is less than cS_p i.e., $3 < 7.26$ therefore *G. perthense* is given a code of 0. The means of the second two adjacent taxa are then compared ($13.7 - 13.4 = 0.3$) the result is compared to the S_p and again $0.3 < 7.26$ therefore *G. wansbeckii* is also given the code 0. The code changes to 1 at *G. pensacole*, as the gap of the adjacent means ($42.7 - 31 = 11.7$) is greater than the cS_p ($11.7 > 7.26$), therefore, it is now in the next gap.

Segment coding

Using this method the range of a character is divided into equal sized segments. The segments (w) are derived by adding S_p to the first taxon mean (Y_i) and then adding S_p to the result until w' is greater than or equal to the value of the last taxon mean.

$$Y_i + S_p = w \quad \text{then } w + S_p = w'$$

The taxa are then coded by comparing the taxon means (Y) to the segments (w). The first taxon is given the code 0. If the next taxon mean falls within the first segment it retains the code 0. If the taxon mean falls outside the first segment it becomes $w + 1$, that is, is given the code 1.

Working through an example using character 11 from the diatom data set (see Table 5.4) The S_p is calculated as 7.26 and using c as 1. The taxa are ranked in order with the taxon with the lowest mean first (see Table 5.4, *G. gibbii*). The first

Table 5.4 Coding for the different methods of coding of character 11 from the diatom data set. Numbers shown in [ ] are the discarded subsets

	Mean	Simple-gap code	Segment coding	Subsets						Generalized Archie
G. gibbii	10.4	0	0	1						0
G. perthense	13.4	0	0	1	2					1
G. wansbeckii	13.7	0	0	1	2	[3]				1
G. turgidum	17.7	0	1		2	[3]	4			3
G. exoticum	21.8	0	1		2	[3]	4	5		5
G. california	25.8	0	2					5	6	7
T. insignis	26	0	2					5	6	[7]	7
G. subtilis	27.9	0	2					5	6	[7]	7
G. sterrenburgii	28.2	0	2					5	6	[7] [8]	7
G. cali	28.3	0	2					5	6	[7] [8] [9]	7
G. murphi	29.1	0	2						6	[7] [8] [9] [10]	8
G. balticum	29.4	0	2						6	[7] [8] [9] [10] [11]	8
P. angulatum	31	0	2						6	[7] [8] [9] [10] [11] [12]	8
G. pensacole	42.7	1	4						6	[7] [8] [9] [10] [11] [12] [13]	10

	Subsets									1st code	Generalized Thiele and Ladiges
G. gibbii	1									0	0
G. perthense	1	2	[3]							0.5	1
G. wansbeckii	1	2	[3]	4						0.5	1
G. turgidum		2	[3]	4						1.5	2
G. exoticum				4	5					2.5	3
G. california					5	6				3.5	4
T. insignis					5	6	[7]			3.5	4
G. subtilis					5	6	[7] [8]			3.5	4
G. sterrenburgii					5	6	[7] [8] [9]			3.5	4
G. cali					5	6	[7] [8] [9] [10]			3.5	4
G. murphi						6	[7] [8] [9] [10] [11]			4.0	5
G. balticum						6	[7] [8] [9] [10] [11] [12]			4.0	5
P. angulatum						6	[7] [8] [9] [10] [11] [12] [13]			4.0	5
G. pensacole						6				14	5.0	6

segment is derived by adding cS_p to the first character mean ($10.4 + 7.26 = 17.66$) this gives us the first segment. All taxa with means less than or equal to 17.66 are given the code 0, that is, *G. gibbii, G. perthense* and *G. wansbeckii*. The next segment is derived from $17.66 + 7.26 = 24.92$ giving the segment of 17.67 to 24.92 and all taxa with means lying in this segment are given the code 1, that is, *G. turgidum, G. exoticum*. This is continued for all taxa.

Generalized gapcoding – Archie 1985

Generalized gap coding was described by Archie (1985) and advocated by Chappill (1989) and Thiele and Ladiges (1988). This method attempts to give a better representation of the characters rather than forcing them directly into arbitrary character states. Generalized gap coding is a two stage process, first subsets are derived from the taxon means and second codes are derived from the subsets. The subsets are found as follows. The cS_p is added to the first taxon mean (Y), and taxa with means less than, or equal to, this value form the first subset. The cS_p is then added to the second taxon mean, all taxa with means less than, or equal to, this value form the second subset. Adding cS_p to each taxon mean continues as above until all taxa are grouped in subsets.

The second stage is to code the subsets. At this stage the method of Thiele and Ladiges (1988) differs from Archies' method. The difference can be summarised as follows:

Generalized gapcoding – sensu Archie (1985) Code the first taxon as 0. Increase the value by one for each successive taxon if it is included in a new subset and by a further 1 if it is no longer in a previous subset. If a subset is fully contained within another previous subset, then the second subset is discarded. This method keeps the differences in magnitude between code values.

Generalized gapcoding – sensu Thiele and Ladiges (1988) Code the first taxon as 0. Increase the value by 0.5 for each successive taxon if it is included in a new subset and by a further 0.5 if it is no longer in the previous subset. The taxon values are then coded as sequential integers $0, 1, 2, 3, 4, 5, \ldots$ removing any differences in magnitude. Ladiges then ordered the states of the generalized gap coded characters from smallest to largest, before running analyses, but did not order the qualitative variables.

Working through an example sensu Archie using character 11 of the diatom data set (see Table 5.4) The $S_p = 7.26$ and c is set to 1. The taxa are ranked in order starting with the lowest mean (see Table 5.4, *G. gibbii*). The cS_p is added to the mean of *G. gibbii* ($10.4 + 7.26 = 17.66$) giving a first subset of 10.4 to 17.66. Then cS_p is added to the next taxon mean i.e. *G. perthense* ($13.4 + 7.26 = 20.66$) giving the second subset as 13.4 to 20.66. Then cS_p is added to the next taxon mean *G. wansbeckii* ($13.7 + 7.26 = 20.96$) giving the third subset as 13.7 to 20.96. The cS_p is added to the next taxon mean, *G. turgidum* ($17.7 + 7.26 = 24.96$) giving the fourth subset of 17.7 to 24.96. *G. gibbii* is given a code of 0 (see Table 5.4). *G. perthense* is given a code of 1 as it starts a new subset, but is still contained in subset 1. *G. wansbeckii* receives a code of 1 as the third subset is discarded as it is totally contained within the second subset, that is, *G. wansbeckii* neither starts a new non-inclusive subset or leaves one

(Table 5.4). *G. turgidum* receives a code of 3 as it leaves subset 1 and starts subset 4. This is continued throughout the data set (see Table 5.4).

Working through an example sensu Thiele and Ladiges using character 11 of the diatom data set (see Table 5.4) The $S_p = 7.26$ and c is set to 1. The taxa are ranked in order starting with the lowest mean (see Table 5.4, *G. gibbii*). The cS_p is added to the mean of *G. gibbii* ($10.4 + 7.26 = 17.66$) giving a first subset of 10.4 to 17.66. Then cS_p is added to the next taxon mean, that is, *G. perthense* ($13.4 + 7.26 = 20.66$) giving the second subset as 13.4 to 20.96. Then cS_p is added to the next taxon mean *G. wansbeckii* ($13.7 + 7.26 = 20.96$) giving the third subset as 13.7 to 20.96. The cS_p is added to the next taxon mean, *G. turgidum* ($17.7 + 7.26 = 24.96$) giving the fourth subset of 17.7 to 24.96. *G. gibbii* is given a code of 0 (see Table 5.4). *G. perthense* is given a code of 0.5 as it starts a new subset, but is still contained in subset 1. *G. wansbeckii* receives a code of 0.5 as the third subset is discarded as it is totally contained within the second subset, that is, *G. wansbeckii* neither starts a new non-inclusive subset or leaves one (Table 5.4). *G. turgidum* receives a code of 1.5 as it leaves subset 1 and starts subset 4. This is continued throughout the data set. The codes are then re-coded $0, 1, 2, 3, \ldots$ every time a new number is encountered regardless of the magnitude of the code change, that is, *G. perthense* at 0.5 has a code of 1 and *G. turgidum* at 1.5 has a code of 2 even though it has shown a greater magnitude of change.

In summary it can be seen each of the different coding methods used here results in a different suite of character codes for character 11 of the diatom data set (see Table 5.4). In all coding methods the taxon with lowest mean value always receives 0 (e.g., *G. gibbii*), whereas the other taxa receive arbitrarily different codes (e.g., *G. pensacole* receives 4 different codes with the 4 different methods). Each method of coding postulates a different hypothesis of homology. The complete data matrices for Pleurosigmataceae and Oxalis showing the 4 different coding methods can be seen in Tables 5.5 and 5.6.

Phylogenetic analyses

Tables 5.2 and 5.3 list the characters used in the analyses and their character states. Character codes are recorded in Tables 5.5 and 5.6. Each data set was analysed using Hennig86 (*ie**) and PAUP (maximum parsimony) under equal weights with all characters unordered.

However, as all characters cannot be assumed to contribute to the same extent in predicting relationships, implied differential weighting was employed using the computer programme Pee-Wee using the options *hold**, *mult*50* to search for trees of the highest fit, performing random addition sequences of 50 replications. All characters were unordered. Replication was followed by tree bisection and branch-swapping. The *amb*-option was used to eliminate the effect of ambiguous support.

The effect on tree length of including the gap-coded characters was to increase it in all cases, as would be expected with increasing the number of characters included in the analysis. Except in the *Oxalis* data set under simple gap coding as all characters received the same code of 0 for all taxa, that is, they did not make any contribution

Table 5.5 Data matrix for the diatom data set with the codes for the 4 different coding methods. 'Original data matrix' is the binary/multistage characters 1–9, the subsequent blocks are the codings of the quantitative characters 10–13 repeated for each coding method

	Original data matrix	Simple	Segment	Generalized Archie	Generalized Thiele
toxonidea	000000000	0000	0222	1784	1463
angulatum	100000000	0000	1221	3863	3542
balticum	111111101	0000	2200	6800	5500
murphii	111111101	0000	2200	6800	5500
subtilis	111121101	0000	2210	6731	5421
gibbii	110100100	0001	0023	0076	0054
california	111121100	1000	1200	5710	4410
sterreburgii	111131101	0000	1200	5700	4400
exoticum	111121110	0000	0120	25A0	2371
perthense	111100110	0000	1010	5141	4131
wansbeckii	110100100	0000	0022	11B4	1183
cali	111121101	0000	3200	8700	6400
turgidum	111121110	0000	2110	6330	5220
pensacole	111121101	1100	3400	8A00	6600

Notes: In the character codes A = 10; B = 11.

Table 5.6 Data matrix for the *Oxalis* data set with the codes for the 4 different coding methods. 'Original data matrix' is the binary/multistage characters 1–8, the subsequent blocks are the codings of the quantitative characters 9–20 repeated for each coding method

	Original data matrix	Simple	Segment	Generalized Archie	Generalized Thiele
debilis	00000111	0000000000??	1111111121??	44354475A9??	332433647???
intermedia	00100110	0000000000?	21222010101?	D9698241433?	96465231333?
eggersii	00100110	0000000000??	0000000010??	11011110 32??	11011110 22??
primavera	10101110	000000000000	222321112111	CA8BC7549866	875785436656
gregarii	20001010	000000000000	000010100000	211230600010	211220500010
nelsonii	21101100	00000?000000	21312?022120	9894B?1BCB83	65637?189963
magnifica	21101100	0000000000??	122012 0222??	6B516A1AED??	48314717AA??
lasiandra	21101100	00?00000 00??	12?0220221??	8D?1A927B6??	59?1662585??
macrocarpa	20001100	00?0?????? ??	21?2??????? ?	B6??????????	74?5????????
drummondii	10001110	000000000000	100110222111	621442987455	421332765445
divergens	11000110	000000000000	000001011100	210116136A22	210114124822
lunulata	20011110	000000000000	000000000000	000001001101	000001001101

Notes: In the character codes A = 10; B = 11; C = 12; D = 13; E = 14.

to the analysis. In the diatom data set the ci and ri are decreased with each coding method. Whereas in the *Oxalis* data set the ri is decreased with all coding methods (except simple gap coding as the characters are not used in the analysis) but the ci varies unpredictably, it decreased with segment coding but increased with generalized gap coding. This may be caused by the higher number of autapomorphic states for these coding methods, which increase ci but decrease ri (Table 5.7).

Table 5.7 Tree statistics for each of the coding methods for each of the data sets

	Original	Simple	Segment	Generalized Archie	Generalized Thiele
Diatom data set					
Trees	9	30	8	12	135
Length	12	16	28	40	41
ci	91	87	78	87	85
ri	95	91	84	84	81
Oxalis data set					
Trees	1	1	4	72	72
Length	14	14	59	112	112
ci	64	64	57	90	90
ri	77	77	62	66	66

Discussion

Coding overlapping morphological characters is fraught with methodological problems. The difficulties are various:

- Simple gap coding is problematic as the addition of new taxa can have a huge effect on codes for all taxa in the data set. The major problem is that taxa which may be very different can be given the same code as if they are part of a long series of closely spaced taxa, as is the case with the diatom data set character 11. Swiderski *et al.* (1998: 511) dismissed simple gap coding as it 'misrepresents the amount of overlap when distributions are skewed or otherwise deviate from normality'.
- Segment coding results in the loss of slight differences in character states between populations. A slight shift in the mean could mean that a gap will disappear, for example, by increasing the population size. Segment coding overcomes the problems of simple gap coding in that it recognises the differences in long series of closely spaced taxon means. But it can put taxa with values of 9.999999 and 10.000001 into different codes if the segment boundary is 10.0.
- Generalized gap coding claims to maintain the relative differences between taxa, but only roughly does so. It can lead to one code per taxon and has a very high number of codes. This can cause problems when using Hennig86 and Pee-Wee as the programmes only accept character codes less than 10. Chappill only accepted characters where gaps separated more than one taxon, that is, she would not use any character that coded as automorphies as they are always fully consistent with the tree.

Variables affecting character codes and resulting phylogenies

When including such data in phylogenetic analyses, tree topology is dependent entirely on the coding method employed. The same is true of all types of characters but is 'hidden' in qualitative characters, usually in the character description or not mentioned at all. In many cases, each of the 4 coding methods used here gave different homology hypotheses for each character, which had direct impact on the resulting phylogenetic trees. The variables that affect character codes in the diatom and higher plant data sets include:

- *Sample choice*: Taxonomic studies often include many specimens of one taxon and only a few representatives from another. It is imperative that as many specimens as possible are examined for each taxa to gain a realistic representation of the variation within the taxon.
- *Character choice*: Phylogenetic analyses assume independence of characters.
- *Measurement error*: This can depend greatly on the skill of the person making the measurements. Factors such as how long they have been looking down a microscope that session can influence the accuracy of the measurements taken, that is, operational error. Variables need to be carefully prescribed so that measurements are taken from homologous landmarks.
- *The choice of constant* c: The choice of value for c is arbitrary. However, c is recommended as 1 and a lower value, for example, 0.5 should be used only when a character shows statistically significant variation between populations yet it does not exhibit a gap between means.
- *Treatment of raw data*: Chappill (1989) chose to log transform her data so that proportional differences would be recognised. Such transformation is not required if data has a normal distribution, that is, it is only needed if the variances are not equal.
- *Scaling data*: A number of workers advocate the use of scaling coded multistate characters to unit range (Colless 1980; Goldman 1988; Chappill 1989; Cranston and Humphries 1988; Thiele and Ladiges 1988). Thiele and Ladiges (1988: 27) stated this was so that 'many state characters did not dominate the analysis' and weighted to give each character the 'same weight as a binary character'. Farris (1990: 91) pointed out that rescaling characters was illogical and should not be undertaken a priori.
- *Statistical values used*: Thorpe (1984: 247) recommended the use of the mean rather than the median for ranking character data as it is a better estimate of the 'centrality of the distribution as it derives its value from the character states of all the samples'. All workers using these coding methods in our review have implemented this recommendation.
- *Missing data*: None of the coding methods used here deal adequately with the problem of missing data. The data set for *Oxalis* had a large number of characters that were unable to be measured due to lack of appropriate specimens. Inclusion of these specimens at a later date would have an effect on the overall patterns of coding.

Other solutions: Swiderski *et al.* (1998) recommended the use of Almeida and Bisby (1984) graphical displays of data. In this method, the ranges of the taxa are plotted out with the median values marked along with the first and third quartet. The diagram is then used to visually divide the measurement data into classes. This method is flawed in that different workers could see a different number of partitions. Even if they see the same number of groups, they may see different members in that group. Gift and Stevens (1997) clearly show the problems with character delimitation by different individuals from graphic displays. Almeida and Bisby's (1984) method worked under the assumption that character states should be discrete or almost so. This does not help, however, in the case where most characters are continuous, which, due to their very nature, show no discrete boundaries. Swiderski *et al.* (1998) considered the problem to be a practical rather than theoretical.

The main question addressed in this chapter is whether it is possible to resolve overlapping variables into homologues. The 4 coding methods above raise both theoretical and practical issues. Each method allocates codes on the basis of taxon means and does not account for a taxon range that may span more than one character state. Taxon means are ranked from lowest to highest prior to coding. When new taxa are introduced into the analysis, or taxa that have missing character values are added, the order of character means can drastically change and alter the character codes. All methods rely on calculation of the cS_p, a value based on all individual data points for all taxa. If new taxa or new data points are introduced, the value of cS_p will almost certainly change and alter the resulting codes. These methods, therefore, are forcing non-homologous states onto the characters.

In summary:

- The assumption that overlapping variables can be resolved into homologues needs reconsideration.
- Gaps created by all these methods are artefacts.
- There still does not appear to be a general solution to the problems of coding continuous characters.
- We need to have a coding method that accurately reflects hypotheses of homology.

Acknowledgements

We would like to thank Chris Humphries and Dave Williams for their advice and encouragement. Ian Kitching for his help and comments on the manuscript.

References

Almeida, M. T. and Bisby, F. A. (1984) 'A simple method for establishing taxonomic characters from measurement data', *Taxon*, 33, 405–409.

Archie, J. A. (1985) 'Methods for coding variable morphological features for numerical taxonomic analysis', *Systematic Zoology*, 34, 326–345.

Boufford, D. E., Crisci, J. V., Tobe, H. and Hoch, P. C. (1990) 'A cladistic analysis of *Circaea* (Onagraceae)', *Cladistics*, 6, 171–183.

Chappill, J. A. (1989) 'Quantitative characters in phylogenetic analysis', *Cladistics*, 5, 217–234.

Cranston, P. S. and Humphries, C. J. (1988) 'Cladistics and computers: a chironomid conundrum?', *Cladistics*, 4, 72–92.

Crisp, M. D. and Weston, P. H. (1993) 'Geographic and ontogenetic variation in morphology of Australian Waratahs (Telopea: Proteaceae)', *Systematic Biology*, 42, 49–77.

Colless, D. H. (1980) 'Congruence between morphometric and allozyme data for *Menidia* species: a reappraisal', *Systematic Zoology*, 29, 288–345.

Farris, J. S. (1990) 'Phenetics in camouflage', *Cladistics*, 6, 91–101.

Freudenstein, J. V. (1994) 'Character transformation and relationships in *Corallorhiza* (Orchidaceae:Epidendroideae). II. Morphological variation and phylogenetic analysis', *American Journal of Botany*, 81, 1458–1467.

Gift, N. and Stevens, P. F. (1997) 'Vagaries in the delimitation of character states in quantitative variation – an experimental study', *Systematic Biology*, 46, 112–125.

Goldman, N. (1988) 'Methods for discrete coding of morphological characters for numerical analysis', *Cladistics*, 4, 59–71.

Kelly, L. M. (1997) 'A cladistic analysis of *Asarum* (Aristolochiaceae) and implications for the evolution of herkogamy', *American Journal of Botany*, 84, 1752–1765.

Ladiges, P. Y., Humphries, C. J. and Brooker, M. I. H. (1987) 'Cladistic and biogeographic analysis of western Australian species of *Eucalyptus* l'her. (informal subseries *Amydalininae*, subgenus *Monocalyptus*) and the description of a new species *E. willisii*', *Australian Journal of Botany*, 31, 565–584.

Ladiges, P. Y., Newnham, M. R. and Humphries, C. J. (1989) 'Systematics and biogeography of the Australian "Green Ash" Eucalypts (Monocalyptus)', *Cladistics*, 5, 345–364.

Ladiges, P. Y., Prober, S. M. and Nelson, G. (1992) 'Cladistic and biogeographic analysis of the "Blue Ash" Eucalypts', *Cladistics*, 8, 103–125.

Linder, H. P. and Mann, D. M. (1998) 'The phylogeny and biogeography of *Thamnochortus* (Restionaceae)', *Botanical Journal of the Linnean Society*, 128, 319–357.

Luckow, M. and Hopkins, H. C. F. (1995) 'A cladistic analysis of *Parkia* (Leguminosae: Mimosoideae)', *American Journal of Botany*, 82, 1300–1320.

Mickevich, M. F. and Johnson, M. S. (1976) 'Congruence between morphological and allozyme data in evolutionary inference and character evolution', *Systematic Zoology*, 25, 260–270.

Morton, C. M. and Kincaid, D. W. T. (1995) 'A model for coding pollen size in reference to phylogeny using examples from the Ebenaceae,' *American Journal of Botany*, 82, 1173–1178.

Pimentel, R. A. and Riggins, R. (1987) 'The nature of cladistic data', *Cladistics*, 3, 201–209.

Pleijel, F. (1995) 'On character coding for phylogeny reconstruction', *Cladistics*, 11, 309–315.

Rae, T. C. (1998) 'The logical basis for the use of continuous characters in phylogenetic systematics', *Cladistics*, 14, 221–228.

Stevens, P. F. (1991) 'Character states, morphological variation, and phylogenetic analysis: a review', *Systematic Botany*, 16, 553–583.

Swiderski, D. L., Zelditch, M. L. and Fink, W. L. (1998) 'Why morphometrics is not special: coding quantitative data for phylogenetic analysis', *Systematic Biology*, 47, 508–519.

Thiele, K. (1993) 'The holy grail of the perfect character: the cladistic treatment of morphometric data', *Cladistics*, 9, 275–304.

Thiele, K. and Ladiges, P. Y. (1988) 'A cladistic analysis of Angophora cav. (Myrtaceae)', *Cladistics*, 4, 23–42.

Thorpe, R. S. (1984) 'Coding morphometric characters for constructing distance Wagner networks', *Evolution*, 38, 244–255.

Wilkinson, M. (1995) 'A comparison of two methods of character construction', *Cladistics*, 11, 297–308.

Williams, S. E., Albert, V. A. and Chase, M. W. (1994) 'Relationships of *Droseraceae*: a cladistic analysis of rbcL sequence data and morphological data', *American Journal of Botany*, 81, 1027–1037.

Chapter 6

Comparability, morphometrics and phylogenetic systematics

Donald L. Swiderski, Miriam L. Zelditch and William L. Fink

ABSTRACT

Although qualitative descriptions of shape are commonly used in phylogenetic systematics, there are numerous objections to using quantitative descriptions of these same features. Previously, we argued that no earlier discussions of this issue provided a general argument that supports exclusion of all morphometric data from phylogenetic studies. Rather, we concluded that data from at least some morphometric methods can support hypotheses of shape homology, but data from other methods cannot. In this chapter, we explore the reasons for this difference. We then present a general criterion for determining whether any morphometric method produces descriptions that are suitable for phylogenetic analysis: the comparability of the shapes described by the variables that a particular method produces. Then we examine several morphometric methods and evaluate whether they meet this criterion. Some methods (like elliptical Fourier and eigenshape analyses) cannot be used in phylogenetic systematics because they have no mechanism to ensure comparability. These methods commonly employ large numbers of points on a given specimen and the points on one specimen need not have a one-to-one correspondence to points on another specimen. Consequently, the variables computed from the coordinates of the points need not refer to the same region of the specimen. Other methods use smaller numbers of points that are judged to be comparable among specimens on a one-to-one basis (i.e., landmarks). Methods that produce variables from the coordinates of landmarks at least have the potential to refer to comparable regions of the specimens. Although the use of landmarks is an important step toward insuring comparability, it is neither necessary nor sufficient. Methods that do not use landmarks may still produce comparable descriptions if other information is used to specify what anatomical feature is described by the variable. Conversely, methods that do use landmarks may not produce comparable descriptions if the landmarks are not adequate to constrain the morphometric description. This latter situation may arise when the landmarks are few and far apart (in analyses using shape coordinates or extended eigenshape analysis) or when additional manipulations of the data are used to generate new variables from combinations of the original ones (as in principal components analysis). Thus, we conclude that morphometric descriptions can be used in phylogenetic systematics if due caution is taken at every stage of the analysis to insure that variables refer to comparable features.

Introduction

Shape differences are often used to infer phylogenetic relationships. For example, a recent analysis of relationships among anteaters (Mammalia: Myrmecophagidae) included such features as 'shape of premaxilla in ventral view' and 'curvature of the basicranial axis' (Gaudin and Branham 1998). An analysis of *Aramigus* weevils (Coleoptera: Curculionidae) used such features as 'convexity of the eyes' and the 'constriction of the pronotal base' (Normark and Lanteri 1998). Analyses of flowering plants may include both the general proportions of the leaf (lanceolate *vs.* cordate) and smoothness of the leaf margin (dentate *vs.* entire) (e.g., Swenson and Bremer 1997). Some systematists have expressed doubts about the reliability of phylogenetic inferences based on shape and other morphological features (Hedges and Maxson 1996; Givnish and Sytsma 1997). However, their principal argument is that there is a high risk of homoplasy (i.e., unrecognized independent origins of similar features) because morphological traits are often convergent. In other words, they are concerned about the reliability of the information obtained from morphological comparisons, not the legitimacy of making those comparisons.

Although the use of morphological traits in phylogenetics is generally considered acceptable if the features are qualitatively described, there is considerable opposition to using quantitative morphometric descriptions of those same features (Pimentel and Riggins 1987; Cranston and Humphries 1988; Mickevich and Weller 1990; Crowe 1994). The foundation of this opposition is the perception that morphometric descriptions do not convey the information needed to formulate and test a hypothesis that two shapes are homologous. For example, Mickevich and Weller (1990: 145) claim '...the simplicity of a morphometric variable renders it less tractable' and Pimentel and Riggins (1987: 208) assert that '...dimensions are indirect measures of a feature, and sufficiently vague to question what is appropriate for tests of homology'.

We addressed some of these issues in a previous paper (Zelditch *et al.* 1995). We agreed with many of the objections to variables generated by commonly used morphometric variables, but we also pointed out that those criticisms do not necessarily apply to all variables generated by all methods. In particular, we concluded that some methods of analyzing locations of landmarks sensu Bookstein (1991, i.e., points judged to be comparable among forms) do produce results that can be used in phylogenetic systematics. We reached this conclusion because the comparability of the landmarks provides information crucial to judging whether two shapes are homologous. In contrast, some methods of describing outlines do not include landmarks and do not employ any other mechanism to insure that the same shape feature is comparable across specimens. Without this information, it is not possible to judge whether two shapes are homologous. Therefore, we concluded that the results of these methods of analyzing outlines cannot be used in phylogenetic systematics. However, there are outline methods that use information from landmarks to improve the fit between the original outline and its morphometric reconstruction (Bookstein and Green 1993; MacLeod 1999). Because the included landmarks could also be used to define comparable regions or segments of the outline, our earlier conclusion regarding outline-based studies may not apply to these methods.

In view of the continuing development of these and other morphometric methods, we reconsider the question of what kind of morphometric data can be used in phylogenetic analyses. We begin by examining what information a phylogenetic analysis requires of morphometric descriptions and the role that this information plays in the inference of phylogenetic relationships. Our goal is to present a set of general criteria that can be used to evaluate any method of morphometric analysis. Then we use those criteria to evaluate several specific methods. Included in this survey are two methods that produce descriptions of the relative locations of landmarks (shape coordinates and partial warps), and three methods that produce descriptions of outlines (elliptical Fourier analysis, eigenshape analysis, and extended eigenshape analysis). To illustrate the methods, we use each to analyze the same data set: a small collection of mandibles from several species of terrestrial squirrels (Sciuridae, Marmotini).

Requirements of phylogenetic analysis

Phylogenetic systematics is rooted in the concept of descent with modification (Hennig 1966). Descendant organisms acquire traits transmitted genetically from their ancestors. These traits may be modified by mutation and recombination, and passed on to subsequent descendants. Accordingly, taxa sharing a recent common ancestor are expected to share derived traits that they inherited from that ancestor. These traits will not be found in other taxa because only the descendants of that ancestor have the novel modifications. Similarly, modifications that occurred in a more distant ancestor will be shared by a more inclusive set of taxa that are the descendants of that earlier ancestor. That expectation is not always met because parallelism and convergence produce misleading similarities, that is, features that have multiple, independent origins. Consequently, phylogenetic analysis is not a simple matter of reconstructing the nested sets of monophyletic taxa, but a process of testing and re-testing hypotheses of homology and monophyly. In other words, the existence of homoplasy forces systematists to adopt an approach that tests hypotheses about the origin of similarities.

The concept of descent with modification is the key to understanding what kinds of information morphometric descriptions must convey if those descriptions are to play a role in phylogenetic inference. The traits used to infer phylogenies must be features that can be transmitted from one generation to the next. That is, they must be intrinsic features of individual organisms. As Pimentel and Riggins (1987) observe, statistical parameters and axes computed by principal components analysis (PCA) do not meet this requirement because these kinds of variables describe properties of aggregates. Individuals do not have means or variances, or principal components. An individual can be scored in terms that describe its distance from the mean, in units of variance, in directions described by principal component axes (and can even be scored on the principal components of a sample that did not include that individual); but this description does not necessarily refer to attributes that the individual could have inherited from an ancestor. Instead, this description specifies the location of the individual in a morphometric space that is defined by an array of individuals and measurements. A change in any component of that array, addition or removal of any individual or measurement, would change every component of the description of the individual because the space, itself, would be different. Therefore, we agree with Pimentel and Riggins (1987) that variables defined by the diversity of the sample that happens to be on hand at the moment are not appropriate variables for an analysis of phylogenetic relationships.

If the traits used in phylogenetic inference must be intrinsic features of individual organisms, then the first requirement of any description is that it must convey information that connects the trait to the organism. This criterion applies to qualitative descriptions as much as it does to quantitative descriptions. We cannot judge whether 'red' is homologous if we do not know what feature is red. Only the specification that 'red' refers to 'feathers' indicates which 'red' and 'not-red' objects are comparable. This information that connects the trait to the organism (the specification of 'feathers') necessarily entails another hypothesis of homology. In this example, that hypothesis is the homology of 'feathers' at some level that includes all the taxa in the current study. If that hypothesis is wrong, the question of which state of the character 'feathers' diagnoses a monophyletic group becomes non-sensical because there is no group to divide. In practice, the hypothesis that the character diagnoses a monophyletic group often goes untested because the study is focussing on the more recent divergence diagnosed by the states of the character (colour of feathers, not origin of feathers). Even so, the character whose states are to be compared should be described in enough detail that the hypothesis of its homology can be evaluated should other data indicate such an analysis is warranted (e.g., discovery of numerous congruent differences in feather structure).

Although we agree with Pimentel and Riggins (1987) that the traits used in phylogenetic inference must be intrinsic features of individual organisms, we do not agree with their claim that all measurement data must be rejected because measurements are abstractions, not anatomical parts. The fact that 'parts' are also abstractions is readily apparent in the empirical studies detailing complex webs of anatomical connections and functional interactions (van der Klaauw 1948–1952; Dullemeijer 1958; Liem 1973; Lombard and Wake 1986; Schaeffer and Lauder 1986; Wake 1993). These kinds of analyses represent a clear challenge to the conventional notion that anatomical parts can be delimited by tracing boundaries between different types of tissues. In fact, Moss and colleagues (Moss and Young 1960; Moss 1962; Moss and Saletijn 1969) take a novel approach to delimiting parts – their functional matrix paradigm – and demonstrate just how different the list of parts can be. Furthermore, the particular kind of abstraction that is the focus of Pimentel and Riggins' (1987) objection, the measurement of a length, underlies the qualitative assessment of 'long' *vs.* 'short' as much as it does the quantitative assessment of 'greater than 10 cm' *vs.* 'less than 10 cm'. Because systematists routinely engage in such abstractions in the course of qualitative description, they cannot count as a reason to reject quantitative descriptions.

The requirement that phylogenetic inference be based on intrinsic features of organisms only means that the description of a measurement must specify a unique location on the organism. This is not an onerous requirement or even a particularly remarkable one. It is nothing more than a requirement to avoid vagueness by specifying which distance was measured among the myriad similar measurements that are possible. For example, skull width could be measured at a number of points, and thus an individual could have several different values for skull width. This ambiguity can be avoided by specifying that the width should be measured between particular end points, such as the distal tips of the mastoid processes. This specificity also helps to insure that the trait passes Patterson's (1982) test of conjunction. A trait fails this test, and the hypothesis of homology is rejected, if different states occur in the same organism. If the endpoints of a measurement are carefully defined, the only way for the measured length to fail the test is for the endpoints to fail the test. In the example above, skull width at the mastoid process would fail the test only if a taxon had two pairs of mastoid processes.

Patterson's other tests of homology are similarity and congruence. As de Pinna (1991) points out, similarity usually is not used as an explicit test, but as a criterion for proposing a hypothesis of homology. Conjunction can also be considered an element of similarity because the test is essentially an evaluation of topographic similarity. This leaves congruence as the only actual 'test' of hypotheses of homology. Because the test of congruence is applied to hypotheses of homology – that are based on judgements of similarity and differences – the testing of congruence does not concern us here. The focus of this chapter is the justification for submitting a trait to the test of congruence in the first place. Thus, we are concerned only with the evaluation of similarity, and the question of whether morphometric description can provide a legitimate basis on which to formulate a hypothesis of homology.

It is important to bear in mind that similarity is a relative description, both in general usage and in phylogenetic systematics. Characterizing two things as 'similar' is meaningless; one must also indicate what would be 'dissimilar'. In phylogenetic systematics, 'similar' means sharing a derived trait and 'dissimilar' means lacking that trait. In other words, 'similar' is shorthand for 'sharing a similar divergence from a more primitive condition'. Again, one must know what the more primitive condition is. However, the inference that one condition is primitive and the other is derived, ultimately depends on the structure of the phylogenetic tree.

A preliminary hypothesis could be based on outgroups or ontogenetic sequences, but it is not unusual for these hypotheses to be rejected. In any event, the inferences of ingroup and outgroup, or primitive and derived, are phylogenetic hypotheses that are beyond the scope of any morphometric analysis and beyond the scope of this volume. The issue addressed in this chapter is whether morphometric methods can provide the information needed to judge similarity according to the relevant criteria. This does not mean all cases must be unambiguous. It does mean that it must be possible to provide an unambiguous description of the alternative states, and formulate an explicit hypothesis of common ancestry that can be falsified by incongruence with the phylogeny. In this regard, 'length greater 10 cm' is better than 'long' as a character-state description because it more clearly delineates the criterion for judging similarities and differences.

There are other objections to using morphometric variables, but most of them concern an expectation that homoplasy in these traits is even more common than homoplasy in qualitatively described traits (Felsenstein 1988; Mickevich and Weller 1990; Garland and Adolph 1994). This expectation is usually justified by referring to studies that show such traits are determined by multiple loci, or by demonstrating that different combinations of contributing elements could achieve the same net effect, as in ratios. However, this is not an argument against quantitative evaluation, it is an argument against the homology of the kinds of traits that are commonly subjected to quantitative analysis. If the same limb length can be accomplished by different combinations of limb segment lengths, then perhaps limb length is not homologous. It is also possible that the limb length in question is primitive and the proportions of the segments were modified while the total length remained constant. The only way to choose between these alternatives is to test whether limb length or limb proportion is congruent with the phylogeny.

Pimentel and Riggins (1987: 201) remarked that 'Good cladistics like all good science is dependent upon good data'. Their implication is that good characters are dependent on good description. We concur, but would also point out that the requirements for good description are the same whether the traits are described qualitatively or quantitatively. There must be information in the description that unambiguously specifies the location of the trait on the organism. That information supplies the basis for comparing the observed values and formulating hypotheses of transformation and homology.

Mandibles and landmarks

Before comparing the abilities of different morphometric methods to anchor shape descriptions to anatomical features, we present here a brief introduction to squirrel mandibles. We have also included a brief discussion of the landmarks used in some of the analyses.

As is typical for rodents, the lower jaw holds a large incisor and a relatively small number of cheek teeth – three molars and a premolar (Figure 6.1). Anterior to the cheek teeth, the bone is little more than a sheath for the incisor. The root of the incisor curls below the cheek teeth and up into the articular process. At the posterior end of the articular process is the condyle, that articulates with the skull at the posterior end of the zygomatic arch. The articular process projects posteriorly more than dorsally, so that the condyle is not far from the occlusal plane of the cheek teeth. Rising between the molars and the articular process is a rather short and thin coronoid process. The small size of this process reflects the small size of the temporalis muscle, the jaw closing muscle that inserts on it. This muscle originates on the side of the brain case and pulls up and back. Below the articular process is a large angular process that provides some of the insertion area for the masseter muscle, another jaw closing muscle, that also occupies most of the area below the cheek teeth. One of the principal portions of the masseter originates on the zygomatic arch and pulls almost straight

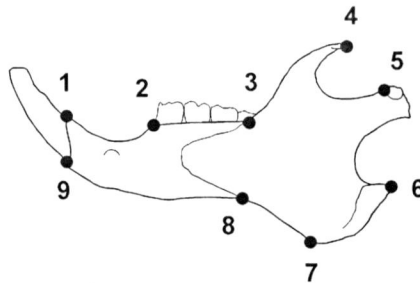

Figure 6.1 Outline of *Spermophilus variegatus* mandible with locations of landmarks.

up. Another portion originates on the side of the rostrum and pulls the jaw forward and up.[1]

For the landmark-based analyses, we identified nine topographical features that could be located reliably in all specimens in the study. We did not use the posterior corner of the articular process, behind the condyle, because a distinct projection in this direction is not found in all taxa. Also, we did not use points on the teeth because their locations are affected by the amount of wear, which differs according to the abrasiveness of the diet. There are additional points that could have been used as landmarks, such as the mental foramen. To increase the comparability of analyses that use landmarks and those that do not, we used only landmarks that are located on the outline of the jaw.

Because the purpose of landmarks (and the reason they are called landmarks) is to represent comparable locations on the anatomy of the organism, we chose points that can be described as unique locations on distinctive anatomical features. For example, landmark 5 is on the elliptical condyle of the jaw joint at the anterior end of the ellipse. Perception of the location of this point may be influenced by the orientation of the specimen relative to the viewer (so care was taken to place all specimens in the same orientation for digitizing), but the definition of the point is unambiguous. To insure that this point is comparable, we verified that all the squirrels examined for this study have an elliptical condyle with the long axis approximately parallel to the anteroposterior axis of the animal. Based on their uniformity, we infer that the elliptical shape and its orientation are primitive for these squirrels, that is, homologous for a group that includes at least these squirrels. Thus, this landmark is a distinctive anatomical feature, one that is present in all taxa in the study and one which we infer was present in the most recent common ancestor of these taxa. Accordingly, we regard landmarks of this type as comparable features in their own right and not as abstract geometric conventions for the representation of more conventional anatomical features (contra MacLeod 1999).

Similar arguments apply to all of the landmarks used in this study. Landmarks 2, 3 and 8 are at the intersections of conventional anatomical structures; these structures and their intersections are found in all specimens in the study. Landmarks 4, 6 and 7 are at the tips of processes and can also be recognized as points of maximal curvature on those processes. These latter three points also have mechanical and anatomical significance as the limits of the attachment areas for their respective muscles. The consistent relationship between the musculature and the bony processes is evidence in support of the homology of the processes, and provides an additional motivation for including landmarks on these processes in the morphometric analysis, but they are not immediately relevant to the inference that these landmarks are comparable. That inference depends only on the consistent shapes of those processes. In fact, the

---

1 Additional details about muscle attachment areas and orientations in squirrels can be found in Ball and Roth (1995) and in Thorington and Darrow (1996). Previous studies of marmotines (ground squirrels, marmots and prairie dogs) suggest there has been an evolutionary transition from a diet that includes primarily fruit and nuts to a diet that includes primarily grasses, herbs and their seeds (Howell 1938; Bryant 1945; Black 1963). In other mammals, similar shifts in diet are associated with such changes in jaw morphology as general elongation, enlargement and posterior shift of the tooth row, and expansion of muscle attachment areas (Greaves 1978; Janis and Ehrhardt 1988).

inconsistency of the bending of the coronoid process is the primary reason we did not recognize other landmarks on that structure (between 3 and 4, or between 4 and 5) even though that information would be valuable for understanding the relationship between the shape of this structure and its function.

The two landmarks whose comparability might seem most dubious are 1 and 9 on the opening of the incisor alveolus. These landmarks are points at the dorsal and ventral extremes of the aperture, but they are not simply points on a diameter across a circular aperture. If they were, there would be legitimate questions about their independence as well as about their comparability (Bookstein 1991). However, the medial wall of the alveolus is flattened where it contributes to the mandibular symphysis (the articulation of the right and left jaws). Consequently, the landmarks are actually corners in the wall of the alveolus and can be recognized independently. If the coordinates of these points are correlated, it is because the evolution of these features is correlated, not because one point is defined in relation to the other.

In summary, all of the landmarks described in this section meet the conditions for comparability set out in the previous section. All are defined with reference to specific anatomical features that a systematist familiar with the group can easily identify. In fact, all of the landmarks are specific anatomical features, themselves. Furthermore, this information about these anatomical features, our inferences regarding their homology can be tested, either by analyzing a more complete set of taxa or by analyzing a more inclusive set of taxa.

Comparison of morphometric methods

The focus of these analyses is to determine which methods are suitable for phylogenetic analysis, not to provide a complete analysis of the evolutionary history of mandibular shape in marmotine squirrels. Accordingly, the specimens in this study represent a small number of functionally and morphologically differentiated taxa. Based on previous systematic studies (Bryant 1945; Black 1963), one of the ground squirrels (*Spermophilus variegatus*) represents the sister group to the monophyletic group containing all of the other marmotines analyzed here. Therefore, the focus of the comparisons will be on identifying features shared by some of these marmotines that distinguish them from *S. variegatus*. However, because these marmotines represent only a small part of the morphological diversity in the group, inferences about evolutionary transformations based on these data should be regarded as preliminary hypotheses and not definitive statements about marmotine evolution.

Bookstein shape coordinates

Bookstein shape coordinates (Bookstein 1991) are the location of one landmark with respect to a line defined by two other landmarks (the baseline). Each specimen is rescaled and rotated so that the baseline points have coordinates (0, 0) and (1, 0). After this rotation, the x- and y-coordinates of every other landmark represent distances away from one end of the baseline in each direction, in multiples of baseline

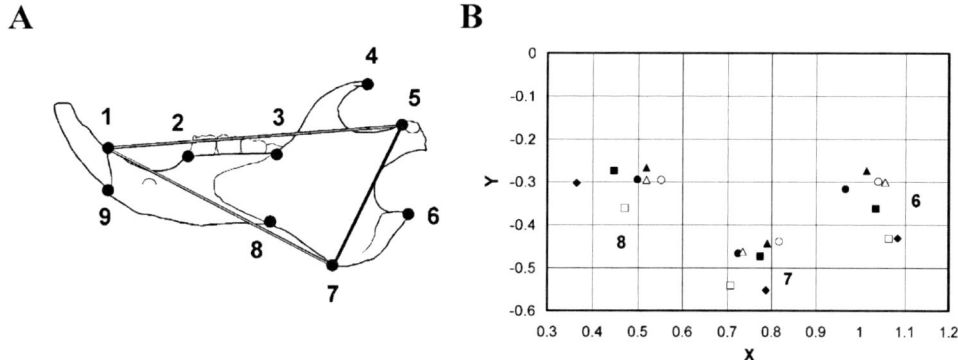

Figure 6.2 Analysis of shape coordinates, (A) Triangle described by shape coordinates of landmark 7, (B) Scatter-plot of shape coordinates for landmarks 6, 7 and 8 in all seven marmotines. Symbol legend: filled circle – *S. variegatus*, open circle – *S. franklini*, filled triangle – *S. tridecemlineatus*, open triangle – *S. spilosoma*, filled square – *S. columbianus*, open square – *C. ludovicianus*, filled diamond – *M. flaviventris*.

length. Differences between specimens in the location of these landmarks relative to the baseline can be displayed in a bivariate plot of the shape coordinates. Each landmark can also be viewed as the apex of a triangle whose base is the baseline. The shape coordinates describe the shape of that triangle, so differences between two specimens in the shape coordinates of a landmark represent a difference in the shape of that triangle.

Figure 6.2A shows the triangle representing the location of landmark 7, on the angular process, relative to the baseline connecting landmarks 1 and 5. The shape coordinates for this landmark (Figure 6.2B) indicate that in *Cynomys ludovicianus* (black-tailed prairie dog) and *Marmota flaviventris* (yellow-bellied marmot) this corner of the angular process is more ventrally located relative to the length of the baseline than it is in *S. variegatus* or any of the other four species of *Spermophilus* (ground squirrels). Based on this information, we could hypothesize that the relatively ventral location of the medial corner of the angular process is a homology shared by a group that includes the prairie dog and marmot and excludes the five ground squirrels. However, before doing that it is important to remember that shape coordinates are not the absolute position of the landmark, but its position relative to the baseline. There is a considerable portion of the jaw between the baseline and landmark 7. The shape coordinates do not indicate whether this corner is lower because the angular process is deeper, or because the body of the jaw is deeper (between landmarks 3 and 8), or perhaps because of some combination of general deepening of the entire jaw and relatively greater deepening of the angular process.

Combining information from several landmarks can provide additional information that can be used to more narrowly circumscribe the region of shape change. In the case of landmark 7, the locations of landmarks 6 and 8 should be particularly informative because they are also on the angular process. Figure 6.2B shows that in *C. ludovicianus*, all three landmarks are more ventral relative to baseline length than they are in *S. variegatus*. Therefore, the body of the jaw, not the angular process, is relatively deeper in *C. ludovicianus* than in *S. variegatus*. In *M. flaviventris*, landmark 8 does not move down with landmarks 6 and 7; instead, landmark 8 moves forward relative to the baseline. Thus, the deepening relative to *S. variegatus* occurs in the angular process of *M. flaviventris*, not in the body of the jaw. Similarly, comparison of the shape coordinates of these three landmarks in other taxa indicates that in three ground squirrels (*S. franklini*, *S. spilosoma* and *S. tridecemlineatus*) the angular process is relatively longer than in *S. variegatus*. Furthermore, the locations of landmarks 6 and 7 indicate that the posterior part of the process is more elongate in *S. spilosoma*, whereas the anterior is more elongate in the other two.

This analysis demonstrates that one triangle of landmarks is insufficient to pin a shape difference to a particular region on a form. Therefore, shape coordinates for a single triangle cannot be directly translated into a character; they do not provide enough information to support a hypothesis about the homology of the observed shapes. However, examination of several shape coordinates does allow detection of the location and extent of a shape difference. This more complete description can be used to construct characters for phylogenetic analysis. In the example above, we described four shape differences distinguishing various taxa from *S. variegatus*. Based on one of these differences we could hypothesize that the relatively elongate shape of the angular process shared by *S. tridecemlineatus* and *S. franklini* is a homologous feature that indicates these two taxa share a recent common ancestor not shared by the other taxa. Each of the other three shape differences distinguishes only one species from *S. variegatus*. These three taxa may have autapomorphic shapes, or they may share these derived shapes with taxa that were not included in this analysis.

Partial warps

In the analysis using shape coordinates, we inferred the localized shape changes from comparison of changes in the relative locations of several landmarks. Thin-plate spline analysis provides a means of precisely describing specific geometric patterns of relative landmark displacements (Bookstein 1989, 1991; see also Bookstein 1996; Rohlf 1996). In thin-plate spline analysis, every shape is described in terms of its differences from a reference form. More technically, each observed arrangement of landmarks is a point on a complexly curved multidimensional surface called Kendall's shape space. The number of dimensions of this surface is a function of the number of landmarks. For two-dimensional forms, the number of dimensions of Kendall's shape space is $2p - 4$, where p is the number of landmarks. Shapes described by different numbers of landmarks exist in different spaces with different numbers of dimensions. Because Kendall's shape space is curved, it has a non-Euclidean geometry. This means that many commonly used methods of analyzing shape variation are not immediately available. The process of defining components of shape and shape variation is equivalent to defining axes of a coordinate system. In Kendall's shape space, as on any curved surface, straight lines that are initially parallel will eventually converge. Consequently, the same amount of shape difference at different places on the surface will have different descriptions in terms of the coordinate system. Thin-plate spline analysis is one way of projecting locations of shapes in Kendall's shape space onto a Euclidean space that is tangent to Kendall's shape space. The reference form defines both the point of tangency and the axes of the tangent space.

In thin-plate spline analysis, the axes of the tangent space are computed by eigenanalysis of the bending-energy matrix, which is a function of the interlandmark distances among all the landmarks of the reference form. Eigenanalysis is a commonly used technique for computing the principal axes of a distribution. The interpretation of those axes depends on the kind of data contained in the matrix that was analyzed. In PCA the matrix under analysis represents observations of many traits in many individuals; therefore, the result is a new description of the diversity in that sample of individuals. The original variables described dimensions of variation; therefore the new variables – the principal components axes – also describe dimensions of variation. Principal components scores represent the projections of specimens (the original specimens or new specimens) onto those axes. In thin-plate spline analysis, the matrix under analysis is a function of the relative positions of landmarks in the reference form; therefore the axes represent potential rearrangements of those landmarks. Scores denoting the projections of other specimens onto these axes represent of descriptions of the arrangements of landmarks in those specimens as rearrangements of the reference form. Because the thin-plate spline axes are functions of the shape of the reference form and are not dependent on an analysis of variability, arguments against using scores on principal components axes in phylogenetic systematics (Zelditch et al. 1995) do not apply to scores on thin-plate spline axes.

As we mentioned above, the axes produced by the thin-plate spline analysis are computed from the relative positions of landmarks in one form and represent patterns of possible change in the relative positions of landmarks. These axes are called partial warps. The patterns of relative landmark displacement they represent can be illustrated by multiplying the pattern by an arbitrary vector. Figure 6.3 illustrates one such pattern computed using the specimen of S. variegatus as the reference form. The scores obtained by projecting a second shape onto the partial warps would indicate the observed contribution of that pattern of relative landmark displacement to the total difference in relative landmark positions between the reference form and the second form. Because the landmarks have x- and y-coordinates, the partial warps scores have x- and y-components. Thus, partial warps scores are the x, y coefficients of the vector that would be multiplied by the partial warp pattern to illustrate that component of the actual difference between the two specimens (Slice et al. 1996).

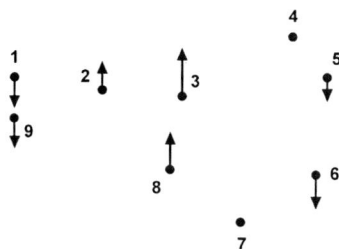

Figure 6.3 Diagrammatic representation of a partial warp for the reference form, S. variegatus.

Thin-plate spline analysis can also be described by analogy to the bending of an idealized thin steel plate (hence the terms 'bending energy' and 'warps'). Steel plate tends to bend in a manner that minimizes both the total amount bending and the energy required for that bending. In the analogy, the reference shape is a set of points that can be pushed up or down orthogonal to the plane of the undeformed plate. The difference in relative landmark positions between the reference shape and the second shape represents the distances that the points on the plate will be pushed up or down. The patterns of bending that can be detected using the vertical displacements of the points representing the landmarks of the reference form are a function of the number of landmarks and the distances between them. These patterns are called principal warps. The number of principal warps may seem rather small, but all possible sets of relative landmark displacements can be described by the sum of a combination of principal warps, each multiplied by the appropriate magnitude. This is similar to the way in which simple sine functions are summed to produce more complex waveforms.[2]

As discussed above, partial warps can be computed for any reference, including a random scatter of points. To insure that descriptions are framed in terms that are relevant to the biological issues under investigation, there needs to be a criterion for choosing a reference form that is appropriate to those issues. Earlier in this chapter we argued that the characters in a phylogenetic analysis must refer to features of an organism. If we are to claim that partial warps, or combinations of partial warps, represent features of an organism, then the reference should be a form that represents an observed morphology. A reference form that is a mean of several species may not resemble any observed form, especially if the species are quite different. Although we would argue against using the mean of several species as reference shape for phylogenetic analysis when there are large gaps between the species, we do support the using a mean computed from a more restricted array of specimens (i.e., representatives of a single species or of an age group within a species). This may seem inconsistent with our opposition to methods dependent on the axes of variation in a sample, but it is not. The partial warps of the mean shape are computed without reference to the variation around that mean. Differences of other specimens from the mean do not enter the analysis until the computation of partial warps scores, that represent those differences projected onto the partial warps defined by the mean. The argument can also be made that the mean is a hypothetical construct not a direct observation, regardless of the diversity in the sample. Although this is technically correct, when the sample size is large and the variance is small, it is likely that several individuals will have shapes that are indistinguishable from the mean. In addition, it can be argued that the variation of such a sample represents random noise and therefore the features of the mean actually are the features of each individual in the sample. Of course, one can always side-step all of these issues by using the individual closest to the mean as the reference form. If the reference is a single specimen, then there is no question that the partial warps refer to features of an organism.

---

2 In some older literature (e.g., Bookstein 1991; Swiderski 1993), the term principal warps was applied to both the three-dimensional patterns of the bending steel plate and the two-dimensional patterns of landmark displacements now called partial warps, and partial warps scores were simply called partial warps; mathematically, the two-dimensional partial warp is a projection of the three-dimensional principal warp.

One of the more attractive features of thin-plate spline analysis is that the interpolation function used to compute the bending of the thin steel plate can also be used to compute deformed grid pictures in the style of D'Arcy Thompson (1917) to illustrate differences between shapes. Figure 6.4 shows deformed grids illustrating differences in jaw shape between *S. variegatus* and the other six marmotines. All shapes are rescaled to the same centroid size (the square-root of the sum of the squared distances of the landmarks from their centroid) and each second form is rotated to an optimal position with respect to the reference form, *S. variegatus* (one that minimizes the square-root of the sum of the squared distances between corresponding landmarks). Grids for two of the ground squirrels, *S. franklini* and *S. tridecemlineatus*, exhibit very little deformation. There is some uniform stretching along the anteroposterior axis, but there is a similar amount of stretching in the grids of other species. Of the remaining four species, *S. spilosoma* appears to be distinct from the other three. In *S. columbianus*, *C. ludovicianus* and *M. flaviventris*, the gridlines are compressed between landmarks 3, 4 and 5, which indicates that the tips of the coronoid and articular processes are relatively closer to the back of the tooth row. In addition, the gridlines are dilated between landmarks 6, 7 and 8, which indicates expansion of the angular process. In *S. spilosoma*, the only posterior landmarks that undergo much relative displacement are 5 and 6. Another feature unique to *S. spilosoma* is a dorsoventral expansion between landmarks 1 and 9, reflecting the relatively thicker incisor of this species.

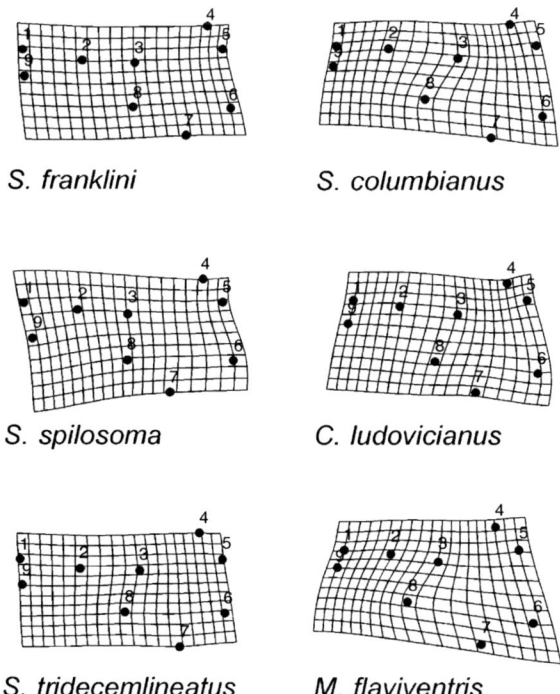

Figure 6.4 Deformed grids illustrating changes in relative landmark locations, comparing each specimen to *S. variegatus*.

80 Donald L. Swiderski et al.

Although *S. columbianus*, *C. ludovicianus* and *M. flaviventris* share some general similarities in the deformations of the posterior region of the jaw, closer examination also reveals several differences. Three warps account for most of the transformations in this region (Figure 6.5). The first is the largest scale warp (partial warp 1), that describes contrasting displacements of landmarks in the middle of the form relative to landmarks closer to the ends of the form. In *M. flaviventris*, the partial warp scores for this feature indicate a large anteroposterior component, a gradient of relative posterior expansion. (A similar, but smaller, expansion is also found *S. spilosoma*). Partial

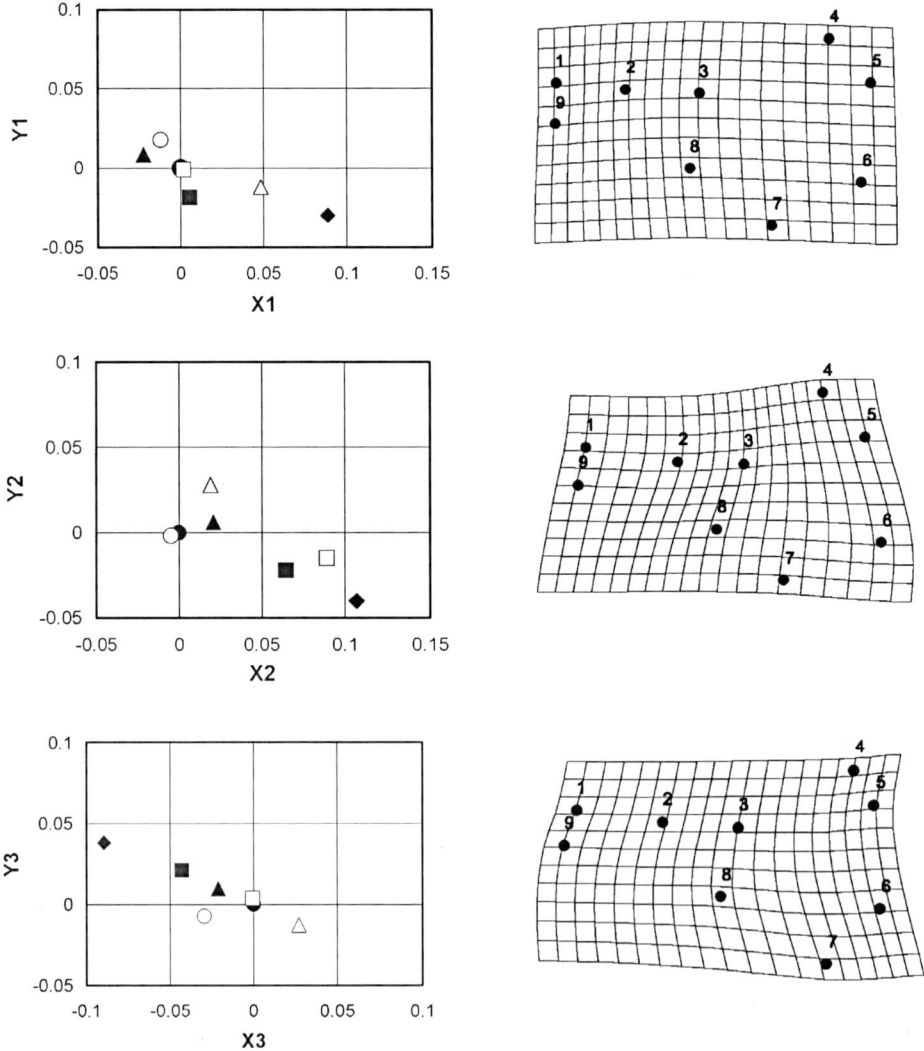

Figure 6.5 Scatter plots of partial warp scores for the three largest scale warps, and deformed grids illustrating that component of the deformation of *M. flaviventris*. Symbol legend: filled circle – *S. variegatus*, open circle – *S. franklini*, filled triangle – *S. tridecemlineatus*, open triangle – *S. spilosoma*, filled square – *S. columbianus*, open square – *C. ludovicianus*, filled diamond – *M. flaviventris*.

warp 2, the next largest scale feature, also accounts for much of the transformation in this region. This feature is a little more complex: the landmarks on the tooth row and the landmark at the posterior end of the angular process move in one direction, the landmark at the anterior end of the angular process and on the tips of the coronoid and articular processes move in the opposite direction. For this feature, *S. columbianus*, *C. ludovicianus* and *M. flaviventris* have large x-scores, which can be most simply described as a transformation of this rectangle of landmarks into a trapezoid. The dorsal landmarks (2, 3, 4 and 5) move closer together, the ventral landmarks (6, 7 and 8) move farther apart. In addition, displacements of landmarks 2 and 3 relative to 1 and 9 at the base of the incisor indicate a relative elongation of the anterior-most region of the jaw. The third feature accounting for substantial transformation in this region is partial warp 3. This feature is similar to partial warp 2, but with the corners of the rectangle closer together (landmarks 4, 6, 7 and 8). Partial warp 3 also involves contrasting displacements of landmarks 1 and 9 at the base of the incisor. The scores for this component indicate that it accounts for a substantial portion of the transformation of *M. flaviventris*, and a smaller proportion of the change in *S. columbianus*.

One possible interpretation of these three partial warps is that they represent three distinct characters to be analyzed separately just as one would analyze differences in forelimbs and hindlimbs separately. Using *S. variegatus* as the outgroup to all the others, as we did when analyzing the shape coordinates, scores on partial warp 1 indicate a small difference may be shared by *S. franklini* and *S. tridecemlineatus*. A much larger difference in the opposite direction may be shared by *S. spilosoma* and *M. flaviventris* (with *M. flaviventris* having a much greater divergence in this direction). The $-y$ component of this change may also be shared by *S. columbianus*. Scores on partial warp 2 indicate there are relatively small differences in this feature among four of the marmotines, and a much larger divergence from *S. variegatus* that may be shared by *S. columbianus*, *C. ludovicianus* and *M. flaviventris*. Scores on partial warp 3 indicate *S. spilosoma* diverges from *S. variegatus* in the opposite direction of the other taxa. Interestingly, most of the taxa fall on a line running from *S. spilosoma* to *M. flaviventris*, which may indicate that the only taxon diverging from *S. variegatus* in a new direction is *S. franklini*.

As we argued before (Swiderski *et al.* 1998), many of the commonly used coding protocols can be used, with minor modifications, to assign character state codes to taxa based on these scatter plots. The number of states recognized for each character would depend on the variation within each species and coding protocol that is used. We have not assigned explicit states in this example because all methods require more than one specimen per species, and because our purpose is not to continue the discussion of coding methods, but to show that the coding would be based scores that refer to specific shape differences in well-defined anatomical regions.

One of the arguments against treating each partial warp as a separate character is that this decomposition of shape differences is biologically arbitrary, that is, they have no necessary relationship to the patterns of variation and evolutionary change (Rohlf 1998). As Rohlf points out, most partial warps describe transformations of broadly overlapping regions and most specimens have scores on several overlapping warps. Naturally, these observations raise concerns about the independence of the characters the partial warps represent. The problem of character independence is not new to systematists; the journals and textbooks contain many discussions of this problem and methods of addressing it. However partial warps present a much more complex problem. Because of their overlap, they can sum to quite different transformations. As Figure 6.5 shows, high scores on partial warps 2 and 3 can represent large changes in angular process associated with lesser changes elsewhere in the jaw. Figure 6.6 shows combinations of these two warps as vectors of landmark displacement for three of the specimens. In *M. flaviventris*, that has high scores on both warps, the entire angular process appears to be relatively longer than in *S. variegatus*. In contrast, *C. ludovicianus* has a high score only on partial warp 2, and only the posterior part of the process appears to be relatively longer than in *S. variegatus*. In *S. spilosoma*, that has negative scores on partial warp 3, the anterior of the process is actually relatively shorter than in *S. variegatus*. Furthermore, these three changes in the shape of the angular process appear to be associated with quite different changes in the rest of the jaw. Thus, the partial warps are not necessarily equivalent to familiar anatomical features, and differences in partial warp scores cannot automatically be translated into hypotheses of homologous shapes. However, decomposition of the total deformations into their component warps can be used as a tool for investigating possible relationships among anatomical regions because the warps refer directly to the specific features of anatomy archived as landmarks. Therefore, hypotheses of homologous shapes can be based on the results of a partial warps analysis.

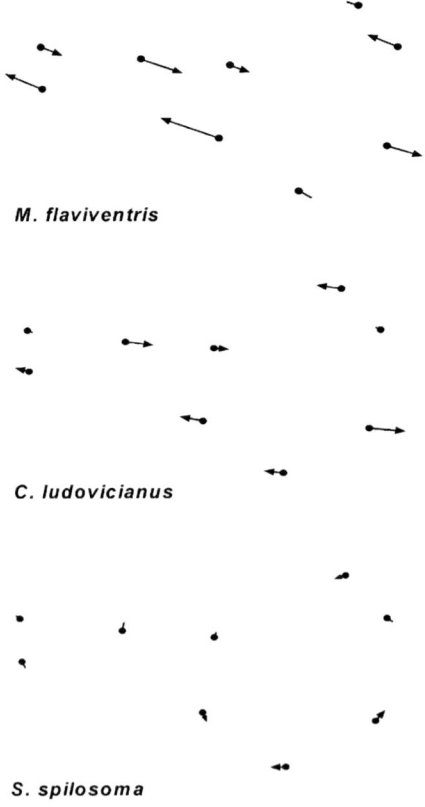

Figure 6.6 Vectors of relative landmark displacement representing the sum of partial warps 2 and 3.

On occasion, a principal warp may, fortuitously, correspond closely to a familiar anatomical feature. In this study, such a feature is partial warp 4, which describes changes at the base of the tooth row relative to the rest of the jaw (Figure 6.7). Both S. columbianus and C. ludovicianus have large scores on the x-axis reflecting similar elongation of the tooth row relative to the rest of the jaw. There also are noticeable displacements of landmarks 4, 5 and 6 relative to one other, but these are small in comparison to the relative displacements of these landmarks implied by other warps. Furthermore, no other warp or combination warps accounts for the relative elongation of the tooth row seen in the total deformations (Figure 6.4). Under these circumstances, it seems reasonable to treat the individual warp as the description of that specific anatomical difference.

The inferences of which taxa share derived features based on the partial warps are somewhat different from the inferences based on the shape coordinates. This difference does not represent a conflict between the methods, nor does it represent a flaw in either method. Instead, the differences between these results reflect the fact that they are describing different features. No warp refers to only a single triangle of landmarks. Each warp refers to relative displacements of all nine landmarks. This information described by the partial warps is present in the shape coordinates, but it is not easily detected from an inspection of them. For example, the shape coordinates for landmark 3 indicate it is located more posteriorly in *S. franklini*, *S. tridecemlineatus*, *S. columbianus* and *C. ludovicianus*, but shape coordinates for landmark 2 indicate it also is located more posteriorly in *S. franklini* and *S. tridecemlineatus* (Figure 6.8). Thus,

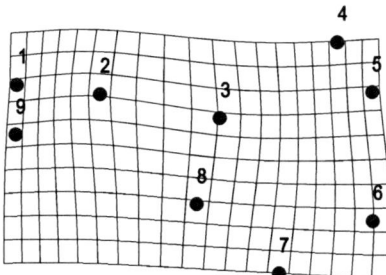

Figure 6.7 Deformed grid for partial warp 4 in *C. ludovicianus*, illustrating relative elongation of tooth row.

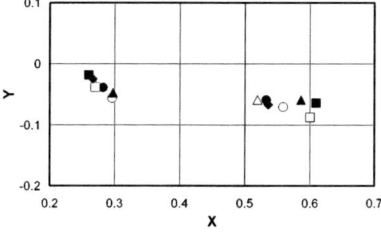

Figure 6.8 Scatter plot of shape coordinates for the landmarks at each end of the tooth row in all seven marmotines. Symbol legend: filled circle – *S. variegatus*, open circle – *S. franklini*, filled triangle – *S. tridecemlineatus*, open triangle – *S. spilosoma*, filled square – *S. columbianus*, open square – *C. ludovicianus*, filled diamond – *M. flaviventris*.

simultaneous examination of the two sets of shape coordinates reveals the relative elongation of the tooth row in *S. columbianus* and *C. ludovicianus*, and also a posterior displacement of the entire tooth row in *S. franklini* and *S. tridecemlineatus*. Our descriptions of the shape coordinates and warps describing changes of the angular process do not match so closely because no single warp closely corresponds to this anatomical region.

More important than the differences between the shape features reported by these two analyses is the similarity in the way landmarks were used to specify the anatomical features to which the scores refer. Because the shape changes described in each analysis could be pinned to specific anatomical features, we could be certain that we were evaluating comparable features. For this reason we conclude that partial warps and shape coordinates can be a basis for formulating hypotheses of homology and monophyly.

Fourier analysis

The methods described above evaluate only the relative positions of a small number of points. Numerous morphometricians have pointed out that other shape information, such as the curvature of the outline or the texture of the surface, is not included in those landmark-based analyses (Ehrlich *et al.* 1983; Read and Lestrel 1986; Bookstein and Green 1993; Lestrel 1997a; Pesce Delfino *et al.* 1997; MacLeod 1999). It may be possible to interpolate changes in the curvature of the coronoid process or the incisor alveolus from the thin-plate spline deformation grids, but information about the actual differences in outline curvatures were not used to compute the deformation grids.

One approach to describing an outline (or any curved line) is to decompose it into a Fourier series, which is a trigonometric series of the general form:

$$f(\theta) = A_0 + \sum_{n=1}^{N} a_n \cos n\theta + \sum_{n=1}^{N} b_n \sin n\theta, \tag{6.1}$$

(Detailed expositions of Fourier methods can be found in Tolstov 1962, Davis 1989, Lestrel 1997b, and references therein). Values of θ range from 0 to 2π, and values of n are positive whole numbers ranging from 1 to a maximum of N, which is $\frac{1}{2}$ the number of digitized points. Each outline is described independently, and comparisons among outlines are based on the values of the coefficients a_n and b_n. When $n = 0$, $f(\theta)$ is the constant A_0, which represents the best fitting straight line (open curves) or circle (closed outlines). The goodness-of-fit criterion is the minimum of the sum of the squared distances of the digitized points from the reconstructed curve.

Each increment of n represents the frequency of a periodic function, commonly called a harmonic, that is added to the constant to improve the fit of the reconstructed curve to the digitized points. For example, when $n = 1$, the function completes one cycle between the starting point and ending point of the curve. The amplitude of the *n*th harmonic is $\sqrt{a_n^2 + b_n^2}$, and represents the contribution of that harmonic to the description of the total form (specifically, the proportion of the squared deviations of the digitized points from the constant that is explained by the harmonic). The coefficients a_n and b_n are also determined by the least-squares criterion. Each harmonic with an amplitude greater than zero produces some improvement of the fit of the reconstruction to the digitized points, but there need not be a steady decline in the increment of improvement of fit. If the digitized outline is highly corrugated (like the suture line of an ammonite), the harmonics with the highest frequencies could be among the ones with the highest amplitudes. In addition to amplitude, the coefficients a_n and b_n can be used to compute the phase angle ($\tan^{-1}(a_n/b_n)$), which represents the position of the starting point of the harmonic relative to the starting point of the digitized outline. If the shapes being compared have been rescaled to the same size and rotated to the same orientation, the phase angle provides a means of checking whether a given harmonic refers to the same anatomical feature on the outline.

Two forms of Fourier analysis are commonly used to describe closed outlines. In the older method, the outline is digitized in equal angular increments around a central point and points on the outline are described in polar coordinates. This method cannot be used to analyze complex forms in which a radius intersects the outline at more than one point. Furthermore, if the outline is digitized in large angular increments, narrow projections could be excessively smoothed, or missed entirely. In Elliptical Fourier Analysis (Kuhl and Giardina 1982; Rohlf and Archie 1984; Ferson *et al.* 1985), points on the outline are described in the more familiar *x*- and *y*-coordinates, and each coordinate is computed by a Fourier series:

$$x(t) = A_0 + \sum_{n=1}^{N} a_n \cos nt + \sum_{n=1}^{N} b_n \sin nt, \tag{6.2}$$

$$y(t) = C_0 + \sum_{n=1}^{N} c_n \cos nt + \sum_{n=1}^{N} d_n \sin nt, \tag{6.3}$$

A_0 and C_0 are the coordinates of the outline's centroid; t ranges from 0 to 2π. The name Elliptical Fourier Analysis refers to the first term of the series, which describes an ellipse; successive terms account for deviations from that ellipse. Although Elliptical Fourier Analysis uses twice as many coefficients to describe the same form, this disadvantage is outweighed by the relative ease of computing those coefficients, and by the greater variety of forms that can be analyzed.

The coefficients produced by Elliptical Fourier Analysis are sensitive to the position of the first point digitized on the outline, and to the size, location, and orientation of the outline in the coordinate space used for digitizing. The effects of this sensitivity on comparative studies can be minimized by performing a few simple operations that are determined by the ellipse described by the first term of the Fourier series (Kuhl and Giardina 1982; Ferson *et al.* 1985). Size is standardized by dividing the coordinates of the outline points by the square root of the area of the ellipse. Location and orientation are standardized by translating and rotating each form so that the ellipse is centered on the origin and the long axis of the ellipse is aligned with the *x*-axis. Starting position is standardized by changing the start of the sequence of points on the outline to a point at the end of the ellipse in the positive *x*-direction. Ferson *et al.* (1985) note that these procedures have the consequence of making the alignment of outlines dependent on their shapes.

For our analysis of marmotine jaws, we digitized 170 points on each outline. As in the analyses using shape coordinates and partial warps, the incisors were not digitized. Figure 6.9 illustrates outlines computed from 1, 2 and 50 harmonics for three of the specimens. Typically, elliptic Fourier analysis produces very accurate reconstructions

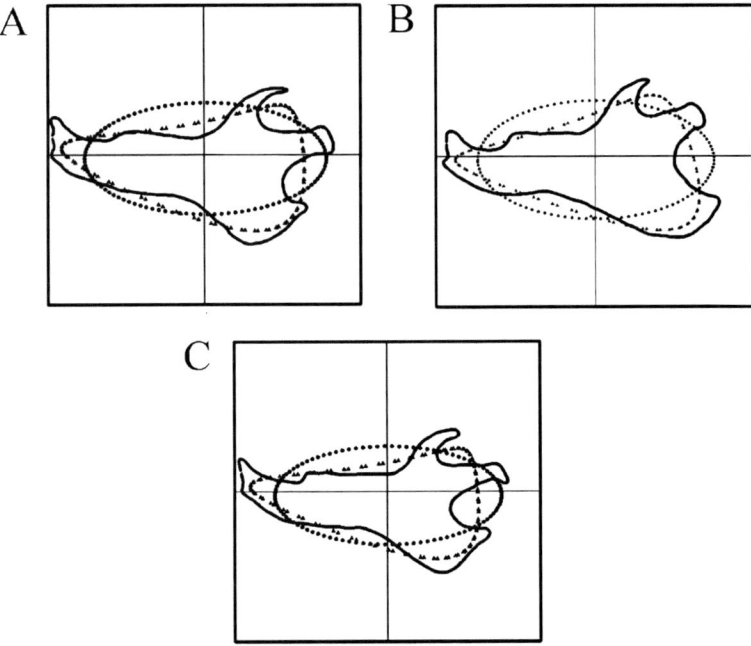

Figure 6.9 Outlines of mandibles reconstructed by Fourier harmonics, (A) *S. variegatus*, (B) *M. flaviventris*, (C) *S. tridecemlineatus*. Legend: circles – one harmonic, triangles – two harmonics, solid line – 50 harmonics.

of the original outline with substantially fewer harmonics than the maximum. This proved to be the case for all of our 50 harmonic reconstructions (well below the maximum of 85). Previous studies have taken advantage of this property to eliminate digitizing error or to remove artifacts introduced by digital imaging (e.g., Ferson *et al.* 1985; Lestrel and Huggare 1997). In these studies, least-squares analysis of the difference between the computed outline and the digitized data is used to produce an objective goodness-of-fit criterion to determine how many harmonics are sufficient to describe the outlines.

In each of the three cases shown, the outline has a different orientation with respect to the coordinate grid. This is because the ellipse described by the first harmonic has a different relationship to the outline. In all three forms, the negative end of the *x*-axis passes near the lower edge of the opening of the incisor alveolus. The positive end of this axis is near the lower edge of articular process in *S. variegatus* and *S. tridecemlineatus*. In *M. flaviventris*, the positive end of the axis is about midway between the articular and angular processes. This means that the ellipse is describing a different aspect of shape in each form, especially in *M. flaviventris*. The mandibles of both *S. variegatus* and *M. flaviventris* may be deeper than the mandible of *S. tridecemlineatus*, but they are deeper in different directions.

The differences in the orientations of the first harmonic ellipses may seem rather small, but they are quite important because they have implications for each subsequent harmonic. The second harmonic refers to the differences between the outline and ellipse that can be described by a trefoil. Because the ellipses have different relationships to the outline, so do the trefoils. The ellipses for S. *variegatus* and S. *tridecemlineatus* cover much of the articular process, leaving the coronoid process projecting above and the angular process projecting below. The triangular shape representing the sum of the first two harmonics is a better fit to the coronoid and angular processes (and to the body of the mandible), and a poorer fit to the articular process. In contrast, both the coronoid and articular processes of M. *flaviventris* project above its first harmonic ellipse, and consequently, this corner of the triangle projects farther above the ellipse than it does in the other two cases. Similarly, the aspects of shape remaining to be described by subsequent harmonics also differ among the specimens. These differences are most noticeable at the lower right corner of the triangle, where different portions of the angular process project beyond the triangle in each case.

The analyses shown here demonstrate that a given Fourier harmonic can refer to a different aspect of outline shape for each specimen in a study. This lack of comparability occurs because nothing in their computation anchors them to specific digitized points. Fourier analysis is a procedure for using a series of progressively more complex trigonometric functions to describe a smooth curve that passes through all of the digitized points. Those functions are not associated with any particular set of points; the coefficients are the values that minimize the sum of squared deviations of all the points. Because there is no link between a harmonic and the locations of a specific set of digitized points, the same harmonic can refer to different anatomical features in different specimens even if all the points are landmarks in the strictest sense. Given the lack of correspondence between Fourier harmonics and anatomical features, we cannot envision any legitimate use of the coefficients of individual harmonics in a phylogenetic study. Because the harmonic cannot be tied to the anatomy, it cannot support a hypothesis about the homology of an anatomical feature. This means that an individual harmonic cannot be interpreted as a character. If a harmonic cannot be interpreted as a character, the values of its coefficients cannot support hypotheses about the homology of the character's states.

The only circumstances in which Fourier decompositions could be used in a comparative analysis would be when the shape is treated as a whole without concern for the homology of its components. Then, one might use a plot of amplitude against harmonic number to sort objects into different shape classes, for example, forms with three lobes and those with four lobes. This kind of analysis could only be used if there is no reason to suspect that any of the individual lobes are homologous. Even under these conditions, the amplitude profile cannot be used to automatically classify shapes because an outline with one large lobe and two small ones might be grouped with outlines that have four lobes rather than with other outlines that have three lobes.

Eigenshape analysis

Zahn and Roskies (1972) observed that a complex curve can be described as a series of steps in which the direction of the step is related to the direction expected of a step around a circle. When a specimen is described with steps of equal length, size (perimeter length) is represented by step length, and shape is represented by the set of angles (ϕ^*) describing the deviation of each step from the expected direction. When each specimen is described by the same number of equal length steps, size differences are described by the differences in step length, and size independent shape differences are described by the differences in the sets of ϕ^* values. Eigenshape analysis (Lohmann and Schweitzer 1990) is a procedure for using the sets of ϕ^* values to analyze patterns of shape variation. The core of eigenshape analysis is a singular value decomposition of the covariance or correlation matrix computed from the descriptions of the individual specimens. The result is a set of orthogonal shape functions (eigenshape functions) that describe patterns of shape variation in the sample, in order of decreasing contribution to the total variance of the sample. These functions are vector multiples of the eigenvectors of the original covariance or correlation matrix. Because the original variables are a series of angles describing step directions relative to a circle, the eigenshape functions can be used to construct outlines that represent canonical patterns of shape variation, called eigenshapes.

We have two objections to using the results of an eigenshape analysis in a phylogenetic analysis. Our first objection concerns the nature of the eigenshape functions. Like principal components, eigenshape functions represent the axes of variation and covariation in the sample at hand. Because the descriptors are features of the sample distribution, not features of individuals, they cannot be used as a basis for inferring homologies. Therefore, eigenshape functions cannot be interpreted as characters and the scores of individuals on these axes cannot be interpreted as character states.

Our second objection concerns the use of a series of equal length steps to describe the outlines of individual specimens. Specifically, we are concerned with the ability to link a specific step to a specific anatomical feature. Below, we illustrate this problem by comparing 50-step Zahn and Roskies functions for outlines of the lower jaws of *M. flaviventris* and *S. franklini* (Figure 6.10). To make the description of step direction more accessible, we have replaced ϕ^* with ϕ, which relates the direction of each step to the direction expected of a step along a straight line. In plots of ϕ against step number, flat regions of the outline are represented in consecutive steps in the same direction; in plots of $\Delta\phi$ against step number, these regions are represented by consecutive steps with no change of direction.

In the two outlines, and in their corresponding ϕ and $\Delta\phi$ plots, the first five steps describe the curvature of the diastema between the incisor and the premolar. After this series, there is an abrupt change of direction as the outline turns to follow the base of the tooth row. From this point, the outline turns upward to follow the anterior edge of the coronoid process. Although the top of the coronoid is reached by step 14 in both outlines, different numbers of steps are used to describe all other anatomical features. Consequently, the same numbered step describes a different anatomical feature in the two outlines. The lack of correspondence between step number and anatomical feature is most evident in the ϕ and $\Delta\phi$ plots between steps 15 and 30. The peaks and troughs in these plots describe the abrupt changes of direction at the tips and bases of the three posterior processes. In this region, a step number that refers to the tip of a process in one organism refers to the base of process in the other organism.

Using the same number of steps to describe both outlines may be a convenient method of partitioning their descriptions into separate size and shape components, but it has several inconvenient consequences for comparisons of shape differences. One consequence is that the number of steps used to describe a given feature in the outline will reflect the proportion of the outline occupied by that feature. This might seem

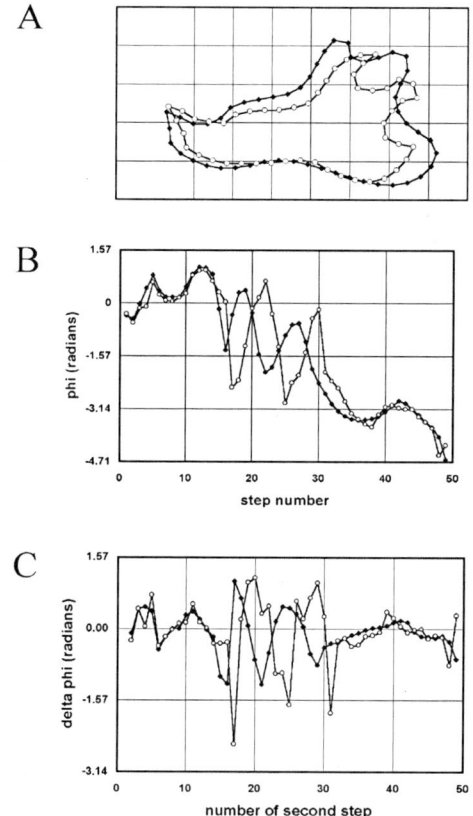

Figure 6.10 Information obtained from the Zahn and Roskies function, (A) Outlines, (B) Step directions, (C) Changes in step directions. (The step from the centroid to the outline has been omitted.) Legend: circles – *S. franklini*, diamonds – *M. flaviventris*.

like a positive consequence because it implies that information about the relative size of the features can be extracted from the number of steps. The number of steps seems to suggest that *M. flaviventris* has relatively smaller coronoid and articular process than *S. franklini*. The reality is that this information is not independent of the fact that *M. flaviventris* also has a relatively larger angular process. Because more steps are needed to describe the angular process, fewer are available to describe the rest of the jaw. In other words, a change in the number of steps required to describe one part of the outline may actually reflect a change in a different part of the outline.

Use of the same number of steps may also impede the description of other aspects of shape. In both specimens, the transition between the tooth row and coronoid process occurs at the same step number. This gives the misleading impression that coronoid process is the same relative distance from the incisor alveolus; however, the base of the coronoid is actually relatively closer to the alveolus in *M. flaviventris*. The true distance between these features is obscured by both the larger number of steps occupied by the angular process and by the relatively deeper curvature of the diastema in *M. flaviventris*. Differences in the shapes of the coronoid and articular process are also obscured by the smaller number of steps allocated to their descriptions in *M. flaviventris*. Because so few steps are used to describe the coronoid process of *M. flaviventris*, it appears to lack the sharp, recurved tip seen in *S. franklini*. In reality, the tip only appears to be blunted in *M. flaviventris* because the tip occurs in the middle of a step. A similar effect occurs at the posterior tip of the angular process of *M. flaviventris*, even though more steps are used to describe this process. Thus, even when step number is ignored and comparisons are based on the set of steps that refer to the same feature, the ϕ values do not necessarily represent equivalent descriptions of that feature.

Based on the lack of correspondence between anatomical features and steps in the Zahn and Roskies function, we conclude that individual ϕ or ϕ^* values, and any variables that might be computed from them, cannot be used alone as a basis for inferring homologies. This lack of correspondence arises because the same number of steps is used to describe each outline. Consequently, any correspondence between anatomical features and the points digitized on the outline is lost. Therefore, additional information is needed to determine whether a particular step, or sequence of steps, describes a comparable portion of the outline in all specimens. Even with that information, descriptions of individual features may not be equivalent; but without that information, data derived from the Zahn and Roskies function cannot be used in phylogenetic systematics.

Extended eigenshape analysis

Extended eigenshape analysis (MacLeod 1999) is designed to address some of the limitations inherent in conventional eigenshape analysis. One difference between the two methods is that extended eigenshape analysis uses the ϕ form of the Zahn and Roskies function not the ϕ^* form. The more important difference between the two methods is that extended eigenshape analysis uses a set of landmarks on the outline to constrain computation of the Zahn and Roskies function so that all cases have the same number of steps between a given pair of landmarks. The number of steps between a particular pair of landmarks is determined by complexity of that portion of the outline (the number of changes in direction between digitized points). As shown in Figure 6.11, the incorporation of the landmark data produces better alignments of the outlines and of the Zahn and Roskies functions.

The premise that underlies the use of landmarks to constrain the Zahn and Roskies function is that a section of outline between comparable points is, itself, a comparable anatomical feature. For example, landmarks 3 and 4 are the anatomical references used to identify the anterior edge of the coronoid process as a comparable feature of all marmotine mandibles in this study. Accordingly, the same set of terms in the constrained Zahn and Roskies function describes this feature in all specimens. This is an important improvement in outline description because it provides a rationale for shifting attention from the outline as a whole to the components of the outline, which may or may not be evolving independently.

Although the use of landmark data to constrain the Zahn and Roskies function produces closer correspondence of descriptive terms to anatomical features, the description of long or complex segments can still be problematic. Figure 6.12 shows

Phylogenetic systematics 91

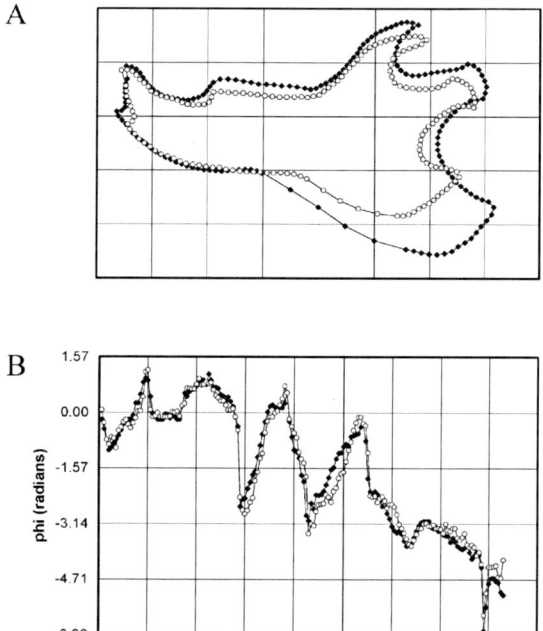

Figure 6.11 Information obtained from the Zahn and Roskies function, using landmarks to delimit comparable segments of outline, (A) Outlines, (B) Step directions. Legend: circles – *S. franklini*, diamonds – *M. flaviventris*.

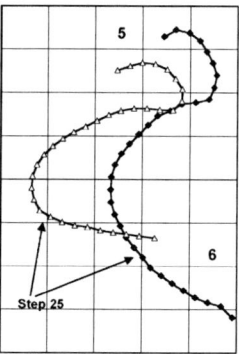

Figure 6.12 Reconstruction of the outline between landmarks 5 and 6, showing lack of correspondence between steps, even when the Zahn and Roskies function is constrained by landmarks.

the reconstructed outline between landmarks 5 and 6 for *M. flaviventris* and *S. spilosoma*. This complex curve covers the articular condyle, posteroventral edge of the articular process, and dorsal edge of the angular process. Among the seven taxa, there are differences in the relative sizes of each of these components. In addition, the region of the articular process immediately below the condyle is expanded in various degrees and in various directions. We did not treat the corner of this process as a landmark because we are uncertain about its homology. We also did not recognize any point in the curve between the articular and angular processes as a landmark because this curve is usually gradual and lacks distinctive anatomical features, so no point along the curve stands out as distinct and comparable. Because of this combination of variability and lack of landmarks, step 25 (of 33) is close to the junction of the articular and angular processes in *S. spilosoma* and considerably further out on the angular process of *M. flaviventris*. Thus, even when landmarks are used to constrain the number of steps for a particular stretch of outline, individual steps in the Zahn and Roskies function do not correspond from specimen to specimen except in the immediate vicinity of each landmark.

The lack of a one-to-one correspondence between step number and anatomical feature is important because the ϕ values in individual steps are used as input for the next stage of the analysis. Thus, there is an assumption that each step of the Zahn and Roskies interpolation is comparable. In contrast, no information about features between landmarks is used at any stage of the thin-plate spline analysis. The thin-plate spline interpolation can be used to draw the outline of the coronoid process of the reference shape before and after a deformation, but the picture of the outline after deformation is completely dependent on the descriptions of the relative positions of the landmarks. No information about the outline of the coronoid process in the second specimen is used to draw the picture of the deformed outline of the reference form. Thin-plate spline analysis uses an interpolation function to describe observations; eigenshape analysis (conventional and extended) uses an interpolation function to define observations.

Our objection to using results of conventional eigenshape analysis in phylogenetic analyses is based on two features of eigenshape analysis: the use of an interpolation function to describe shapes of individual specimens and the use of axes of variation in a sample as a basis for describing differences between shapes. The principal distinction of extended eigenshape analysis is that it employs a better version of the interpolation function. This improvement results in more accurate descriptions of the outlines, but it still does not meet the requirements for a phylogenetic analysis.

Other shape indices

Our rejection of the constrained Zahn and Roskies function should not be construed as a rejection of the idea that segments of outline can be judged to be comparable. We are only opposed to the idea of further dividing a segment in the absence of concrete anatomical evidence to support that division. Instead, we suggest that the segments deemed comparable should be treated as wholes. In this section, we present some suggestions for implementing this approach.

Figure 6.13 shows plots of points digitized along the stretch of outline between landmarks 3 and 4. This segment of outline covers the anterior edge of the coronoid process, which is the area of insertion for the temporalis muscle. The coordinates of these points were transformed to shape coordinates relative to a baseline between landmarks 3 and 5 (articular condyle). Qualitatively, the anterior edge of the coronoid process can be described as having an S-shape. The lower limb of the S becomes progressively more vertical, the upper limb becomes horizontal again, and in some

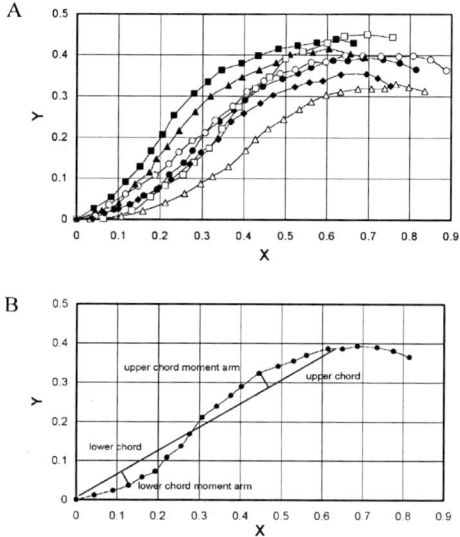

Figure 6.13 An alternative approach to comparing sections of outline bounded by two landmarks, (A) Outlines of the anterior edge of the coronoid process (between landmarks 3 and 4) reconstructed from shape coordinates using landmarks 3 and 5 as the baseline. Symbol legend: filled circle – *S. variegatus*, open circle – *S. franklini*, filled triangle – *S. tridecemlineatus*, open triangle – *S. spilosoma*, filled square – *S. columbianus*, open square – *C. ludovicianus*, filled diamond – *M. flaviventris*, (B) Diagrammatic representation of some of the measurements used to analyze differences between outlines.

taxa tilts below horizontal. The four taxa shown in Figure 6.13 differ in the height of the tip of the process above the baseline, the tightness of the bends in the S, and the slope of the middle of the S (where the direction of curvature is reversed). The taxa also appear to differ in the symmetry of the S and in the smoothness of the curves. Judging from the differences among the taxa, curvature (or sinuosity), symmetry, and smoothness appear to be independently evolving features of the shape of this curve. The combination of curvature, symmetry and chord lengths account for the height of the process.

Each aspect of the outline described qualitatively above, can also be evaluated quantitatively. Table 6.1 lists results a few of the possible variables. Smoothness was computed as the sum of squares of the signed changes in step direction divided by the net change in step direction. Symmetry was evaluated by finding the point at which the direction of curvature is reversed and drawing a line from the origin, through that point, to its intersection with the reversed curve and then comparing the chord lengths (their difference divided by their sum) (Figure 6.13B). Sinuosity was computed as the net curvature of the two limbs. Curvature for a single bend can be computed as the length of the moment arm, which is the 'depth' of the curve, divided by the length of its chord (Swartz 1990). To compute sinuosity, we used the sum of the moment arms divided by the sum of the chord lengths. When asymmetry is low, sinuosity is close to the average of the curvatures of the two limbs; as asymmetry increases, sinuosity approaches the curvature of longer limb.

Table 6.1 Variables describing the shape of the anterior edge of the coronoid process (See text and Figure 6.13 for descriptions of the measurements.)

Taxon	Smoothness	Chord asymmetry	Sinuosity	Curvature – lower limb	Curvature – upper limb
S. variegates	2.42	0.108	0.108	0.128	−0.092
S. franklini	1.21	0.009	0.068	0.052	−0.084
S. spilosoma	2.76	−0.211	0.134	0.107	−0.176
S. tridecemlineatus	2.01	0.307	0.125	0.110	−0.132
S. columbianus	3.31	−0.062	0.082	0.065	−0.101
C. ludovicianus	2.57	0.008	0.135	0.132	−0.137
M. flaviventris	2.38	0.001	0.090	0.092	−0.087

The variables we describe above would evaluate many of the same features evaluated by a Fourier analysis. In a Fourier analysis, net curvature and sinuosity would be described by low frequency harmonics, asymmetry by intermediate harmonics, and smoothness by high frequency harmonics. The crucial difference is that the coefficients of the Fourier harmonics are not independent of each other. Each Fourier harmonic explains the deviation of the digitized points from the curve described by the sum of all lower frequency harmonics. In contrast, each of the variables we used was evaluated independently. It may be troublesome to some that our measures of smoothness and asymmetry do not refer to the residuals from a fitted shape, but we view this as a positive attribute because a curve can be uneven or asymmetric whether the underlying shape is a C, an S, or some more complex shape.

Some of the measurements we used also have attributes in common with the Zahn and Roskies function. Our computation of smoothness entails measurements of each change in step direction. However, these computations do not require the assumption that specific points on different specimens are comparable simply because they have the same position in the sequence of points digitized on comparable curves. Our computations of chord asymmetry and sinuosity use the inflection point to demarcate the two limbs of the S-curve, and those limbs are measured separately. However, the chord lengths and moment arms of those limbs represent nothing more than intermediate steps in the computation of a value that refers to an attribute of the entire segment of outline between the two landmarks. Furthermore, this analysis does not treat the inflection point as anything other than a geometric attribute of the S-curve. If we had evidence that the inflection point was biologically comparable (e.g., if we had evidence that different components of the temporalis muscle inserted on either side of this point), then we would have another landmark, which would justify treating the upper and lower limbs of the S as separate characters. If that were the case, we would evaluate the curvatures of the separate limbs, not the sinuosity of the entire S.

Another useful attribute of our computations of chord asymmetry and sinuosity is that they return a single value for each specimen. This attribute eliminates the temptation to use a fitted power series or Fourier series to decompose the curve into a set of terms describing features that may not be comparable. It also eliminates the need to compute principal component axes or other *n*-dimensional vectors to specify the locations of specimens in the measurement space. Consequently, there is also no need to rely on coefficients of similarity based on the distances between specimens in the morphometric space. Instead, the values of smoothness, chord asymmetry and sinuosity can be evaluated directly for evidence of evolutionary divergence; such variables can be a basis for formulating hypotheses of monophyly.

Based on the values reported in Table 6.1, we would infer that *S. spilosoma* and *S. columbianus* share an increase in smoothness and *S. franklini* has a unique decrease. The S-curve is slightly asymmetric in *S. variegatus*, with the upper limb slightly longer than the lower. Three taxa share a reduction of this asymmetry, but in *S. tridecemlineatus* it has increased. Two taxa, *S. spilosoma* and *S. columbianus*, share reversed asymmetry in which the lower limb is longer than the upper. With regard to sinuosity, *S. variegatus* has the median value; *S. spilosoma*, *S. tridecemlineatus*, and *C. ludovicianus* have higher values (larger deflections relative to net length); *S. franklini*, *S. columbianus*, and *M. flaviventris* have lower values. Based on these results, we would infer that there are two derived traits shared by *S. spilosoma* and *S. columbianus* (increased smoothness and reversed asymmetry), and one that conflicts with this pairing (sinuosity). These hypotheses about transformations of the anterior margin of the coronoid process can be coded and combined with any other characters we might wish to use to infer the phylogenetic relationships of these taxa. Similar analyses can be used to evaluate the features of any other stretch of outline judged to be comparable among specimens.

We recognize that there are other ways of quantifying the attributes of a curve, and that there may be other attributes that could be described. Our intention was only to present some variables that do not imply claims of comparability beyond those that can be based on the landmarks. If comparability is based on anatomy, a hypothesis of homology can be justified. This criterion will be met by any variable that can be described in terms that designate the specific anatomical domain that is evaluated.

Conclusions and recommendations

Based on our analyses of marmotine mandibles, we conclude that it is possible to use information about outline shape in phylogenetic analysis, if certain conditions are met. Thus, our previous statement (Zelditch *et al.* 1995) opposing the use of all outline data in phylogenetic systematics was too broad. The methods that are commonly used to describe outlines do have problems, but the problems are specific to the ways in which variables are computed. In Fourier analysis, the first term can refer to a different aspect of shape in each specimen. Because each subsequent term refers to an aspect of shape not described by the previous terms, it is possible that no term in the description describes a comparable feature of all specimens. In the Zahn and Roskies function, lack of correspondence arises because step length is a function of outline length, not anatomy, so steps with the same number can fall on different anatomical features. Eigenshape analysis exacerbates this problem by computing principal axes of variation from the Zahn and Roskies descriptors. If a variable refers to measurements taken on different parts of the outline, its variance is meaningless and the covariances of such variables are also meaningless. (The same criticism applies to principal components computed from Fourier coefficients.) Thus, we do not withdraw our earlier rejection of these methods. Rather, we now leave open the possibility that methods lacking these flaws may produce variables that can be used in phylogenetic systematics.

If morphometric descriptions of outlines are to be used in phylogenetic analyses, the variables they produce must refer to features of the outline that are comparable. In other words, a variable must refer to a region of outline that is anatomically defined. This does not mean that every digitized point on the outline must be a landmark. It does mean that the coordinates of points that are not landmarks and the steps between such points, cannot be compared between specimens on a one-to-one basis. Rather, the set of points describing a particular part of an outline must be treated as related bits of information about the larger whole. Furthermore, each variable computed from this information must describe some aspect of the entire curve, not an aspect of a region within the curve that is not comparable between specimens.

The requirement that the region of outline must be comparable does not mean that the curve must be bounded by landmarks. Certainly, they are useful when they are available, but they are not necessary. Other anatomical references can be used to define the comparable region. For example, a region of outline might simply be defined as the curve joining the articular and angular processes. This curve can be described quantitatively using variables like the ones we used to describe the anterior edge of the coronoid process. There is an additional complication that must be addressed when the curve is not delimited by landmarks: care must be taken to insure that the description does not depend on the location of the endpoints. However, this is still a generous requirement that permits the computation of several descriptive variables.

Using landmarks and other anatomical references to identify comparable features of outlines solves the problem of using information from outlines in phylogenetic analyses. If the features are comparable, the tests of hypotheses about the homology of curve shapes are both reasonable and legitimate. The principal problem that remains to be solved is evaluation of the evolutionary independence of traits prior to mapping them onto a phylogenetic tree. The problem arises whether we are talking about different partial warps or different sections of outline. It is not unique to morphometric analyses (e.g., Vermeij 1973; Swiderski 1991; Wake 1993), but it assumes a somewhat greater importance in these studies because morphometric variables often describe overlapping regions or different aspects of the same region. Principal components analysis has been used to investigate these relationships among quantitative variables (Lessa and Stein 1992; Auffray et al. 1996; Rohlf et al. 1996), but it is not a valid solution because it describes sample variances and covariances, not comparable features of individuals. What is needed are methods that examine combinations of shape variables without losing the comparability that morphometricians have worked so hard to incorporate in their measurement data.

Acknowledgements

We thank Norm MacLeod and Peter Forey for the opportunity to participate in the symposium, and for funding that made our participation possible. We also thank Norm and an anonymous reviewer for their constructive evaluations of an earlier draft of this contribution. Our research was supported by National Science Foundation grant DEB-9509195 (MLZ and WLF).

References

Auffray, J.-C., Alibert, P. and Latieule, C. (1996) 'Relative warp analysis of skull shape across a hybrid zone of the house mouse (*Mus musculus*) in Denmark', *Journal of Zoology*, 240, 441–455.

Ball, S. S. and Roth, V. L. (1995) 'Jaw muscles of New World squirrels', *Journal of Morphology*, 224, 265–291.

Black, C. C. (1963) 'A review of the North American Tertiary Sciuridae', *Bulletin of the Museum of Comparative Zoology*, 130, 109–248.

Bookstein, F. L. (1989) 'Principal warps: thin-plate splines and the decomposition of deformations', *IEEE Transactions on Pattern Analysis and Machine Intelligence*, 11, 567–585.

Bookstein, F. L. (1991) *Morphometric tools for landmark data: geometry and biology*, New York: Cambridge University Press.

Bookstein, F. L. (1996) 'Combining the tools of geometric morphometrics', in Marcus, L. F., Corti, M., Loy, A., Naylor, G. J. P. and Slice, D. E. (eds) *Advances in morphometrics. NATO ASI series. Series A*, New York: Plenum Press.

Bookstein, F. L. and Green, W. D. K. (1993) 'A feature space for edgels in images with landmarks', *Journal of Mathematical Imaging and Vision*, 3, 321–361.

Bryant, M. D. (1945) 'Phylogeny of the Nearctic Sciuridae', *American Midland Naturalist*, 33, 257–390.

Cranston, P. and Humphries, C. (1988) 'Cladistics and computers: a chironomid conundrum?', *Cladistics*, 4, 72–92.

Crowe, T. (1994) 'Morphometrics, phylogenetic models and cladistics: means to an end or much ado about nothing?', *Cladistics*, 10, 77–84.

Davis, H. F. (1989) *Fourier series and orthogonal functions* (reprint of original 1963 edition), New York: Dover Publications.

de Pinna, M. C. C. (1991) 'Concepts and tests of homology in the cladistic paradigm', *Cladistics*, 7, 367–394.

Dullemeijer, P. (1958) 'The mutual structural influence of the elements in a pattern', *Archives Neerlandaises de Zoologie*, 13, supplement 1, 74–88.

Ehrlich, R., Pharr, R. B., Jr. and Healy-Williams, N. (1983) 'Comments on the validity of Fourier descriptors in systematics: a reply to Bookstein *et al.*', *Systematic Zoology*, 32, 202–206.

Felsenstein, J. (1988) 'Phylogenies and quantitative characters', *Annual Review of Ecology and Systematics*, 19, 445–471.

Ferson, S., Rohlf, F. J. and Koehn, R. K. (1985) 'Measuring shape variation of two-dimensional outlines', *Systematic Zoology*, 34, 59–68.

Garland, T., Jr. and Adolph, S. C. (1994) 'Why not to do two species comparative studies: limitations on inferring adaptation', *Physiological Zoology*, 67, 797–828.

Gaudin, T. J. and Branham, D. G. (1998) 'The phylogeny of Myrmecophagidae (Mammalia, Xenarthra, Vermilingua) and the relationship of *Eurotamandua* to the Vermilingua', *Journal of Mammalian Evolution*, 5, 237–265.

Givnish, T. J. and Sytsma, K. J. (1997) 'Consistency, characters, and the likelihood of correct phylogenetic inference', *Molecular Phylogenetics and Evolution*, 7, 320–330.

Greaves, W. S. (1978) 'The jaw lever system in ungulates: a new model', *Journal of Zoology*, 184, 271–285.

Hedges, S. B. and Maxson, L. R. (1996) 'Re: molecules and morphology in amniote phylogeny', *Molecular Phylogenetics and Evolution*, 6, 312–314.

Hennig, W. (1966) *Phylogenetic systematics*, Urbana: University of Illinois Press.

Howell, A. H. (1938) 'Revision of the North American ground squirrels, with a classification of the North American Sciuridae', *North American Fauna*, 56, 1–256.

Janis, C. M. and Ehrhardt, D. (1988) 'Correlation of relative muzzle width and relative incisor width with dietary preference in ungulates', *Zoological Journal of the Linnean Society*, 92, 267–284.

Kuhl, F. P. and Giardina, C. R. (1982) 'Elliptic Fourier features of a closed contour', *Computer Graphics and Image Processing*, 18, 236–258.

Lessa, E. P. and Stein, B. R. (1992) 'Morphological constraints in the digging apparatus of pocket gophers (Mammalia: Geomyidae)', *Biological Journal of the Linnean Society*, 47, 439–453.

Lestrel, P. E. (1997a) 'Introduction', in Lestrel, P. E. (ed.) *Fourier descriptors and their applications in biology*, Cambridge: Cambridge University Press, pp. 3–21.

Lestrel, P. E. (1997b) 'Introduction and overview of Fourier descriptors', in Lestrel, P. E. (ed.) *Fourier descriptors and their applications in biology*, Cambridge: Cambridge University Press, pp. 22–44.

Lestrel, P. E. and Huggare, J. A. (1997) 'Cranial base changes in shunt-treated hydrocephalics: Fourier descriptors', in Lestrel, P. E. (ed.) *Fourier descriptors and their applications in biology*, Cambridge: Cambridge University Press, pp. 322–339.

Liem, K. F. (1973) 'Evolutionary strategies and morphological innovations: cichlid pharyngeal jaws', *Systematic Zoology*, 22, 425–441.

Lohmann, G. P. and Schweitzer, P. N. (1990) 'Chapter 6: on eigenshape analysis', in Rohlf, F. J. and Bookstein, F. L. (eds) *Proceedings of the Michigan Morphometric Workshop*, Special Publication Number 2, Ann Arbor, Michigan: The Unversity of Michigan Museum of Zoology, pp. 147–166.

Lombard, R. E. and Wake, D. B. (1986) 'Tongue evolution in the lungless salamanders, family Plethodontidae IV. Phylogeny of plethodontid salamanders and the evolution of feeding dynamics', *Systematic Zoology*, 35, 532–551.

MacLeod, N. (1999) 'Generalizing and extending the eigenshape method of shape space visualization and analysis', *Paleobiology*, 25, 107–138.

Mickevich, M. F. and Weller, S. J. (1990) 'Evolutionary character analysis: tracing character change on a cladogram', *Cladistics*, 6, 137–170.

Moss, M. L. (1962) 'The functional matrix', in Krauss, B. S. and Riedl, R. (eds) *Vistas in orthodontics*, Philadelphia: Lea and Febiger, pp. 85–98.

Moss, M. L. and Saletijn, L. (1969) 'The primary role of functional matrices in facial growth', *American Journal of Orthodontics*, 55, 566–577.

Moss, M. L. and Young, R. W. (1960) 'A functional approach to craniology', *American Journal of Physical Anthropology*, 18, 281–292.

Normark, B. B. and Lanteri, A. A. (1998) 'Incongruence between morphological and mitochondrial-DNA characters suggests hybrid origins of parthenogenetic weevil lineages (genus *Aramigus*)', *Systematic Biology*, 47, 475–494.

Patterson, C. (1982) 'Morphological characters and homology', in Joysey, K. A. and Friday, A. E. (eds) *Problems of phylogenetic reconstruction*, Systematics Association Special Volume No. 21, London: Academic Press, pp. 21–74.

Pesce Delfino, V., Lettini, T. and Vacca, E. (1997) 'Heuristic adequacy of Fourier descriptors: methodologic aspects and applications in morphology', in Lestrel, P. E. (ed.) *Fourier descriptors and their applications in biology*, Cambridge: Cambridge University Press, pp. 250–293.

Pimentel, R. A. and Riggins, R. (1987) 'The nature of cladistic data', *Cladistics*, 3, 201–209.

Read, D. W. and Lestrel, P. E. (1986) 'Comment on uses of homologous-point measures in systematics: a reply to Bookstein *et al.*', *Systematic Zoology*, 35, 241–253.

Rohlf, F. J. (1996) 'Morphometric spaces, shape components and the effects of linear transformations', in Marcus, L. F., Corti, M., Loy, A., Naylor, G. J. P. and Slice, D. E. (eds) *Advances in morphometrics. NATO ASI series. Series A*, New York: Plenum Press, pp. 117–129.

Rohlf, F. J. (1998) 'On applications of geometric morphometrics to studies of ontogeny and phylogeny', *Systematic Biology*, 47, 147–158.

Rohlf, F. J. and Archie, J. W. (1984) 'A comparison of Fourier methods for the description of wing shape in mosquitoes (Diptera: Culicidae)', *Systematic Zoology*, 33, 302–317.

Rohlf, F. J., Loy, A. and Corti, M. (1996) 'Morphometric analysis of Old World Talpidae (Mammalia, Insectivora) using partial warp scores', *Systematic Biology*, 45, 344–362.

Schaeffer, S. A. and Lauder, G. V. (1986) 'Historical transformation of functional design: evolutionary morphology of feeding mechanisms in loricarioid catfishes', *Systematic Zoology*, 35, 489–508.

Slice, D. E., Bookstein, F. L., Marcus, L. F. and Rohlf, F. J. (1996) 'Appendix I. A glossary for geometric morphometrics', in Marcus, L. F., Corti, M., Loy, A., Naylor, G. J. P. and Slice, D. E. (eds) *Advances in morphometrics. NATO ASI series. Series A*, New York: Plenum Press, pp. 531–551.

Swartz, S. M. (1990) 'Curvature of the forelimb bones of anthropoid primates: overall allometric patterns and specializations in suspensory species', *American Journal of Physical Anthropology*, 83, 477–498.

Swenson, U. and Bremer, K. (1997) 'Patterns of floral evolution in four Asteaceae genera (Senecioneae, Blennospermatinae) and the origin of white flowers in New Zealand', *Systematic Biology*, 46, 407–425.

Swiderski, D. L. (1991) 'Morphology and evolution of the wrists of burrowing and nonburrowing shrews (Soricidae)', *Journal of Mammalogy*, 72, 118–125.

Swiderski, D. L. (1993) 'Morphological evolution of the scapula in tree squirrels, chipmunks and ground squirrels (Sciuridae): an analysis using thin-plate splines', *Evolution*, 47, 1854–1873.

Swiderski, D. L., Zelditch, M. L. and Fink, W. L. (1998) 'Why morphometrics is not special: coding quantitative data for phylogenetic analysis', *Systematic Biology*, 47, 508–519.

Thompson, D' Arcy W. (1917) *On growth and form: the complete revised edition*. New York: Dover Publications, Inc. (reprinted 1992).

Thorington, R. W., Jr. and Darrow, K. (1996) 'Jaw muscles of Old World squirrels', *Journal of Morphology*, 230, 145–165.

Tolstov, G. P. (1962) *Fourier series*, Englewood Cliffs, New Jersey: Prentice-Hall.

van der Klaauw, C. J. (1948–1952) 'Size and position of the functional components of the skull', *Archives Neerlandaises de Zoologie*, 9, 1–361.

Vermeij, G. J. (1973) 'Adaptation, versatility and evolution', *Systematic Zoology*, 22, 466–477.

Wake, M. H. (1993) 'Non-traditional characters in the assessment of caecilian phylogenetic relationships', *Herpetological Monographs*, 7, 42–55.

Zahn, C. T. and Roskies, R. Z. (1972) 'Fourier descriptors for plane closed curves', *IEEE Transactions on Computers*, C-21, 269–281.

Zelditch, M. L., Fink, W. L. and Swiderski, D. L. (1995) 'Morphometrics, homology and phylogenetics: quantified characters as synapomorphies', *Systematic Biology*, 44, 179–189.

Chapter 7

Phylogenetic signals in morphometric data*

Norman MacLeod

ABSTRACT

Although many of the goals and concepts of qualitative morphological analysis and morphometrics are similar, systematists have largely rejected the use of morphometric methods in phylogenetic analysis on a variety of grounds. This review finds that (1) the concepts of a cladistic character and a morphometric variable are essentially identical, (2) morphometric methods can be instrumental in discovering and documenting new morphological characters and character states, (3) prior objections to the use of morphometric variables because of their continuous nature confuse the issues of variable type with those surrounding the distributions of sets of observations, (4) morphometrics offers the best method of determining whether morphological observations are discontinuous (= can be coded as discrete character states) or continuous (= cannot be coded as discrete character states), (5) constellations of landmark-based morphometric variables represent adequate summaries of putative structure-level homologues for use in phylogenetic analyses, (6) partial warp analyses do not perform well in either simulated or actual phylogenetic systematic analyses because of their inherent instability and lack of adequate spatial localization, and (7) a new method of subdivided relative warp analysis (described herein) performs very well at recognizing simulated morphological character states and recovering a simulated morphological phylogenetic hierarchy. Based on these results it is concluded that the potential of morphometric data analysis methods (especially relative warp-based methods) to contribute to phylogenetic-systematic investigations should be explored further.

Introduction

The fundamental observation of biology is morphology. Morphological data form the basis of virtually all systematic descriptions. Morphological features define the basic units of biology: the species and other monophyletic taxa (Nelson 1989) and are used by all biologists – including geneticists and molecular systematists – to identify those groups in the overwhelming majority of cases. Indeed, even molecular data are morphological insofar as the chemical properties that enable particular molecules to function in biological processes – and so be maintained by natural selection – derive

---

* This chapter is dedicated to F. James Rohlf on the occasion of his 65th birthday.

as much from the arrangement of atoms in each molecule's structure (= its shape) as from its composition.

Morphological data are regarded as being of significance in systematics because morphological variation is believed to be characterized by gaps between taxa. The presence of these gaps makes each taxon uniquely diagnosable and their hierarchical structure reflects action of morphological change superimposed on the evolutionary process of ancestry and descent. These gaps may arise as a result of a number of evolutionary processes (see Otte and Endler 1989 for reviews), but their discovery, description, and interpretation represents the first and most basic task of all systematic research.

Morphometrics is the study of covariances between patterns of morphological variation and patterns of variation in other associated or causal variables (Bookstein 1991; MacLeod in press). As such, morphometrics and systematic biology share a common interest in the analysis of morphology, in assessing the nature of morphological variation, and in studying degrees of covariance with those patterns (e.g., taxonomic covariances, ecological covariances, functional covariances, phylogenetic covariances). Given that morphometrics also invariably incorporates strong elements of quantification and formal hypothesis testing, it would seem natural for biological morphologists to regard morphometric tools as an integral part of their approach to systematics. This, however, has not been the case.

The reasons for the persistent lack of a strong connection between systematics and morphometrics are many. But, for contemporary systematists I believe they can be traced back to a sense of unease within the systematics community over historical connections between the systematic philosophy of phenetics and many morphometric procedures (e.g., Crowe 1994). Indeed, it often seems as though many systematists equate morphometrics with phenetics – even though this is demonstrably not the case (see Bookstein *et al.* 1985; Bookstein 1994) – and regard both as being beyond the bounds of accepted systematic practice. This, perhaps unrecognized, avoidance of morphometrics by systematists is mirrored within the morphometrics community which has, for the most part, avoided taking phylogenetic patterning into consideration in their interspecific data analyses despite many recent and clear demonstrations of the need to do so (Felsenstein 1985, 1988; Harvey and Pagel 1991; MacLeod 2001).

The purpose of this chapter is to explore the past, present, and future of relations between systematics, and morphometrics. In keeping with the theme of the volume, this exploration will be organized around the topic of morphological phylogenetic analysis, though the methods, discussion, and conclusions drawn should be applicable to other areas of systematics (e.g., biogeography, ecology) as well. In particular, it will consider the question of how and why morphometric data should be utilized in the context of descriptive and analytic phylogenetic systematics. These explorations will take the form of both a (re)consideration of systematic concepts and practices, as well as their demonstration via example analyses. By venturing into the 'no-man's land' between systematics and morphometrics this study represents a gathering together of this topic's disparate strands in an attempt to bridge the conceptual divides that presently separate large segments of the systematics community in an area that has traditionally stood at the heart of organism-centered biological investigations.

Characters and variables

Any discussion of the relation between contemporary systematics and morphometrics should begin by examining issues surrounding the concepts of 'systematic characters' and 'morphometric variables'. Farris *et al.* (1970: 172) defined a systematic character (= the transformation series of Hennig 1966), as "a collection of mutually exclusive states which (a) have a fixed state such that (b) each state is derived directly from just one other state and (c) there is a unique state from which every other state is derived." Pimentel and Riggins (1987: 201) defined a 'character' as "a feature of organisms that can be evaluated as a variable with two or more ordered states." Contrast these descriptions with the standard biometric concept of a variable (Zar 1974: 2) as "a characteristic that varies from one biological entity to another." Similarly, Sokal and Rohlf (1981: 11) define a biometric variable as "a property with respect to which individuals in a sample differ in some ascertainable way." If one strips away parts b and c from the Farris *et al.* (1970) definition – which are matters of theory and interpretation – and restricts the Zar (1974) and the Sokal and Rohlf (1981) definitions to morphological characteristics or properties, it can be appreciated that the operational concepts of 'systematic character' and 'morphometric variable' are essentially identical.

There are four basic types of variables: ratio-scale variables, interval-scale variables, ordinal-scale variables, and nominal variables. Ratio-scale variables represent continuous, infinitely divisible, numerical scales in which a unit difference represents the same quantity regardless of its location along the scale (e.g., measured heights, lengths, widths). These variables are typically represented by real numbers. Meristic variables are considered a special class of ratio-scale variables that can take only discrete, integer values (e.g., number of eggs layed in a clutch, number of eyes, number of digits). Interval-scale variables represent continuous numerical scales in which a unit difference represents different quantities depending upon its location along the measurement scale. The classic examples of this variable type are the Fahrenheit and Centigrade (but not the Kelvin) temperature scales (e.g., 100 °C is not twice as hot as 50 °C) since the zero point is set arbitrarily. Circular variable scales are also of this type. Ordinal-scale and nominal variables both represent discrete measurement scales, but along these scales there is no requirement that a unit difference represents the same quantity regardless of its location along the scale. These scales differ depending on whether the attribute can be represented as a ranked (ordinal) or non-ranked (nominal) sequence. Both ratio-scale and interval-scale variables can be transformed into ordinal-scale or nominal variables via application of appropriate rules.

The characters described and discussed in the theoretical systematic literature (e.g., Kitching *et al.* 1998) are almost always nominal-scale variables. In addition to the G, A, T, C nominal variables of molecular phylogenetics, systematic attributes typically treated as nominal variables include such standard examples as eye color (e.g., brown, hazel, blue), hair color (e.g., brown, black, blonde), egg type (e.g., aminote, non-aminote), and body covering (e.g., scales, feathers, hair). A standard symbology for a nominal systematic variable is A, A'. The convention of naming these variables with numerical symbols (0, 1, 2, ...) does not make them interval-scale variables because the difference criterion essential to that variable type is violated. Nominal variables may be converted to ordinal-scale variables through application of a rule (e.g., outgroup comparison) that establishes the variable's order. Hennig's (1966) original concept of the transformation series represents an example of a systematic ordinal-scale variable. However, the common contemporary practice of submitting numerically coded state variables to phylogenetic analysis in the unordered mode means that most systematic character/character-state datasets are composed of nominal variables. This data type contrasts markedly with morphometric datasets, that are almost always composed of ratio-scale variables.

I suspect it is the fundamental differences between these two variable types, along with a desire to work as close to the abstract ideal of a systematic character as possible, that many systematists have in mind when they make categorical statements like the following.

> Continuously varying quantitative data are not suitable for cladistic analysis because there is no justifiable basis for recognizing discrete states among them. (Pimentel and Riggins 1987: 201).

> ... it would be inadmissible to use a [morphometric] length variable that had been arbitrarily divided into to states, one of lengths less than, the other of lengths greater than the median length, since it would be just as reasonable to choose any point along such a continuum at which to delimit states. (Crisp and Weston 1987: 67).

> None of the authors on coding methods has yet faced the question of how we could test for the presence of underlying discrete states. Lacking such a test, there is no reason to discretize [sic] quantitative characters. (Felsenstein 1988: 462).

Such statements would be uncontroversial if each organismal phenotype presented large sets of ratio-scale and nominal variables for analysis in that one could simply identify the variable type and exclude the ratio-scale subset. In practice, however, things are a bit more complicated. Many authors have pointed out that it is a common practice for phylogenetic systematic analyses to be performed on data matrices containing quantitative, ratio-scale variables that have been semantically 'discretized' into nominal variables (= characters, see Simon 1983; Almeida and Bisby 1984; Thorpe 1984; Archie 1985; Baum 1988; Goldman 1988; Chappill 1989; Stussey 1990; Stevens 1991; Thiele 1993; Rae 1998; Strait *et al.* 1996; Swiderski *et al.* 1998). Moreover, even many standard examples of 'good' cladistic (= nominal) variables fail to stand up to even casual scrutiny.

For example, Pimentel and Riggins (1987: 202) refer to flower-petal colour as a cladistic (nominal) variable in their discussion of character-state concepts. Colour, of course, is really a ratio-scale variable – hence the endless list of names for slightly different colours – based on the frequency spectrum of reflected light. The fact that systematists often find it convenient to describe colour as though it were a nominal variable, changes neither the nature of the phenomenon that produces colour nor the arbitrariness of the various nominal scales used to describe the visible light spectrum. Based on the fact that ratio-scale variables *are* routinely used as characters in contemporary phylogenetic analysis, one cannot logically exclude morphometric variables from the list of useful systematic character types or covariates simply on the basis of their continuous measurement scales.

Using morphology to discover character states

Contemporary systematists are able to treat ratio-scale and interval-scale variables as ordinal-scale or, more typically, nominal variables because they regard the pattern(s) of variation they exhibit as being discontinuously distributed along the theoretically continuous measurement scale. In those instances where a discontinuous distribution of size or shape measurements can be documented for a sample there can be no objection to the morphometric definition of subdistribution limits and the relabelling of the subdistributions as nominal classes. This operation conforms exactly to the recognition of discrete states within a larger character (see Eldredge and Cracraft 1980; Nelson and Platnick 1981; Wiley 1981; Smith 1994; Kitching et al. 1998). Under this conceptualization the theoretically continuous variable axis represents the character and the discontinuously distributed clusters of observations arrayed along this axis represent fixed and mutually exclusive state classes. Indeed, explicit demonstration of the distributional discontinuities on which such state boundaries are based, their formal definition, and their illustration should be required of all peer-reviewed, morphological, systematic communications.

For instance, morphological size variables (e.g., length, area) are routinely used to subdivide the variation observed between presumed homologous structures into states. Typically these subdivisions are given qualitative names (e.g., feature size: small, large) and coded as nominal states (e.g., 0, 1), often without any supporting data demonstrating discontinuity or defining state boundaries. Morphometric data analysis methods were originally formulated to provide such demonstrations and definitions. In this context it is difficult to understand how recourse to morphometric variables and analyses has been so consistently ignored by systematists. The fact that these simple principles of demonstration–documentation have not been followed in the past has been responsible for much unnecessary confusion and over the nature of phylogenetic systematic analysis and the validity of particular analytic results.

The salient aspects of this process – and the advantages of the morphometric approach – can be illustrated with a simple example. In their phylogenetic analysis of Silurian encrinurine trilobites from the central Canadian Arctic, Adrain and Edgecombe (1997) employed 40 different morphological characters drawn from six different trilobite character complexes. Of these characters – all of which were nominally coded in the standard manner – 25 represented continuous, ratio-scale morphometric variables (e.g., Height of Eye: not stalked [0], very tall, 'stalked' [1]; Size of Transverse Tubercle Row: subdued [0], prominently expressed [1]; Depth of Doublural Notch: shallow, broad [0], deep [1]). Figure 7.1 shows tracings of the 'Depth of Doublural Notch' character taken from the plates provided by Adrain and Edgecombe (1997) with the character-states assigned to these morphologies marked. Obviously, there is a wide range of morphological variation inherent in this feature. Just as obviously, there are notch outlines that could be described accurately as 'shallow, broad' (e.g., *Avalanchurus simoni*) and others that could be described as 'deep' (e.g., *Struszia epsteini*). Nevertheless, a wide range of intermediate morphologies also exist that do not seem to fit clearly into either nominal class (e.g., *Mackenziurus deedeei, S. dimitrovi, S. onoae*).

Is the variation exhibited by these morphologies distributed continuously or discontinuously? What are the class boundaries of Adrain and Edgecombe's (1997)

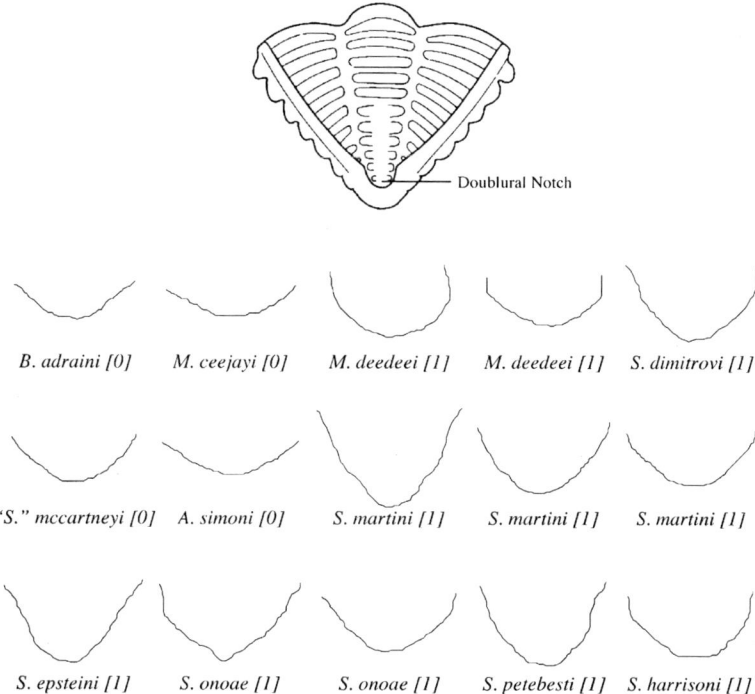

Figure 7.1 Doublural notch and notch outlines for a selection of Arctic encrinurine trilobite pygydia (= tail segments). The doublural notch is located on the underside of the pygydium and accepts the anterior portion of the specimen's glabella when the animal is enrolled. As such, it is related to a cranidial character. Adrain and Edgecombe (1997) used this fundamentally shape-based, morphometric character in their phylogenetic analysis of this group. In that study the notch outlines of *Billevittia adraini*, *Mackenziurus ceejayi*, *"Struszia" mccartneyi*, and *Avalanchurus simoni* were nominally described as 'shallow, broad' [0] while the notch outlines of the remaining species were described as 'deep' [1]. S. = *Struszia*, M. = *Mackenziurus*, A. = *Avalanchurus*, B. = *Billevittia*.

character-state classes? Are the character-state assignments correct? Are other characters lurking in this feature and could be used in phylogenetic systematic contexts? Systematists must have answers to these questions – along with analogous questions for all of the other morphometric characters used in this study – if they are to evaluate the validity of the conclusions reached, attempt to reproduce the analysis on their own, or (perhaps most importantly) attempt to extend this analysis to other groups of encrinurine trilobites in particular, or to trilobites in general. The answers to these questions are very difficult to obtain, explain, or justify through simple, qualitative inspection of these figures.[1] Obtaining answers to these questions is perhaps the most obvious area in which morphometrics can contribute to systematics.

---

1 Adrain and Edgecombe (1997) did not provide a table of illustrations such as Figure 7.1 for this – or any other – character, but illustrated their characters via traditional plates of photographs within which many more sources of variation were present.

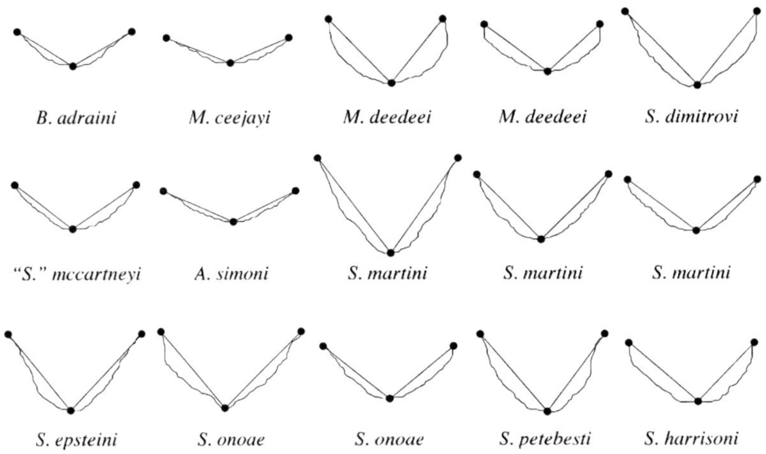

106   Norman MacLeod

B. adraini	M. ceejayi	M. deedeei	M. deedeei	S. dimitrovi
"S." mccartneyi	A. simoni	S. martini	S. martini	S. martini
S. epsteini	S. onoae	S. onoae	S. petebesti	S. harrisoni

Figure 7.2 Landmarks and implied triangles used to geometrically assess doublural notch shape using Bookstein Shape Coordinates. For this analysis the baseline was taken as the chord between the end-point landmarks. The central landmark was taken as the lowest point on the curve nearest the midpoint of the baseline chord. Although this central landmark definition identifies it as a 'constructed' point, the concept effectively quantifies the qualitative descriptors 'shallow' and 'deep' used by Adrain and Edgecombe (1997) to assign these morphologies to nominal character states.

A variety of morphometric approaches to the analysis of this dataset could be selected. Perhaps the simplest geometric analysis method would be to characterize each notch by a triangle of landmarks, with two landmarks representing the endpoints of the structure and the third representing the notch's nadir. Once coordinate values for these landmarks had been obtained the curves could be compared to one another using the Bookstein Shape Coordinate (BSCoord) method (Bookstein 1986). Bookstein Shape Coordinates are well-suited to this analytic situation in that the Adrain and Edgecombe (1997) 'Depth of Doublural Notch' character seems concerned primarily with the depth of the notch nadir relative to the baseline formed by the notch endpoints.

Figure 7.2 shows the landmarks points selected to represent the doublural notch character and Figure 7.3 shows the results of BSCoord analysis. The distribution of the free coordinate (= notch nadir) suggests that, for this small dataset, shape variation is continuous from the shallowest to the deepest notch profiles. Given these results there seems no obvious morphological discontinuity at which to place a line of definition between Adrain and Edgecombe's (1997) depth of doublural notch state classes.

In addition to raising questions about the definition of Adrain and Edgecombe's character states, these results cast doubt on their character-state assignments. Although the four species they assigned to the 'shallow, broad' character-state class do represent the shallowest curves in this dataset, the variation exhibited by these curves along the *y*-shape coordinate axis (= curve depth *sensu stricto*) is greater or equal to the variation between the lower limit of this provisional subgroup and the next grouping of shape coordinates. Based on these results it is difficult to understand why *"Struszia" mccartneyi*'s notch is described as 'shallow, broad' while *Struszia martini*'s and *Mackenziurus deedeei*'s notches are described as 'deep'.

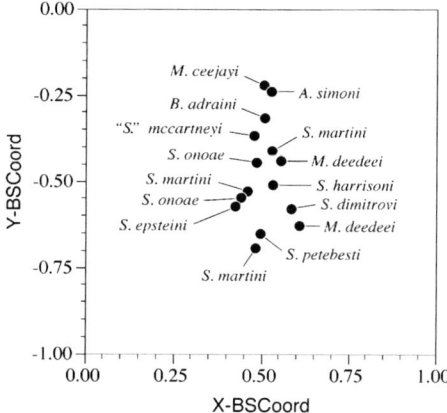

Figure 7.3 Results of a shape coordinates analysis of the trilobite doublural notch landmarks shown in Figure 7.2. This plot represents the distribution of *x* and *y* values of the central landmark for each representative individual after the baseline (see Figure 7.2 caption) had been rescaled to unit length and standardized orientation. Note the lack of a pronounced discontinuity among these shapes along the *y*-shape coordinate. Such a discontinuity would be expected given Adrain and Edgecombe's (1997) subdivision of the doublural notch character into two discrete character states. S. = Struszia, M. = Mackenziurus, A. = Avalanchurus, B. = Billevittia.

Note that this BSCoord-based analysis reduced these somewhat complex curves to triangles of landmark points (Figure 7.2). In some cases the underlying morphology is indeed triangular (e.g., *A. simoni*, *S. martini*) and so would not suffer by representation as a geometric triangle. For the majority of the curves in this small dataset, though, representation as a triangle severely distorts the true nature of the morphology that is being used in the qualitative assessment of their variation.

In addition, to these considerations the reliability of the biological correspondence with which the notch nadir landmark has been placed on the curve is open to question. While various geometric criteria can be advanced to help guide the placement of this landmark, none of these geometric criteria have anything necessarily to do with the underlying biological processes responsible for the notch structure. In the absence of much more biological information about the notch, the physio-chemical processes responsible for its formation, its comparative ontogeny, etc., it is questionable whether this third coordinate represents the same type of observation represented by the baseline landmark coordinates. Given this inherent biological uncertainty, coupled with the distortions imposed on the morphological system as a result of abstraction to just three landmarks, one might suspect that the negative results obtained by the BSCoord analysis may have as much to do with how the morphological variation was measured as with the nature of geometric variation among the original notch curves. Consequently, it seems reasonable to employ an alternative method of morphometric analysis in order to (1) test the BSCoord results for robustness to examination by different geometric methods and (2) determine whether the addition of geometric data to the measurement system supports refinements to the Adrain and Edgecombe (1997) system of morphological descriptors.

Once again, there are a variety of methods that could be employed to analyze these curves and draw conclusions about the manner in which their shapes are distributed. One available candidate is open-curve eigenshape analysis (MacLeod 1999). This method is similar to relative warp analysis (Bookstein 1991), but draws a distinction between landmarks and semi-landmarks (Bookstein 1997). In essence, open-curve eigenshape analysis uses formal landmarks to define the endpoints of the curve and (if necessary) subdivide the curve into segments. These segments are then represented by semi-landmarks that are accorded a sequence-level correspondence to one another between shapes within the sample.

It is important to note that there is no necessary implication of biologically homologous correspondence between semi-landmarks. The nature of their correspondence resides at the level of geometry and sequence order only. This type of correspondence is justified when it is the only level of correspondence assessment available on which to base morphological comparisons. Sequence-level correspondence is no different in principle from the qualitative assessments of between-curve-segment correspondence in the absence of additional biological information that are made routinely by systematists. Such comparisons have been accepted as a basis for the characterization of morphological attributes for centuries. If one can base comparisons on point-to-point correspondences that have some larger biological-phylogenetic significance one should do so (but see discussion of morphological homology below). In the absence of such biological information, morphologies can still be quantitatively compared – and shape-distribution hypotheses tested – using the sequence-level correspondences inherent in the concept of semi-landmarks (Bookstein 1997).

In the present example the fifteen doublural notch outlines shown in Figure 7.1 were digitized and the two end-point coordinates designated as biological landmarks. These semi-landmark-defined outlines were then converted to Zahn and Roskies (1972) angular-deviation shape functions and submitted to a singular value decomposition in the manner that has become standard for eigenshape-based methods (see MacLeod 1999 and references therein). A plot of the notch-curve scores on (= covariances with) the first two eigenshape axes (Figure 7.4) suggests that discontinuities do exist in the distribution of shapes within this dataset. In particular, *A. simoni*, *M. ceejayi*, and *Billevittia adraini* exhibit shape scores along the first eigenshape axis (ES-1) that are atypically low relative to the remainder of the dataset. Since Adrain and Edgecombe (1997) assigned these three species to their 'broad, shallow' character-state class, this grouping is not unexpected. However, "*S*". *mccartneyi* occupies a position along the ES-1 axis on the other side of the morphological discontinuity delimited by *A. simoni*, *M. ceejayi*, and *B. adraini*. Since "*S*". *mccartneyi* was also characterized as exhibiting a 'shallow, broad' doublural notch by Adrain and Edgecombe (1997), these eigenshape-based morphometric results are inconsistent with their qualitative assessment of morphological variation. Similarly, the distribution of shapes along the second eigenshape axis (ES-2) suggests that shape variation within the two *M. deedeei* specimens also exhibits a morphological discontinuity with respect to the remainder of the sample. This distinction in doublural notch morphology was not noticed by Adrain and Edgecombe (1997).

To obtain a better understanding of the types of morphological contrasts implicit in Figure 7.4, assessments of 'pure' shape variation along these two eigenshape axes can be obtained through modeling (Figure 7.5, see MacLeod 1999 for an explanation of

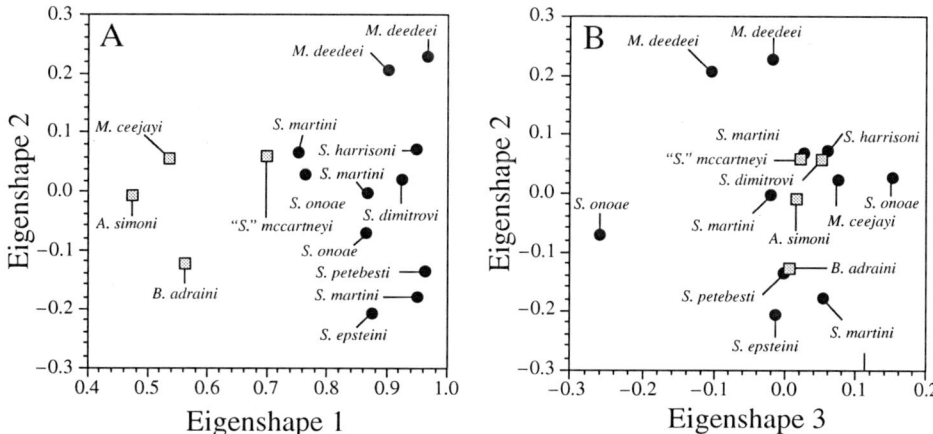

Figure 7.4 Eigenshape results for the doublural notch curves shown in Figure 7.1. A. Ordination of notch curves within the plane formed by eigenshapes 1 and 2. B. Ordination of notch curves within the plane formed by eigenshapes 2 and 3. By taking the geometry of the entire notch into consideration geometric discontinuities between the central cluster of morphologies and *Billevittia adraini*, *Mackenziurus ceejayi*, and *Avalanchurus simoni* are evident along Eigenshape 1. A similar shape discontinuity exists between the two *Mackenziurus deedeei* specimens and the remaining shapes along Eigenshape 2. Qualitative assessments of discontinuities such as these form the basis for traditional systematic character-state assignments. Morphometric methods, such as eigenshape analysis, are available to help systematists evaluate observed shape distributions. Once such results have helped 'sharpen the eyes' of systematists they would be free to re-evaluate shape distributions in a traditional qualitative manner, or to explicitly use results like these to make character-state assignments. See text for additional discussion. *S.* = *Struszia*, *M.* = *Mackenziurus*, *A.* = *Avalanchurus*, *B.* = *Billevittia*.

the modeling method). The series of shape models determined for the ES-1 axis confirm Adrain and Edgecombe's (1997) original distinction between shallow and broad curves (= low scores) and narrow deep curves (= intermediate and high scores). But, contrary to the results of Adrain and Edgecombe's (1997) qualitative analysis, this is not the only shape-variation mode that has potential for defining systematically interesting morphological discontinuities. The series of models determined for the (ES-2) suggest that the contrast between V-shaped notch outlines (= low and intermediate scores) and U-shapes notch outlines (= high scores) also has potential for characterizing *M. deedeei* in a manner consistent with traditional qualitative analyses. Inspection of Figure 7.1 after viewing these results shows that the two *M. deedeei* notch outlines do indeed exhibit characteristically more U-shaped notch profiles – much more steeply sided with broader, flatter bottoms – than the remainder of the dataset.

In this example, morphological character states were defined on the basis of pronounced gaps or discontinuities in the distribution of shapes within a sample of trilobite pygydia. It is important to note that even though the morphometric variables used to ordinate these shapes were continuous, ratio-scale variables, the eigenshape analysis revealed two different discontinuities within the distribution of observations along these continuous variables and that these quantitative gaps were later confirmed by a qualitative (re)inspection of the shapes. Consequently, the specification and analysis of continuous morphometric variables facilitated the recognition of discontinuous

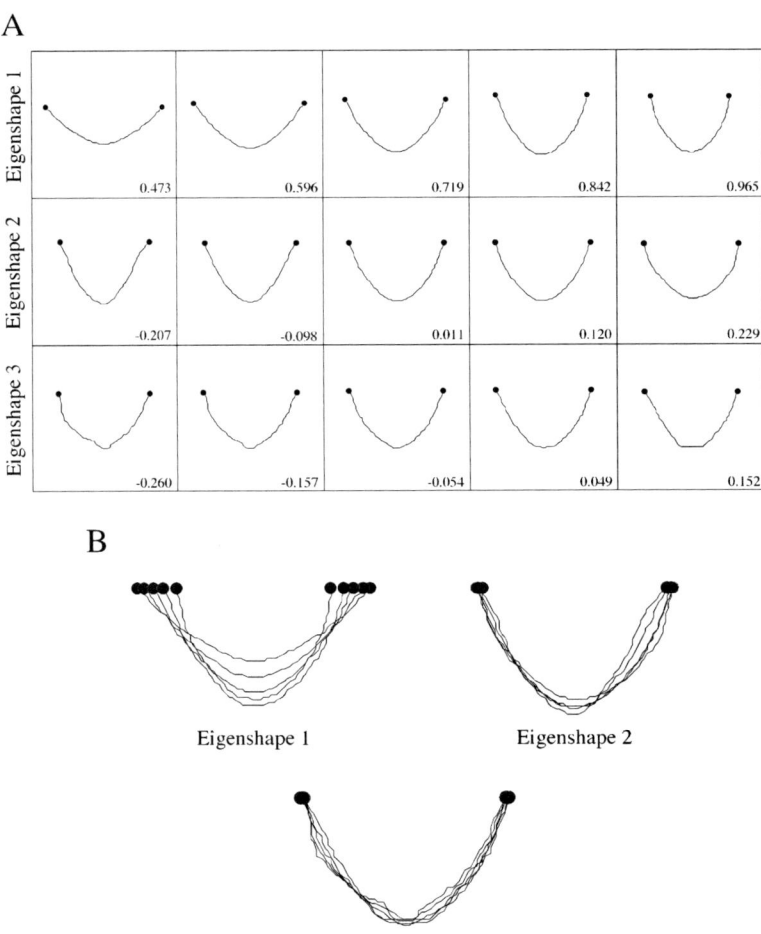

Figure 7.5 Geometric models of the first three eigenshape axes shown in Figure 7.4. A. Model sequences computed at designated locations along the eigenshape 1, 2, and 3 axes. B. Along-axis or superposition-style representations of model sequences. Using these figures it is clear that the shape contrast represented along Eigenshape 1 corresponds to Adrain and Edgecombe's (1997) qualitative states 'shallow, broad' and 'deep'. Eigenshape 2 corresponds to a contrast between U-shaped and V-shaped notch geometries (a distinction missed by Adrain and Edgecombe, 1997). Eigenshape 3 corresponds to a contrast between curve symmetry or irregularity in the region of the nadir. Models such as these can aid in the interpretation of abstract morphometric ordinations – and in the recognition of new characters or character states – by portraying coordinate positions within the shape space as geometric figures that can be compared in a traditional, qualitative manner.

patterns of shape variation that could be retrospectively coded by the traditional method of qualitative inspection. This is consistent with the descriptions of standard contemporary systematic procedure (e.g., Farris *et al.* 1970; Pimentel and Riggins 1987; Felsenstein 1988).

Based on this example analysis it seems hard to argue that morphometric data analysis can, in principle, play no useful role in the systematic study of biotic character-state data. Indeed, these results suggest that morphometric approaches can represent nothing more than a direct — albeit more sophisticated — extension of traditional qualitative morphological analysis methods that simply draw the systematist's eye to patterns that he or she might not have immediately recognized. Once the presence of such discontinuities within the sample has been made obvious by the morphometric analysis, they can be readily understood and utilized by more traditional qualitative inspection. Similar conclusions have been reached by a number of other systematists (Thiele and Ladiges 1988; Thiele 1993; Fink and Zelditch 1995; Zelditch *et al.* 1995; Rohlf 1998; Swiderski *et al.* 1998; Rae 1998), though none has provided a simple example of this process.[2]

Continuity: axes vs observations

Obviously, not all patterns of variation between putative taxonomic groups will be characterized by pronounced discontinuities. For instance, morphological characters that are the result of polygenic suites will exhibit a progressive shift in the mean value for a population under directional selection; even if the selection pressure is intense (Falconer 1981; Felsenstein 1988). While the continuous patterns of variation characteristic of anagenetic evolution should not be confused with the morphological discontinuities implied by cladogenesis (Zelditch *et al.* 1995), it is the case that both modes of evolutionary change may be characterized by zone of between-group morphological overlap that are determined (at least in part) by group-membership (Figure 7.6).

This observation has been used by a number of systematists to argue that continuously-distributed variables should be used, together with discontinuously-distributed variables, to reconstruct phylogenetic patterns. However, continuously-distributed ratio-scale or interval-scale variables cannot be transformed into the nominal or sometimes ordinal variables used by most phylogenetic inference algorithms except through the application of ad hoc rules. Gap coding (Mickevich and Johnson 1976; Archie 1985), segment coding (Colless 1980; Thorpe 1984; Chappill 1989) statistical 'difference between means' tests (e.g., homogeneous subset coding, Farris 1990; Thiele 1993; Rae 1998); and statistical 'overlap analysis' (Almedia and Bisby 1984; Swiderski *et al.* 1998) all represent attempts to devise rule-based systems for subdividing continuous patterns of variation in morphological variables based on a consideration of frequency and/or a priori group-membership.

The difficulty with these rule-based methods is that they — sometimes to a greater extent, sometimes to a lesser — alter the nature of the variable description from that of the individual to that of the group. Figure 7.6 illustrates this problem. Figure 7.6A shows the typical situation for a morphological character measured on a continuous, ratio scale. All values along the scale are possible. Above this scale a hypothetical frequency distribution of a set of morphometric measurements is illustrated. In this

---

2 See also Chappill (1989) and Thiele (1993) for discussions of semantic confusion between the notions of continuous variables and discontinuous patterns of variation that often occur in the systematic literature.

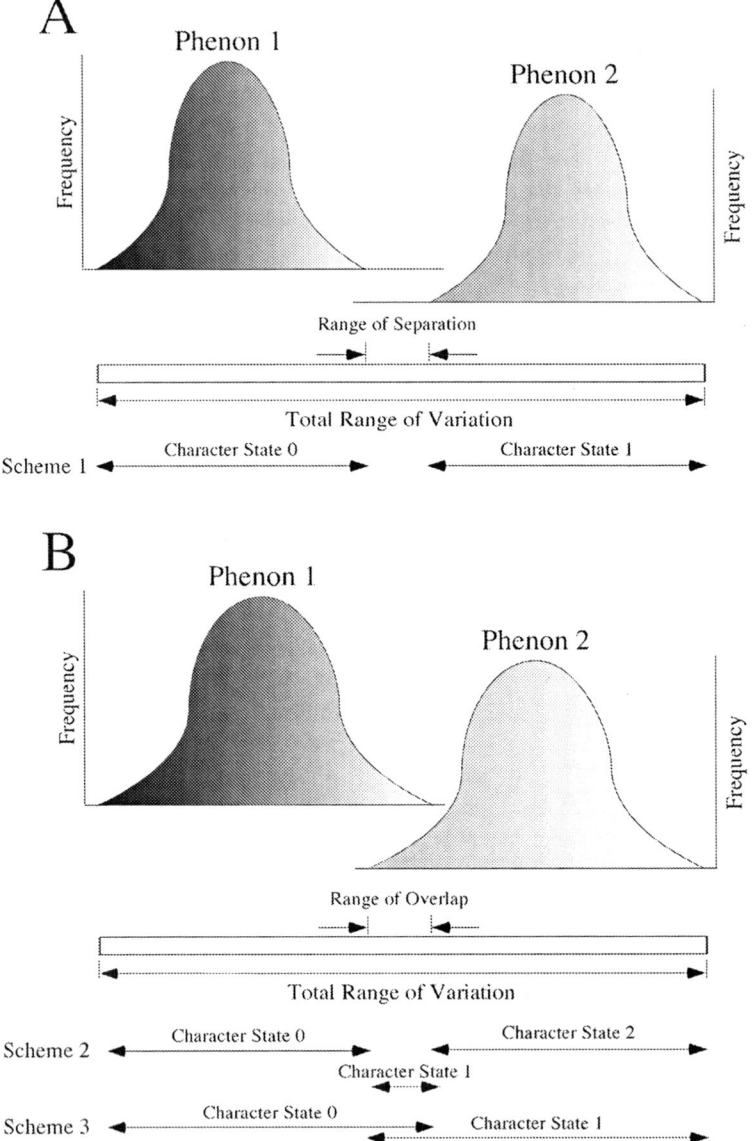

Figure 7.6 Conceptual diagram illustrating the difference between continuous variables and continuous distributions of variable observations. A and B represent sets of frequency distributions along a continuous morphometric variable axis for discontinuously distributed (A) and continuously distributed (B) sets of observations. These observations have been gathered into two putative taxonomic groups, or phena. The gap in A represents a morphological discontinuity that serves to distinguish the phena from one another objectively and independently and that may be used to recognize one, or the other, or both as a monophyletic group. This situation poses no theoretical or practical problems in terms of character recognition of character-state coding, as evidenced by the widespread inclusion of such characters in systematic datasets (Scheme 1). The lack of an inter-phenal gap in B represents several

example the overall distribution is discontinuous with two well-separated subgroups evident along the continuous variable axis. Using morphometric methods it is a simple matter to determine the observed range of separation between the subdistributions (= discovery phase of analysis). Once this is accomplished, the morphological discontinuity discovered by the systematist can be used to define the range of variation for putative state classes of the morphological variable (Figure 7.6A, Scheme 1). Individuals can then be assigned to these state classes without needing to take putative group membership into consideration. As a result, the variable axis (= character) is objectively subdivided into segments (= state classes) by the observations with taxic groups emerging as a result, rather than as an assumption, of the analysis. Character states discovered and described in this manner will be genuinely independent in that they make no a priori reference to group membership.

Contrast this with the situation in Figure 7.6B which shows the same morphological scale above which the frequency histogram for an alternative set of morphological observations is illustrated. In this case, the distribution is continuous and bimodal along the measurement scale. Since there is no morphological discontinuity that can be used to recognize and define phenotypic states unambiguously, partitioning must be made on the basis of group membership. Either the range of the overlap between putative phena is designated as a separate state from the non-overlapping ranges of the distribution (e.g., Archie 1985) or the means of the putative phena are tested for statistically significant separation and the entire phenon assigned to a state on the basis of the test results (e.g., Thiele 1993; Rae 1998). This practice, which lies at the heart of all gap coding, gap weighting, segment coding, subset coding, overlap analysis, etc. procedures, might be described by the generic term 'member coding'.

Member coding leads to the production of questionable character-state definitions. For example, since the definition of the group must be based on other criteria (presumably discontinuously distributed character states), the member-coded character state cannot be regarded as an independent descriptor of morphological variation. Logical consistency is also challenged by member coding. Member-coded character-state definitions cannot be recognized as 'features of organisms' because some proportion of the member specimens exhibit morphologies that are dependently distinguished from (Figure 7.6B, Scheme 2) or identical to (Figure 7.6B, Scheme 3) to those of other groups. Such procedures can even result in group members receiving a single character-state code as if all members of the group were morphologically distinct from all members of another member coded group *regardless of whether this disjunction was actually observed*. Member-coded character states codify assessments of group-based frequency trends or probability statements that are dependent on group diagnosis rather than objective and independent morphological descriptors. As such they represent a fundamentally different type of quasi-morphological variable that should

---

systematic problems. Continuously distributed observations cannot be used to objectively or independently distinguish the phena from one another, because the limiting criterion is irreducibly dependent on the recognition of one of the other phena, which are presumably diagnosed on the basis of other variables. Various schemes have been developed to arbitrarily subdivide such continua (e.g., gap coding, interval coding, see Schemes 2 and 3), but all are dependent on a priori group recognition to work. Since character states defined in these 'member-coding' schemes cannot objectively and independently delineate phena they are of little use in phylogenetic systematics.

not be mixed with real morphological variables (whether continuously distributed or discontinuously distributed) in either phylogenetic or morphometric analyses unless appropriate measures are implemented to take the introduction of such mixed-mode data into consideration.

Rather than spending time trying to rationalize the arbitrary subdivision of continuous morphological data so that these can be used in phylogenetic systematic analyses, systematists would be better advised to acquaint themselves with tools of morphometrics since the use of these tools will greatly improve their chances of recognizing and correctly interpreting the morphological discontinuities that are present in their data. The doublural notch analysis discussed in the previous section contains a practical example of this situation. Despite the fact that Adrain and Edgecombe's (1997) qualitative analysis failed to recognize the distinction between V-shaped and U-shaped notch morphologies, application of a more generalized morphometric procedure was able to reveal its presence in those data. The addition of a new character and set of discontinuous character-state distinctions – such as can be achieved by morphometric methods – is of far more use to systematics than the inconsistent and logically suspect results of member-coding procedures.

Distances, landmarks, and homology

While morphometric variables cannot be rejected for systematic study because of their type – they are no different in this respect from traditional systematic characters, cladistic characters, statements or systematic keys – there may be other problems with this class of morphological descriptors that would limit their utility in some systematic contexts. Pimentel and Riggins (1987: 201) argued that in addition to being able to be ordered and independent, true systematic characters must be "homologous expressions of a feature found in the ingroup and outgroup".

Morphometric variables have long been criticized as being intrinsically non-homologous representations of form on several different grounds. The most common of these is an objection to the reification of (usually multivariate) morphometric variables as organismal 'attributes' in any meaningful biological sense of that term (Pimentel and Riggins 1987). This criticism confuses the methodology of principal components (= eigenanalysis) with the data to which the method has been applied.

Until recently the most common type of morphometric data was linear distances between pairs of landmark points (see Blackith and Reyment 1971; Reyment et al. 1984; Bookstein et al. 1985; Reyment 1991). Eigenanalysis of covariance or correlation matrices derived from such data yield variables that are differentially weighted amalgamations of scalar magnitudes. Pimentel and Riggins (1987) point out correctly that such amalgamations do not correspond to the concept of biological homology, but this deficiency arises as a result of the lack of topological information in the scalar distance matrix rather than arising out of the eigenanalytic procedure itself. For example, if the necessary topological information is restored to the analytical system by keeping track of the relative orientations of the distance variables (e.g., the 'truss analysis' method of Strauss and Bookstein 1982; see also Bookstein et al. 1985) the results of such an analysis can be recombined into a model of morphological deformation that exhibits the properties of topological correspondence and spatial localization required of biological homologues (MacLeod in press). Such topologically-informed eigenanalysis-based procedures could be used, in principle, to identify and interpret discontinuities in shape distributions in a manner conceptually similar to qualitative morphological analysis procedures. Thus, the problem Pimentel and Riggins (1987) refer to in their criticism of morphometrically-defined characters is a problem of the manner in which morphological variation had been portrayed up to that point in time (as topology-free scalar magnitudes) rather than a problem that arises from the eigenanalytic methods used to summarize patterns of variation within those data.

At approximately the same time that Pimentel and Riggins (1987) made their criticism several morphometricians independently began to recognize and understand the origins of the same problem: that the absence of topological information from morphometric datasets severely constrained the interpretability of their analytic results. From 1986 through 1991 F. L. Bookstein, C. Goodall, D. G. Kendall, F. J. Rohlf, and others effectively synthesized and reformulated several disparate data and method-based schools or morphometric analysis into a single, unified 'geometric morphometrics' with topology at its center. This synthesis was achieved by refocusing attention on the coordinate positions of landmark points scattered over a structure and regarding deformations of those, using those coordinate point constellations, as morphometric variables. Operationally, this reformulation of morphometrics involves the submission of landmark constellations – variously adjusted to remove the effects of size and differential orientation – as input into (for the most part) eigenanalysis-based procedures. The results of such analyses produce mathematically elegant summaries of geometrical deformation patterns that could be expressed in either the abstract notation of mathematics (e.g., data matrices, deformation grids, scatterplots) or the geometric representation of morphological variation in traditional qualitative systematics (e.g., Bookstein 1991; MacLeod 1999). This was possible because the magnitudes of Cartesian coordinates preserve information about the relative amount *and* directions – the topology – of landmark displacements between forms.

Since topological information has now been inextricably embedded into the corpus of geometric morphometrics, a reconsideration of Pimentel and Riggin's (1987) claim that morphometric variables cannot express biological homology is necessary. In my view, this question has two aspects, (1) whether the constellations of landmark points defined on the basis of biological structures are homologous to other constellations of landmark points similarly defined (= structure-level homology), and (2) whether individual landmark points are homologous with other landmark points (= point-level homology).

Unfortunately, the morphometric literature makes little distinction between the concepts of geometric homology and biological homology. Landmarks are defined as relocatable coordinate positions on an object in a two-dimensional or three-dimensional Euclidean measurement space (Bookstein 1991, MacLeod in press). Since geometrical homology is defined on the basis of topological correspondence, corresponding landmarks are, by definition, geometrical homologues.

Biological homology begins with topological (= geometrical) homology, but extends its concept to embrace aspects of history and origin. Although the concept of the homologue was known to Aristotle (who used it to infer correctly that porpoises were more closely related to mammals than to sharks), Richard Owen (1843: 374) is responsible for the concept's canonical formulation as "the same organ in different animals under every variety of form and function." Darwin (1859) regarded his theory of common descent as providing a biological explanation for the difference between homologous and analogous structures, after which the former was re-defined as 'similarity due to common ancestry'. Unfortunately this reformulation led to the concept's use in two different senses: (1) a transcendental or transformational sense as a sequence of idealistic modifications (e.g., fish jaw bones changing into mammalian ear ossicles) and (2) a taxonomic or taxic sense (e.g., tetrapods being recognized as a monophyletic lineage based on the fact that all members exhibit four limbs, except for those in which the limb number has been reduced due to secondary loss). Patterson (1982) pointed out this duality, rejected the tranformationalist conceptualization as being non-falsifiable, and equated taxic homology with the Hennigian concept of synapomorphy.

Thus, while the triangular dorsal fins of sharks, porpoises, and goldfish might be regarded as geometrically homologous, these structures are not biologically homologous because the implied taxic grouping is not supported by other morphological characteristics (e.g., developmental patterns, skeletal characteristics, soft anatomy). This stands in contrast to the popular description of these three organismal groups as belonging to the group 'fish' which is a phenetically-defined morphological concept based entirely on topological similarities in gross external morphology.

Since the concept of biological homology is logically tied to organic structures and incorporates the notion of topological similarity (Rieppel 1980, 1994; Patterson 1982), morphometric variables that assess aspects of topological similarity between those structures can be used to delimit taxic groups in a manner consonant with the strictures of biological homology. This is nothing more, or less, than what is done in the qualitative assessment of morphological variation patterns within or between groups of organisms. Contra Zelditch *et al.* (1995), morphometric decompositions of landmark constellations are not biologically homologous *by definition* because (1) morphometrics measure topological similarity and topological similarity is only one aspect of biological homology, and (2) the taxic groups recognized by a morphometrical analysis of one structure may be falsified by other morphometric or qualitative results for other structures (this is also true of 'standard' systematic characters). Caution must be exercised not to confuse the failure of a particular morphometric result to recognize a distinction between complex morphological structures with a failure of such a distinction to exist Morphometrics usually assesses patterns of topological similarity and difference between aspects of organisms shape, not the organisms themselves; see doublural notch example above. Nevertheless, the overall similarity between morphometric and qualitative procedures of analysis at the level of morphological structures (structure-level homology) seems clear and consonant.

The idea that individual landmark points represent biological homologues is logically separate from the issue of structure-level homology. Landmark points were originally described as 'homologous' in order to distinguish them from the geometrically-constructed boundary point locations used in most forms of morphometric outline analysis (Bookstein *et al.* 1986; Bookstein 1990, 1991). Bookstein (1991) identified three classes of biological landmarks: discrete juxtapositions of structures or tissues (Type 1), maxima of curvature (Type 2), or extrema (Type 3). This classification focuses attention on the amount of information necessary to identify or relocate the landmark.

Type 1 landmarks may occur at any point on or within a form so long as that form is composed of different structures or tissue types. While these landmarks are constrained to exist on the boundaries (= outlines) of structural components or tissue-defined regions, their locations are not determined by any characteristics of the overall boundary or outline. Type 2 landmarks lie on the boundaries of single structures or regions and are defined by the nature of the curving surface of that boundary. Type 3 landmarks represent those coordinate locations on single structures (irrespective of whether the structure is composed of various substructures or regions) that represent the extremes of the structure's boundaries. Like Type 2 landmarks these points are constrained to lie on the object's outline.

No consideration has been traditionally given to the nature of any substructure or tissue when locating Type 3 landmarks. Their definition is dependent on the nature of the outline (= by the distribution of adjacent boundary coordinates), on the orientation of the object, and on the number of axes one wishes to locate extrema along. Because the nature of Type 3 landmarks is so variable and dependent on such a wide variety of conditions, Bookstein (1997) revised his 1991 classification and termed this class of landmarks 'semi-landmarks.' The category semi-landmarks includes the former Type 3 landmarks of Bookstein (1991) as well as the boundary coordinates used in outline morphometrics (e.g., Fourier analysis, eigenshape analysis, edgels).

Bookstein's (1997) revised landmark classification describes the range of landmark-based morphometric observations more comprehensively and recognizes fundamental similarities between observational types more consistently. Since the newer landmark taxonomy abandons the older distinction between individual landmarks and boundary coordinates – which formed the rationale for labeling the former as 'homologous' in order to distinguish them from the latter – this appellation no longer serves any purpose. More importantly though, Bookstein's (1997) revised landmark classification recognizes the fundamental unity of all landmark types as relocatable points that correspond across specimens in a geometrical sense. In other words, individual landmarks represent geometrical homologues. But, the ability to represent and summarize topological patterns among biologically homologous structures does not render corresponding landmarks themselves biologically homologous. To make such a conceptual leap is to confuse the idiosyncrasies of a representative with the characteristics of the group being represented.

In order to appreciate this distinction, in your mind's eye visualize a familiar morphological feature. While it is acceptable to describe the alternative forms of a radius bone, a canine tooth, a genal spine, or a pectoral fin as 'long' or 'short', 'elliptical' or 'subquadrate', 'pointed' or 'blunt', etc. – because any reasonable set of morphometric measurements derived from sets of landmarks located on these objects exhibit non-overlapping distributions – it is quite a different matter to claim that the 'geometrical midpoints' or the 'proximal and distal termini' of differently shaped bones or teeth, or spines, etc. correspond with the 'geometrical midpoints' or the 'proximal and distal termini' of differently shaped bones or teeth, or spines, etc. of other specimens in any save a topological sense. Within reasonable limits it is irrelevant whether corresponding landmark points fall on precisely the same point of a feature because the level of topological similarity required by this decision is not part of the biological homology concept. Since, in the great majority of instances, a wide variety of alternative landmark pairs can pass the similarity, conjunction, and congruence tests of biological homologues (see below for an example), true homology – for there can only be one pair of landmark points that define a length on any biological structure that are biological homologues of another pair of landmark points on another homologous structure – cannot be separated from false homology at the level of the mathematical point by the tests available to systematists. Consequently, the entire question of 'biologically homologous landmarks' is moot. Wagner (1994) points out that cases may exist in which the concepts of biological and topological homology coincide (e.g., point-intersections between three skull bones representing a Type 1 landmark). However, such situations are conjectural at present and represent a distinct minority of landmarks currently used for morphometrical analyses. In the absence of highly-detailed developmental and phylogenetic evidence, the notion of mathematical 'point homologies' will likely remain either an assumption or an assertion for the foreseeable future.

Bookstein (1991) described another type of relation that bears on the issue of homology in morphometrics: deformational homology. Tracing the origins of this concept to Thompson (1917), deformational homology begins with a series of geometrically-homologous point-to-point mappings on two forms and postulates sets of smooth deformations implied by a comparison of the forms. Often these deformations can be described by a single or a series of generalized deformational type(s) (e.g., pure inhomogeneous, quadratic, rigid motion involving several landmarks, spiral deformation; see Bookstein 1991 for examples). Unfortunately, Bookstein's (1991) discussion is unclear as to whether he was referring to geometric or biological homology in advancing this concept of Thompsonian deformational homology. Thompson (1917) believed his deformational types (even though he never referred to them in those terms) resulted from the operation of basic physical laws. However, Thompson refrained from discussing his concept in evolutionary-phylogenetic terms because he rejected Darwin's theory (see Mayr 1982).

Regardless of Thompson's opinions on evolutionary theory, his concept of deformational homology underpins much of the 'morphometric synthesis' (Bookstein 1993) because the language of deformations is useful in summarizing and interpreting the results of geometrical morphometric data analyses. It is this concept of deformational homology, however, that Pimentel and Riggins (1987) implicitly refer to when they criticize morphometric variations as being capable of representing only transformational homology. However, as pointed out by Bookstein (1994), Rohlf (1998) and MacLeod (in press) this metaphorical linkage between the morphometric-geometric 'language of deformations' and transformational homology diverts uncautious readers from the main point of systematic morphometrics. The deformational graphic methods used to portray geometric morphometric results (e.g., thin plate splines) are only illustrative conventions; useful for visualizing geometric relationships. Just as written descriptions of transformational homologies can be recast rhetorically as statements of taxic homology (Zelditch *et al.* 1995), so too can ratio-scale morphometric variables – even complex, multivariate variables – be used to quantitatively define taxic groups on the basis of topological similarities or differences among a priori-defined putatively homologous features. The relevant questions, then, are not whether individual landmarks can be declared homologous (they cannot), whether morphometric variables can be used to recognize groups of taxa on the basis of shared topological similarity between putative homologues (the evidence for this is quite overwhelming as demonstrated by over a century of morphometric analyses), nor whether morphometric variables represent some quality different from what is typically represented by a large number of qualitative morphological characters (they do not). Rather the relevant questions are (1) whether morphometrically-defined variables exhibit a hierarchical structure that can be logically represented on a cladogram and used to define congruently nested sets of taxa and (2) whether the taxic groups defined on the basis of morphometric analysis agree with groupings defined on the basis of more traditional morphological analyses.

Partial warp variables as phylogenetic characters: a test of congruence

Extending from their arguments regarding the homology of landmarks, Zelditch *et al.* (1995, see also Fink and Zelditch 1995) have recently advocated the use of partial warp-based morphological-deformation variables in phylogenetic analysis. Partial warps are calculated from principal warps, which are eigenvectors of the bending-energy matrix that express the ways a reference configuration of landmarks can be geometrically deformed (Bookstein 1991). The partial warp scores are computed by projecting the values of a Procrustes-aligned set of landmarks separately onto each of the principal warp vectors.

Zelditch *et al.* (1995) preferred the partial warps approach to all other multivariate methods of morphometric data analysis for systematic studies because they believe such summaries to represent non-arbitrary and spatially-localized features of a geometric-biological system that are unique to individual organisms (as opposed to being arbitrary summaries of populations or samples), and that variables so-defined can be used to recognize morphological characters and define character states in a manner that supports their analysis within a hierarchical data-analysis system (e.g., those used for phylogenetic inference). The primary problems these authors sought to solve by advocating the use of partial warp variables in systematic-phylogenetic contexts were (1) improved ways of discovering of new morphological characters to be used in phylogenetic analysis, and (2) demonstrating that quantitative morphological characters were no different from qualitative morphological characters in systematic contexts.

The Zelditch *et al.* (1995) arguments have not been met with widespread agreement within the morphometrics or systematics communities. Bookstein (1994, see also Monteiro 2000) questioned whether any morphometrically-defined shape variables could unambiguously order shape transformations. The Shape Nonmonotonicity Theorem shows that deformation-based shape variables can be created to support any ordering of end-member shapes. Moreover, Bookstein (1994) argued that geometric deformations per se could not be used as systematic characters because they lack the property of commutativity. Fink and Zelditch (1995) and Zelditch *et al.* (1998), in very brief responses to the Bookstein (1994) article did not dispute any of his geometrical arguments, but attributed Bookstein's disagreements with them over the use of morphometrical variables in systematics to disagreements over 'semantics'.

Lynch *et al.* (1996) attempted to use the method of Zelditch *et al.* (1995) to create systematic characters, but were uncomfortable with the results. These authors recommended that simulation studies be used to validate the Zelditch *et al.* (1995) method. Naylor (1996) conducted such a study using a simulated fish phylogeny that was reported to contain no homoplasy and was based on only a single morphological character-state change per branch. His results showed that although parsimony-based analysis of the entire multistate-coded, partial-warp dataset did recover the correct tree topology, these morphometrically-defined characters exhibited an extraordinarily high degree of homoplasy (RI = 0.48). In addition, Naylor's results indicated that none of the known character-state transformations – most of which could have been captured easily by traditional, qualitative analysis – were represented in the character-state matrix based on partial warp scores. Zelditch et al. (1998) dismissed Naylor's (1996) results claiming that he used a different method from the one they proposed.[3]

Rohlf (1998, see also Monteiro 2000) also criticized several practical aspects of the Zelditch et al. (1995) method. These include (1) the arbitrariness in the Zelditch et al. (1995, see also Fink and Zelditch 1995) advocacy of using a single individual exhibiting an extreme landmark configuration (e.g., representative of an outgroup in the case phylogenetic studies, and early developmental stage in the case of ontogenetic studies) rather than a Procrustes mean landmark configuration as the basis (= tangent point) for the principal warp decomposition; (2) the arbitrariness of using a method that makes no reference to patterns of shape variation present within a sample as a basis for summarizing shape patterns of shape variation within the sample; (3) the well-known sensitivity of the partial warps method to changes in the reference landmark configuration (Bookstein 1991); (4) the notion that partial warp decompositions are uniquely spatially localized (see also MacLeod in press); (5) the inevitability of even simple morphological changes being partialled out into a complex of distinct morphometric variables with consequent loss of interpretability; and (6) the unsuitability of partial warp-based results for subsequent statistical analysis due to lack of independence. Zelditch et al. (1998) did not dispute any of Rohlf's (1998) geometric arguments, but appealed repeatedly to non-specific, putative failures on Rohlf's (1998) part to understand their 'biological logic', 'biological interpretation', or 'biological reasoning'. Variations on these criticisms have also been voiced by Adams and Rosenberg (1998) with a response by Zelditch and Fink (1998).

In order to continue the evaluation of the Zelditch et al. (1995) partial warp method, a comparative strategy was used to determine the level of congruence between a tree obtained from traditional, qualitatively-defined, morphological characters and one based on morphometrically-defined, partial warp characters with both datasets being obtained from a small sample of real organisms. Previous partial warp-based phylogenetic studies of real organisms have either mapped a few character-state transitions onto independently justified cladograms (e.g., Zelditch et al. 1995) or included them along with other qualitatively-assessed morphological data in a single analysis (e.g., Fink and Zelditch 1995). While former cannot distinguish between consistent and coincidental patterns of correspondence between coded morphological and morphometrical variables (Rohlf 1998), the latter cannot measure the unique contribution of the morphometric subset or support direct comparisons between alternative summaries of morphological variation. The more rigorous approach of Naylor (1996), which is similar to the approach used by Cranston and Humphries (1988) to evaluate the

---

3 Since details of the methods advocated by Zelditch and Fink have differed in different studies (e.g., Fink and Zelditch 1995; Zelditch and Fink 1995; Zelditch et al. 1995) the differences Zelditch et al. (1998) are alluding to is unclear. I have been unable to recognize any substantive difference between the method described by Zelditch et al. (1995) and the one used by Naylor (1996).

contribution of quantitative characters to tree resolution, accomplishes both of these tasks and has not, to my knowledge, been attempted previously with this type of morphometric data. The approach described below also avoids the simplicity and artificiality of simulations.

Patterns of hierarchical character-state variation were compared for two datasets collected from images of 13 encrinurine trilobite species illustrated by Adrain and Edgecombe (1997, see Figure 7.7). These authors conducted a traditional, morphology-based parsimony analysis on a larger group of encrinurine trilobites, of which these 13 species form a subset. Between 33 percent and 71 percent of the 40 characters assigned to the character complexes used by Adrain and Edgecombe (1997) to infer phylogenetic relations among these trilobites were morphometric variables whose axes had been semantically subdivided and described-defined as nominal character states. Figure 7.8 shows the position of these 12 species within a maximum parsimony cladogram calculated on the basis of the entire 40-character dataset (8A) and within an agreement subtree formed from the nine equally-parsimonious cladograms calculated from the 12 Adrain and Edgecombe (1997) cranidial characters (8B). While there are differences between these two cladograms, the primary topology of *A. simoni* – *S. harrisoni* → *Struzia* species + *F. bachae* → *Mackenziurus* is stable, as is the unity of the *MacKenziurus* subclade. The topology of these trees, along with their ensemble consistency and retention indices (CI = 0.6593, RI = 0.6643, RC = 0.4380) indicate that a substantial degree of hierarchical structure in morphological characters exists within these two datasets.

In order to determine whether partial warp-based morphometric methods can recover phylogenetically informative characters, these trees were compared with a tree derived from partial warp scores computed in the manner recommended by Zelditch *et al.* (1995). For this analysis a total of 10 landmarks representing consistency relocatable positions on the trilobite cranidium were collected from each specimen (Figure 7.7, upper drawing). Because trilobites are bilaterally symmetrical in dorsal view, these landmarks were confined to the left side of the cranidium. Aspects of the glabella and fixed cheek – both of which supplied characters for the traditional analysis – were quantified by these landmarks.

Principal warps were calculated from the *B. adraini* landmark constellation (the outgroup used in the Adrain and Edgecombe 1997) and used to determine partial warp scores for the 12 ingroup taxa. Translation of these partial warp scores into a series of nominal character states was accomplished using scatterplots of the scores to look for gaps in the score distribution along the partial warp x and y axes (see Zelditch *et al.* 1995; Fink and Zelditch 1995). Figure 7.9 illustrates two examples of these, along with the corresponding thin-plate splines, for warps that represent different spatial scales of deformation.[4]

Interestingly, none of the partial warp plots could be regarded as resembling any of the nominal characters used in the traditional analysis. Whereas the latter are almost always confined to various subregions within the form (e.g., degree of glabellar

---

[4] Although this analysis differs from Zelditch *et al.* (1995) and Fink and Zelditch (1995) in that only a single representative of the species in question was used, this simplification does not change the principles involved; especially insofar as each of the specimens illustrated in Figure 7.7 exhibits the entire range of cranidial characters and character states used in the traditional, qualitative analysis.

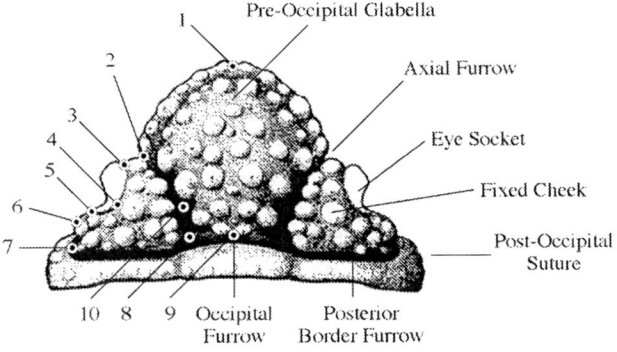
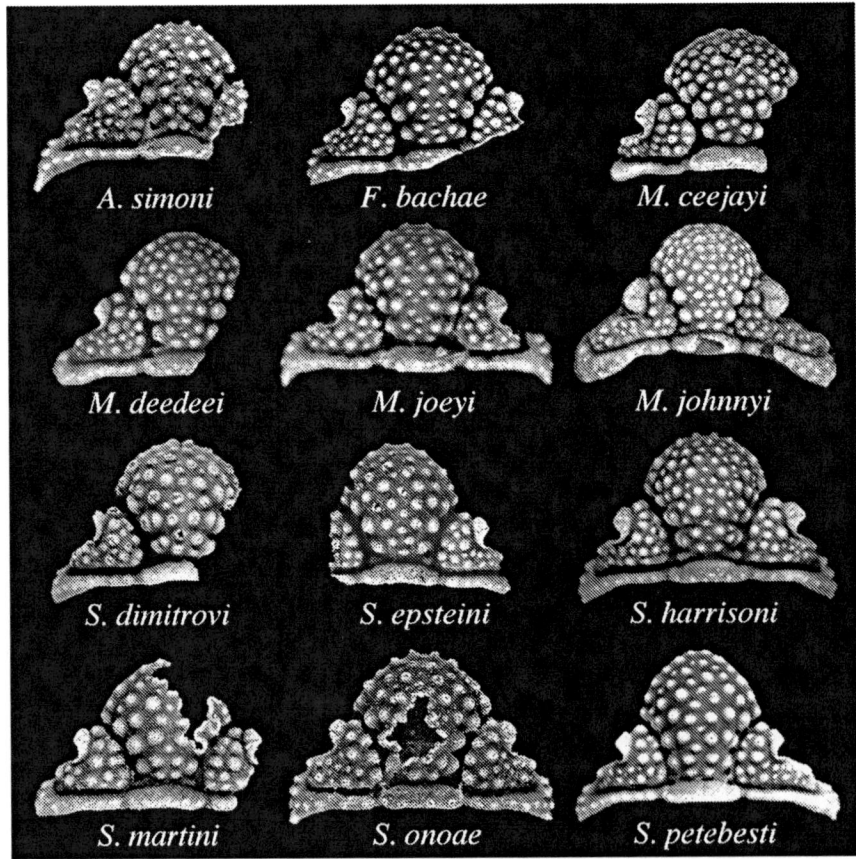

Figure 7.7 Encrinurine trilobite cranidial morphology (upper figure, right), landmarks (upper figure, left) and morphological variation for 12 representative species. Images from Adrain and Edgecombe (1997) and used with permission from Palaeontographica Canadiana.

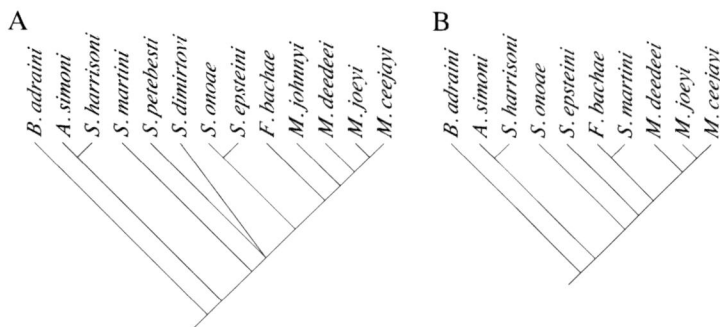

Figure 7.8 A. Strict consensus tree for the two equally parsimonious cladograms that resulted from a branch-and-bound analysis (equal character weighting) of the 40 morphological characters coded for these 13 encrinurine trilobite species by Adrain and Edgecombe (1997). Tree statistics as follows: CI = 0.6593, RI = 0.6643, RC = 0.4380. Note general agreement between phylogenetic and taxonomic groupings, especially for the genus *Mackenziurus*. B. Agreement subtree for the eight equally parsimonious cladograms that resulted from a branch-and-bound analysis (equal character weighting) of the 13 cranidial morphological characters coded for these same 13 trilobite species. Tree statistics as follows: CI = 0.6818, RI = 0.7021, RC = 0.4787. Note general agreement between the cranidial character agreement subtree and total character consensus tree topologies.

elongation, angle formed by the posterior margin of the fixed cheek) that are independent structural units of the carapace, the former – by definition – represent patterns of variation over the entire landmark series irrespective of any structural subdivision. Of course, different landmarks within the series receive different weights within the partial-warp vectors (in the same way that different variables receive different weights on a PCA axis), but these spatially defined weight patterns do not respect obvious structural boundaries. Moreover, the associated scores must be calculated from the entire set of landmark data. This means that the principal warps – and the partial warps of which they are a part – are not truly localized in the same sense that this term is applied in a traditional, qualitative, morphological analysis (see Rohlf 1998 and MacLeod in press for further discussions of this issue).

The agreement subtree derived from these partial warp-based morphometric variables is shown in Figure 7.10. Obviously, the partial warps analysis failed to recover a consistent hierarchical structure in these partial warp-defined variables. As a result, the cladogram based on a qualitative analysis of morphological characteristics (Figure 7.8) was not recovered. These results are consistent with the previous results of Naylor (1996) based on simulated patterns of morphological variation, and the warnings of Bookstein (1996) and Rohlf (1998) regarding the stability and consistency problems inherent in attempting to use partial warp-defined variables as taxonomic characters. The best that can be said of this result is that the partial warps method demonstrated very low discriminatory power for inferring adequately resolved hierarchical patterning from these landmarks. The poor performance of this character-definition method on these morphologies takes its place among the list of similar empirical failures to infer credible phylogenetic patterns from morphometric data (e.g., Cranston and Humphries 1988; Chappill 1989; Crowe 1994).

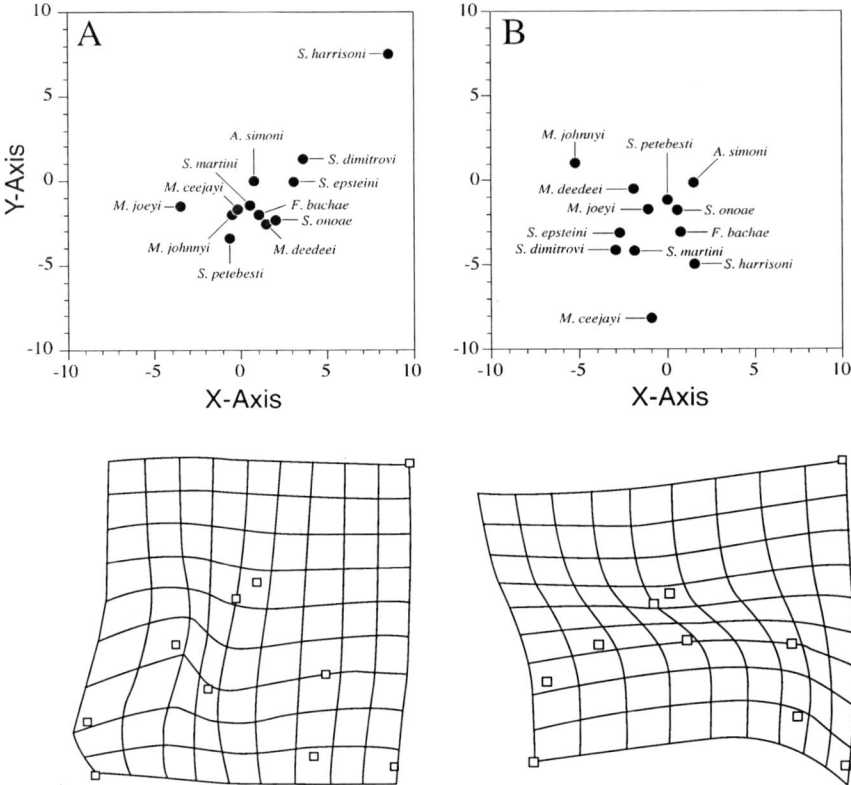

Figure 7.9 Representative partial warps and partial warp score plots for a principal warp analysis of the trilobite cranidial landmarks shown in Figure 7.1. A. Scatterplot of partial warp 2 scores (above) and corresponding principal warp (below). This primarily expresses shape variation in the region of the eye socket and fixed cheek. B. Scatterplot of partial warp 5 scores (above) and corresponding principal warp (below). This primarily expresses shape variation as a transverse antero-posterior compression of the cranidium. Zelditch et al. (1995) suggest that partial warp scatterplot axes such as these can be used as systematic characters. These authors recommend that gaps in the distribution of taxa along partial warp axes be used to define character states. For example, the pronounced gaps between *Struszia harrisoni* and the remaining trilobite species along both partial warps 2x and 2y (A, upper figure), and between *Mackenziurus ceejayi* and the remaining trilobite species along partial warp 5y (B, upper figure), could be coded as separating these three putative characters into two states.

A relative warp approach to quantitative morphological analysis in systematics: congruence, interpretability, and extension

Relative warps in systematics

As an alternative to the partial warp-based approach to morphometric character definition advocated by Zelditch *et al.* (1995), the method of relative warps was evaluated

Figure 7.10 Agreement subtree for the 976 equally parsimonious cladograms that resulted from a branch-and-bound analysis (equal character weighting) of the 14 partial warp-based morphometric characters coded for 12 trilobite species using the methods and conventions described by Zelditch et al. (1995). Tree statistics as follows: CI = 0.8571, RI = 0.7255, RC = 0.6218. Note lack of agreement between this partial warp-based morphometric character agreement subtree and the total character consensus tree and cranidial character agreement subtree topologies. The failure of partial warps to recover a consistent, hierarchically structured, morphological signal from these trilobite data is consistent with the predictions of Bookstein (1994, 1996) and Rohlf (1998) regarding the analytic utility of partial warp methods in phylogenetic contexts.

using the same comparative test. Relative warps differ from partial warps in being based on patterns of shape covariance between objects included within an empirical or reference sample. They are, in their simplest formulation, the results of an eigenanalysis of the reference sample covariance matrix where the objects from which the covariances are calculated represent a series of column vectors of landmark locations, in either 2-space (x, y) or 3-space (x, y, z).

While this relative warps approach has been advocated indirectly by Rohlf (1998, see also Bookstein 1996 and Monteiro 2000), it has, to my knowledge, never been tested empirically. Zelditch et al. (1995) criticized methods that employ an eigenanalysis of covariance-matrices, arguing that such methods (1) employ optimization criteria that do not correspond to putative principles of 'phylogenetic informativeness' (2) are inherently tied to particular samples (and so do not constitute independent descriptions of morphological state, see also Pimentel and Riggins 1987), and (3) and do not produce variables that are spatially localized descriptions of morphological variation to a degree sufficient to conform to the biological concept of homology.

With respect to the first criticism, the purpose of a systematic analysis of organismal morphology is to search for discontinuities or gaps in the patterns of morphological variation. This is true whether the analysis is qualitative – as is the case in traditional systematic investigations – or quantitative. Small discontinuities or gaps will exist between each individual in a sample or population. Nevertheless, to be of systematic utility, the discontinuities or gaps between character states must be greater than the discontinuities or gaps that exist between individuals that exhibit the same character state. If morphological gaps exist between subsets of species in a sample their existence will contribute to the variance of any morphological descriptors (e.g., shape functions) that are sensitive to the presence of the gap. Since covariance-based eigenanalysis aligns multivariable vectors (= axes) with the directions of maximum variance within a dataset, it likely that such procedures will be of great use in locating any gaps that are of genuine systematic interest gaps within morphological datasets. In other words, we would expect that the discontinuities we seek as systematists would be reflected in specimen ordinations that are variance-optimized over the entire sample. This is precisely the sort of pattern we look for when we qualitatively assess morphology. It cannot be logical to accept such evaluation procedures in qualitative contexts, while, at the same time, denying their relevance in quantitative contexts. In addition, exactly the same type of eigenanalysis-based optimization is employed by partial warp analysis (where a hypothetical bending-energy matrix is substituted for the covariance matrix) to define the principal-partial warps whose use Zelditch *et al.* (1995) advocate.

With respect to the second criticism, this too is inconsistent with accepted contemporary phylogenetic practice. While theoretical treatments of cladistic characters often make reference to the desirability of non-relative character definitions (e.g., Pimentel and Riggins 1987), in practice, the use of relative, sample-referenced, character definitions is commonplace (Chappill 1989; Stevens 1991; Thiele 1993). Moreover, Zelditch *et al.* (1992, 1995, 1998), Zelditch and Fink (1995), and Fink and Zelditch (1995) employ sample-referenced methods to obtain the 'mean forms' that they then use as the basis (= tangent point) for their partial warp analysis. Given the extreme instability of partial warps in the face of changes in the reference shape, their criticism of covariance-based eigenanalysis methods – which are, on the whole, much more robust to variation in appropriately constructed samples than partial warps – seems erratic.

This also holds for the Zelditch *et al.* (1995, as well as Pimentel and Riggins' 1987) criticism that any alteration of a sample's composition might significantly perturb the orientation of a variance-optimised morphometric variable axis such as those determined by PCA or relative warp analysis. Certainly this is true if a strongly atypical individual is included in the sample. But in appropriately selected samples the need to deal with substantially outlying individuals should be minimized. Inclusion of a 'typical' individual in a sample would not necessarily produce strongly divergent results (see below). Moreover, inclusion of a strongly contrasting individual in a parsimony-based morphological phylogenetic analysis or a likelihood-based molecular analysis also has the ability to perturb the results (e.g., Zelditch *et al.* 1998), yet this is obviously not regarded by most systematists as grounds on which to preclude the use of such methods in phylogenetic contexts. Indeed, in the same way that Zelditch *et al.* (1992, see other references above) employed a small, but representative, reference sample to establish the basis for their morphometric analysis of ontogeny, a small, but representative reference sample could be used to establish a basis for any covariance-eigenanalysis morphological analysis with additional individuals – that were not part of the sample used to estimate the population eigenvectors – being projected into the eigenvector-defined reference space (see MacLeod and Rose 1993; MacLeod in press for examples and discussion). While this issue deserves a much more in-depth treatment, once again, it should not be the case that particular morphometric methods are dismissed for being sensitive to certain analytic situations while, at the same time, the implications of such sensitivities for other methods are casually accepted elsewhere.

With respect to the third criticism, MacLeod (in press) has shown that the interpretation of partial warps is no more, and no less, spatially localized than the interpretation of relative warps (= coordinate point eigenshapes) or PCA axes. As a result, these latter constructs provide as good a fit to the concept of biological homology as partial warp axes; which is to say, not a very good fit at all (see Distances, Landmarks, and Homology section above).

Interestingly, in the canonical examples of partial warp-based character-state descriptions the arrays of partial warp axes are calculated for a constellation of landmarks scattered over the entire body of the organism under investigation. This stands in striking contrast to the normal, qualitative systematic practice of subdividing a complex organic structure into a number of putatively homologous structures and then treating these structures as independent units of morphological variation. From their discussions of the partial warps method it is clear that Zelditch and Fink regard the eigenanalytic decomposition of the reference form's bending-energy matrix as being the mathematical equivalent of the traditional systematist's qualitative disassembly of an organism into quasi-independent character complexes. Nevertheless, the partial warps are never truly localized in that spatial information from all parts of the morphology (via the landmark positions) is used to compute all partial warp scores.

To achieve an analytic procedure that is closer to accepted contemporary practice in comparative morphology it would be necessary to first subdivide the organism's body into landmark-defined substructures, carry out separate morphometric analyses on these substructures, use those results to search for discontinuous patterns of variation, code those patterns in the standard manner, and then compute the results of a parsimony-based phylogenetic analysis. Of course, such a subdivided data analysis strategy could be undertaken for any morphometric procedure, including partial warps. However, the fact that no examples of this analytic variant currently exist, along with the emphasis on the putative homology of partial warps that underpins these canonical examples, suggests that the original partial-warps-as-systematic-characters concept does not include such a subdivided data analytic strategy.

The Naylor simulated fish phylogeny revisited

To test the proposition that a subdivided relative warp morphometric data analytic approach can recover systematically useful character states better than a canonical partial warps approach the former was applied to the Naylor (1996) fish morphology simulations. Those results were then compared to the 'phylogenetic' derivation of the simulation models and the results of Naylor's partial warp analysis of these same data. Naylor's simulations are regarded as the most relevant set of test data for these comparisons because the 'phylogeny' is known and because the fish simulations represent a level of morphological complexity commensurate with the canonical examples of the partial warp approach.

Figure 7.11 shows the fish simulations used in the relative warps test. These images represent new drawings based on scans of the original Naylor (1996, figure 2) figures. Since these are not the original figures there may be some variance between the originals and these new drawings; especially insofar as the originals were reproduced at such a small size that the precise original landmark locations were very difficult to discern. Inferred landmark locations on the new drawings were quantified as pairs of Cartesian coordinates in the usual way.

In order to test the fidelity of the new drawings and the reproducibility of the Naylor (1996) results, a partial warps analysis was conducted on the new drawings

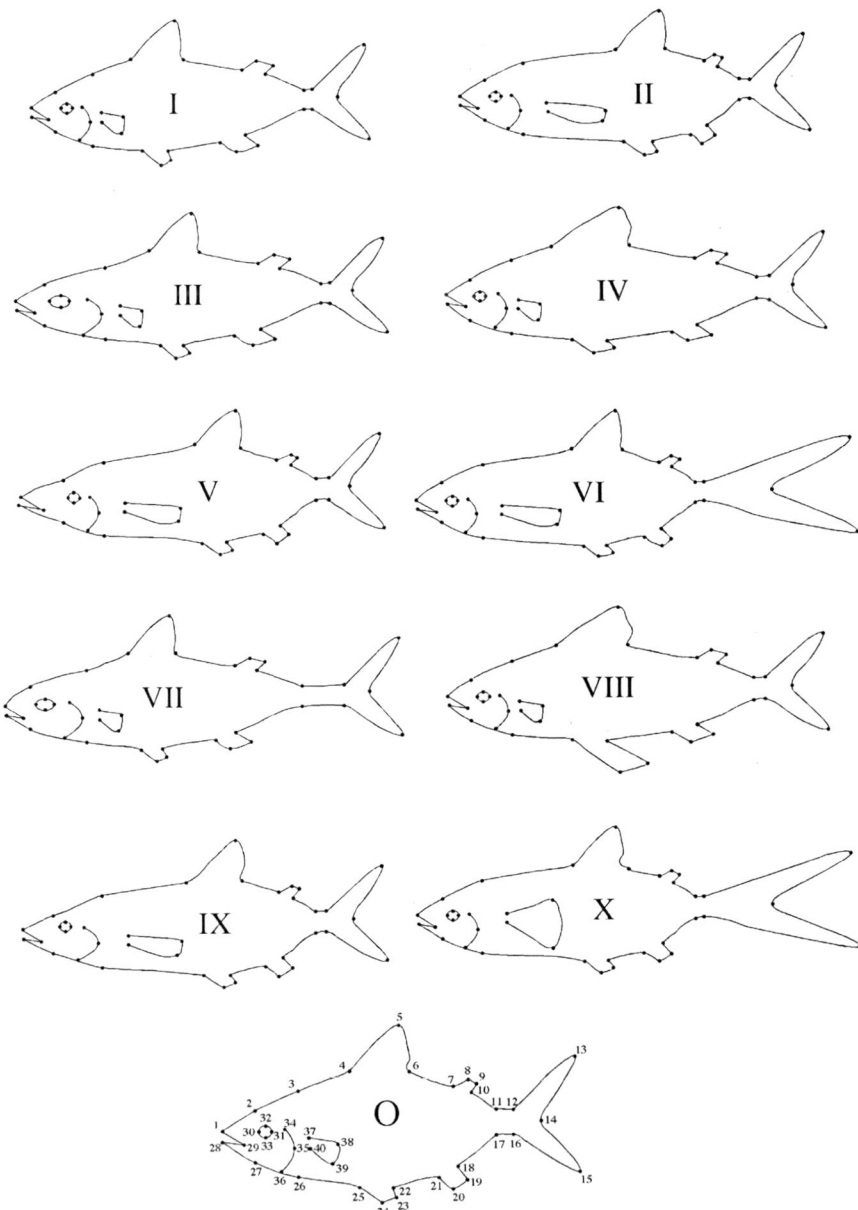

Figure 7.11 Morphological variation in a series of fish morphology simulations used by Naylor (1996) to exemplify the ability of partial warp-based morphometric methods to recover phylogenetic signals. Landmarks are the same as those used by Naylor, except for the four eye landmarks (Naylor used a single landmark located in the eye center) which were used to represent eye shape as well as (relative to other landmarks) eye location. Morphological simulation figures redrawn from Naylor (1996).

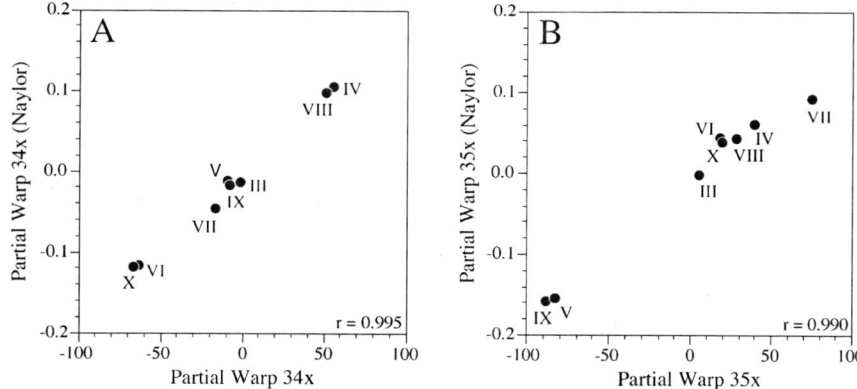

Figure 7.12 Comparison of two representative partial warp score distributions for this study (*x*-axis) and the original Naylor (1996) study (*y*-axis). A. Partial Warp 34x. B. Partial Warp 35x. The near identity of the partial warp score patterns confirms that the redrawn simulations are virtually identical to the original Naylor (1996) simulations.

and the resulting warp score distributions compared to Naylor's figure 3 and table 1.[5] Although the magnitude of the scores differed (presumably as a result of modifications to the tpsSpln program used to compute the partial warp scores between 1995 and 2000), the relative patterning of scores was virtually identical (Figure 7.12). In addition to supporting the fidelity of the new drawings with respect to Naylor's originals, this near identiy makes a useful point about the nature of landmarks. Since it is very unlikely that exactly the same (= homologous) landmarks were chosen on the original and new drawings in all cases, the similarity of the old and new partial warp results demonstrates that there is no need for landmarks to be absolutely homologous in either biological or geometrical senses in order to represent the gross shapes of landmark-defined objects. The concept of a landmark as nothing more that a relocatable reference whose purpose is to locate the approximate relative positions of gross structural elements is sufficient to achieve remarkably consistent results provided all other aspects of the analysis remain constant.[6]

Prior to the relative warp analysis, these landmarks were combined into a series of 13 groups (Table 7.1) that effectively subdivided the overall morphology into a set of quasi-distinct, but biologically homologous, characters. This step parallels the subdivision of complex morphologies that is universally applied in systematic practice. While this subdivision was arranged to reflect the characters used to construct the Naylor (1996) simulation, a complete measurement of the form required the inclusion of additional characters (e.g., dorsal fin shape, pelvic fin shape).

---

5 For this analysis a single 'eye' landmark located in the middle of the eye ellipse was used (instead of the four eye landmarks shown in Figure 7.11) in order to render the landmark system comparable to that used by Naylor (1996).

6 This result should not, however, be taken as supporting a claim that partial warps *per se* are robust to changes in the reference or basis shape, see Rohlf (1998).

Table 7.1 Characters and defining landmarks used in the relative warp analysis of Naylor's fish icons. See Figure 7.11 for landmark positions

n	Characters	Landmarks
1	Mouth Region	1-2-29-28-27
2	Orbital-Branchial Region	3-2-27-26
3	Pectoral Region	4-3-26-25
4	Abdomen	11-10-7-6-22-21-18-17
5	Caudal Peduncle	12-11-17-16
6	Tail	16-15-14-13-12
7	Dorsal Fin	4-5-6
8	Adiopose Fin	7-8-9-10
9	Anal Fin	18-19-20-21
10	Pelvic Fin	22-23-24-25
11	Eye	32-30-31-33
12	Gill	34-35-36
13	Pectoral Fin	37-38-39-40

Once these subsidiary datasets had been assembled each set of landmark constellations was registered (= oriented and scaled) using the Generalized Least Squares (GLS) algorithm (Rohlf 1990). These registered coordinate data were then used as input for a series of thirteen separate relative warp analyses. Representative patterns of shape variation on the two most important relative warp shape difference axes for six of these characters is shown in Figure 7.13. In each instance the relative warp results separated the nine shapes into a series of mutually exclusive groups; usually into two groups, but in two cases (eye shape and pectoral fin shape) into three. Each of these taxic groupings save one was consistent with the pattern of morphological changes used by Naylor (1996) to construct the simulation.

This single exception was the Orbital-Brachial Region Shape (Figure 7.13E). Naylor's (1996) change (3) – 'lengthening of region containing gill and eye' was purported to be shared by simulations III and VII in his simulation. The relative warp results for this character groups together simulations III, VII, and IX as exhibiting orbital-brachial region landmark constellations that are similar to one another and distinctly different from the remaining substructure (= character) constellations of the other simulations by a considerable degree along the most important shape difference axis (Relative Warp 2, RW-2). However, inspection of the fish shape simulations in Figure 7.11, and Naylor's (1996) original figure 2, shows that simulation IX does indeed exhibit an antero-posteriorly lengthened orbital-brachial region that is strongly reminiscent of the states for this character found in simulations III and VII. This interpretation is also supported by Character 12 (gill shape, Figure 7.13F), that suggests simulation IX is apomorphic for this orbital-brachial region-related attribute.

Although the relative warp analysis was quite successful in finding the same taxic groupings implied by Naylor's simulated phylogeny, this does not necessarily mean that character-state assignments made of the basis of these morphometric data exhibit a consistent hierarchical structure that could be used to recover the simulated pattern of sister-group relationships. To conduct this test the distributions of taxa for the thirteen characters were divided into subgroups based on the presence of unambiguous

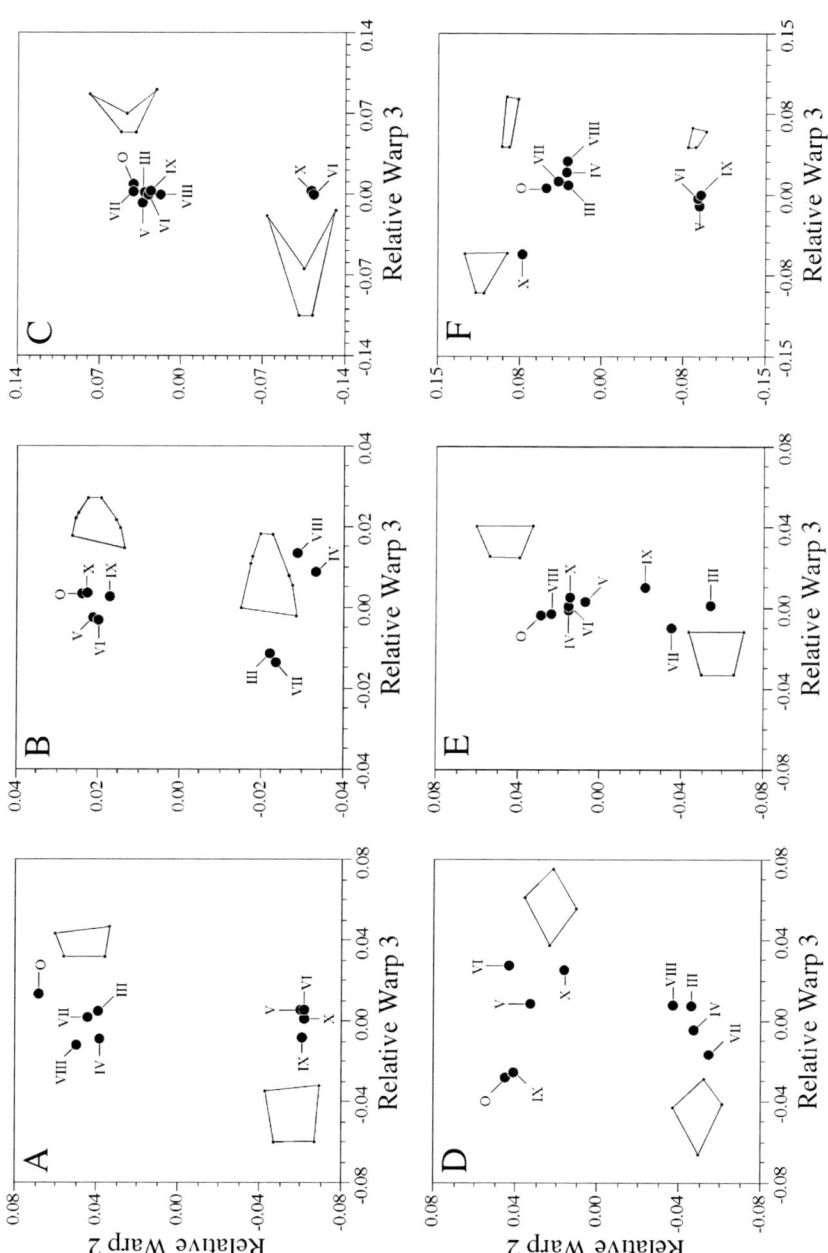

Figure 7.13 Representative results of a relative warp-style analysis of morphological variation in four different character complexes based on the redrawn Naylor (1996) fish simulations. A. Pectoral fin region shape. B. Abdomen shape. C. Tail shape. D. Pelvic fin shape. E Optic-branchial region shape. F. Pectoral fin shape. Inset landmark constellations illustrate representative character state morphologies for the morphometrically-defined taxic subgroups. These scatterplots represent patterns of shape variation among the outgroup form and the eight crown simulations of the simulated phylogeny within the plane of the two most important shape-discrepancy axes (relative warps 2 and 3). In each case a clear morphological discontinuity oriented along the single most important shape discrepancy axis (RW-2) was revealed by the analysis. It is suggested that these axes represent putative morphological characters, and the gaps represent the divisions between different states of these characters, that can be useful in phylogenetic analysis. See Figure 7.11 for simulation morphologies.

discontinuities in the patterns of shape variation along either the RW-2 or RW-3 axes. These taxic subgroups were then assigned nominal character-state labels with state '0' being assigned to the group containing the outgroup morphotype. The resultant morphometric character-state matrix is shown in Table 7.2.

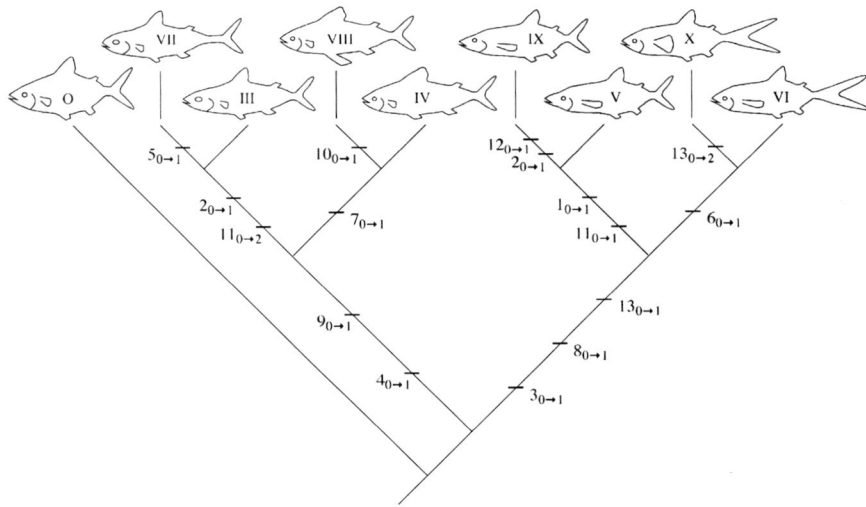

Figure 7.14 Maximum parsimony cladogram resulting from a branch-and-bound analysis (equal character weighting) of the 13 relative warp-based morphological characters coded for the nine Naylor (1996) simulated fish morphologies (see Figure 7.13 for examples). Tree statistics as follows: CI = 0.9286, RI = 0.9474, RC = 0.8797. Naylor's (1996) previous partial warps analysis of these simulations succeeded in recovering this tree but yielded a much lower tree Retention Index (0.48). This suggests that the partial warps approach creates substantial amounts of homoplasy in morphometrically-based systematic datasets. See text for discussion.

Table 7.2 Relative warp-based character/character-state matrix. See Figure 7.11 for simulation morphologies. See Table 7.1 for character definitions and Figure 7.13 for character-state groupings

Simulations	Characters												
	1	2	3	4	5	6	7	8	9	10	11	12	13
III	0	1	0	1	0	0	0	0	1	0	2	0	0
IV	0	0	0	1	0	0	1	0	1	0	0	0	0
V	1	0	1	0	0	0	0	1	0	0	1	0	1
VI	0	0	1	0	0	1	0	1	0	0	0	0	1
VII	0	1	0	1	1	0	0	0	1	0	2	0	0
VIII	0	0	0	1	0	0	1	0	1	1	0	0	0
IX	1	1	1	0	0	0	0	1	0	0	1	1	2
X	0	0	1	0	0	1	0	1	0	0	0	0	0
Outgroup	0	0	0	0	0	0	0	0	0	0	0	0	0

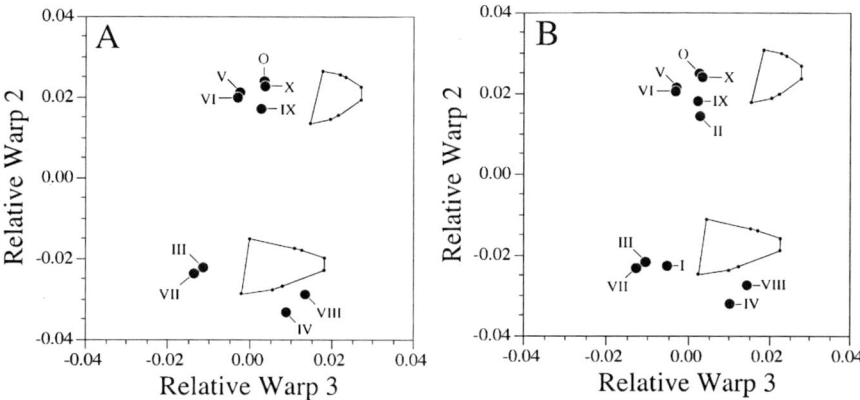

Figure 7.15 Effect of including new morphologies on the Naylor (1996) fish simulation relative warp results. A. Relative warp ordinations within the plane of the two most important shape-discrepancy axes (relative warps 2 and 3) for simulations III through X and the ancestral form (A). B. Relative warp ordinations within the plane of the two most important shape-discrepancy axes (relative warps 2 and 3) for simulations I through X and the ancestral form (A). Inset landmark constellations illustrate representative character state morphologies for the morphometrically-defined taxic subgroups. Note that inclusion of the internal node simulation morphologies I and II did not appreciably change the separation between character-state groupings or the intra-state ordinations of morphotypes within this plane through the shape space. While the inclusion of additional (typical) individuals to a sample will engender some small alternation the geometry of the sample's shape space, this alteration would not be expected to automatically obscure the presence and recognition of real morphological discontinuities. Moreover, it is possible to project individuals into the space defined by a statistically representative sample of morphotypes without altering the nature of the eigenvector-defined shape space at all (see MacLeod and Rose 1993 for an example). Accordingly, objections to the use of eigenanalysis-based methods in morphological systematics because of their so-called inherent instability in the face of additions to the reference sample (e.g., Zelditch *et al.* 1995) may, in many instances, be irrelevant to systematic practicalities. Geometric analysis should be able to duplicate any result obtained by traditional, qualitative methods. See Figure 7.11 for simulation morphologies.

Analysis of this character-state matrix eliminating simulations I and II (as ancestral internal nodes) to repeat the conditions of the Naylor partial warp analysis resulted in the location of one maximally parsimonious trees (Figure 7.14). Although the orientations of the relative warp axes between topology of this tree is identical to that of the original Naylor (1996) simulation, of the 13 relative warp characters used in this analysis, only one (Orbital-Brachial Region Shape, see above) shows any homoplasy and this homoplasy represents a single character-state reversal. This reversal is regarded as a mistake in Naylor's original coding of his morphological simulations. Taken as a whole, these relative warp-based characters represents a perfect match in terms of topology and tree descriptive indices to the Naylor (1996) simulated phylogeny. This relative warp-based tree also correctly predicts the character states of simulations I and II (the ancestral forms withheld from the phylogenetic analysis). Finally, contrary to the predictions of Zelditch *et al.* (1995), this relative warp-based character analysis is robust to the inclusion of additional specimens in sample (Figure 7.15).

While one successful analytic result – especially of a simulated dataset – does not definitively prove the case for the utility of morphometric approaches to character analysis, this relative warp-based result is by far the most successful example of morphometric variables being used to locate and define systematic-phylogenetic characters currently known. In the most successful partial warps analysis to date (Fink and Zelditch 1995) the tree based on only the morphometrically-defined characters differed from the tree produced by the combined morphometric + qualitative character dataset suggesting that these two datasets disagree. Tree statistics for the relative warp analysis of the Naylor (1996) simulations are overwhelmingly better than Naylor's (1996) partial warps results and substantially better than for the (Fink and Zelditch 1995) partial warps characters.

Summary

Goals of qualitative and quantitative morphological analysis in systematics are the same. Both are concerned primarily with representing patterns of morphological variation in organisms and relating these patterns to other patterns of variation in other variables. Systematists are particularly concerned with the documentation of morphological discontinuities that exist between groups of individual organisms because these discontinuities (may) reflect cladogenesis.

Many of the morphological patterns that interest systematists represent geometries that can, in principle, vary continuously. These are best expressed as ratio-scale variables; the same types of variables that are routinely used in morphometric analysis. Such variables are compatible with the standard systematic concept of the 'character' as defined by Farris *et al.* (1970) and by Pimentel and Riggins (1987). Systematists have often been confused by the fact that sets of observations or measurements along continuous, ratio-scale, morphometric variables can exhibit either continuous or discontinuous (= clustered) patterns of variation. Sets of observations or measurement that exhibit the latter offer no particular difficulties in terms of devising objective rules for transforming morphometric ratio-scale variables (= characters) into the nominal variables (character states) required by contemporary phylogenetic analysis algorithms. Sets of observations or measurement that exhibit the former offer no hope of being able to delineate groups of taxa logically, consistently, or objectively, irrespective of various 'member coding' procedures that have been devised.

The fact that a large number of systematic studies employ characters that are ratio-scale morphometric variables that have been subdivided arbitrarily into discrete states emphasizes the practical and uncontroversial use of morphometric data in systematic contexts, as well as the routine systematic observation of discrete distribution of observations or measurements along such variable axes. Explicit morphometric analysis methods, however, offer systematists practical means of discovering, assessing, and describing the morphological gaps on which taxic diagnoses are based as well as more consistently coding observations or measurements along variable (= character) axes for systematic analysis.

Traditional objections to the use of morphometric data in systematic contexts because they do not conform to the concept of biological homology derive from a misunderstanding of morphometric variables (especially landmark coordinates), an inconsistent approach to the specification of traditional, qualitative observations in systematics, and a lack of appreciation for the spatial limits implicit in the concept of biological homology. As has been pointed out by a pantheon of systematists stretching from Richard Owen to Colin Patterson, homologues are structures that exist at particular spatial scales and are most often recognized by either internal or external topological similarity with other such structures. Landmarks may represent homologous structures for the purposes of assessing topological similarity, but landmarks – as geometric points – are not intrinsically homologous (see MacLeod 1999 for examples of non-homologous systematic landmarks). Rather they represent abstractions of structures that may (or may not) represent true homologues.

Finally, the recent suggestion that morphometric partial warps of landmark configurations can be used as a new source of morphological characters and character-states was evaluated via comparison of trees resulting from a traditional qualitative, and partial warps-based quantitative analysis of trilobite cranidial characters for a selection of 12 encrinurine species from the Canadian Arctic. Results showed that the partial warps-based tree differed strongly from the tree derived from qualitative morphological analysis and that none of the characters used for the latter was 'discovered' by the former. These results are consistent with the theoretical and practical results obtained by others on both simulated and real taxa. In aggregate, they raise serious concerns regarding the applicability of using partial warps in systematic contexts. Alternatively, a new procedure for using landmark-based morphometric analysis based on relative warps is described and applied to a simulated phylogeny. Results show that this new method (1) correctly 'discovered' the distinctions between simulated morphologies, (2) produced exactly the same tree as would have been produced by a qualitative analysis of the simulated morphologies, (3) did not introduce elevated levels of homoplasy to the phylogenetic analysis, and (4) resolved heretofore unsuspected ambiguities in the simulation.

Reliance on qualitative methods to recognize and document morphological discontinuities in systematic datasets has led, in many cases, to needless confusion and controversy over the validity of phylogenetic systematic results and appropriateness of interpretations based on those results. The descriptive and analytic rigor that would result from introducing morphometric methods into phylogenetic systematics would have substantial and positive implications for both fields. A period of active experimentation with these methods is now needed to further explore their compatibility.

Acknowledgements

I wish to thank the following for informative discussions regarding morphometric procedures and/or their use in systematic contexts: Fred Bookstein, Jim Rohlf, Chris Humphries, David Polly, and Donald Swiderski. This essay was reviewed by Peter Forey and Andrew Smith both of whom submitted comments and criticisms that greatly improved the arguments presented.

References

Adams, D. C. and Rosenberg, M. S. (1998) 'Partial warps, phylogeny, and ontogeny: a comment on Zelditch and Fink (1995)', *Systematic Biology*, **47**, 168–173.

Adrain, J. M. and Edgecombe, G. D. (1997) 'Silurian encrinurine trilobites from the central Canadian Arctic', *Palaeontographica Canadiana*, **14**, 1–109.

Almeida, M. T. and Bisbey, F. A. (1984) 'A simple method for establishing taxonomic characters from measurement data', *Taxon*, 33, 405–409.

Archie, J. W. (1985) 'Methods for coding variable morphological features for numeric taxonomic analysis', *Systematic Zoology*, 34, 326–345.

Baum, B. R. (1988) 'A simple procedure for establishing discrete characters from measurement data, applicable to cladistics', *Taxon*, 37, 63–70.

Blackith, R. E. and Reyment, R. A. (1971) *Multivariate morphometrics*, London: Academic Press.

Bookstein, F., Chernoff, B., Elder, R., Humphries, J., Smith, G. and Strauss, R. (1985) *Morphometrics in evolutionary biology*, Special Publication 15, Philadelphia: The Academy of Natural Sciences of Philadelphia.

Bookstein, F. L. (1986) 'Size and shape spaces for landmark data in two dimensions', *Statistical Science*, 1, 181–242.

Bookstein, F. L. (1990) 'Analytic methods: introduction and overview', in Rohlf, F. J. and Bookstein, F. L. (eds) *Proceedings of the Michigan Morphometrics Workshop*, Ann Arbor, MI: The University of Michigan Museum of Zoology, Special Publication 2, pp. 61–74.

Bookstein, F. L. (1991) *Morphometric tools for landmark data: geometry and biology*, Cambridge: Cambridge University Press.

Bookstein, F. L. (1993) 'A brief history of the morphometric synthesis', in Marcus, L. F., Bello, E. and García-Valdecasas, A. (eds) *Contributions to morphometrics*, Madrid: Museo Nacional de Ciencias Naturales 8, pp. 18–40.

Bookstein, F. L. (1994) 'Can biometrical shape be a homologous character?', in Hall, B. K. (ed.) *Homology: the hierarchical basis of comparative biology*, San Diego, CA: Academic Press, pp. 197–227.

Bookstein, F. L. (1996) 'Combining the tools of geometric morphometrics', in Marcus, L. F., Corti, M., Loy, A., Naylor, G. J. P. and Slice, D. E. (eds) *Advances in morphometrics*, New York: Plenum Press, pp. 131–151.

Bookstein, F. L. (1997) 'Landmark methods for forms without landmarks: localizing group differences in outline shape', *Medical Image Analysis*, 1, 225–243.

Chapill, J. A. (1989) 'Quantitative characters in phylogenetic analysis', *Cladistics*, 5, 217–234.

Colless, D. H. (1980) 'Congruence between morphological and allozyme data for Menidia: a reappraisal', *Systematic Zoology*, 29, 288–299.

Cranston, P. S. and Humphries, C. J. (1988) 'Cladistics and computers: a chironomid conundrum', *Cladistics*, 4, 72–92.

Crisp, M. and Weston, P. (1987) 'Cladistics and legume systematics, with an analysis of the Bossiaeeae, Brongniartieae and Mirbelieae', in Stirton, C. (ed.) *Advances in legume systematics, part 3*, Kew: Royal Botanical Gardens, pp. 65–130.

Crowe, T. M. (1994) 'Morphometrics, phylogenetic models and cladistics: means to an end or much ado about nothing?', *Cladistics*, 10, 77–84.

Darwin, C. (1859) *On the origin of species by means of natural selection, or the preservation of favoured races in the struggle for life*, London: John Murray.

Eldredge, N. and Cracraft, J. (1980) *Phylogenetic patterns and the evolutionary process*, New York: Columbia University Press.

Falconer, D. S. (1981) *Introduction to quantitative genetics*, London: Longman.

Farris, J. S. (1990) 'Phenetics in camouflage', *Cladistics*, 6, 91–100.

Farris, J. S., Kluge, A. and Eckhardt, M. (1970) 'A numerical approach to phylogenetic systematics', *Systematic Zoology*, 19, 172–189.

Felsenstein, J. (1985) 'Phylogenies and the comparative method', *American Naturalist*, 125, 1–15.

Felsenstein, J. (1988) 'Phylogenies and quantitative characters', *Annual Review of Ecology and Systematics*, 19, 445–471.

Fink, W. L. and Zelditch, M. L. (1995) 'Phylogenetic analysis of ontogenetic shape transformations: a reassessment of the Piranha genus *Pygocentrus*', *Systematic Biology*, 44, 343–360.

Goldman, N. (1988) 'Methods for discrete coding of morphological characters for numerical analysis', *Cladistics*, 4, 59–71.

Harvey, P. H. and Pagel, M. D. (1991) *The comparative method in evolutionary biology*, Oxford: Oxford University Press.

Hennig, W. (1966) *Phylogenetic systematics*, Urbana: University of Illinois Press.

Kitching, I. J., Forey, P. L., Humphries, C. J. and Williams, D. M. (1998) *Cladistics: the theory and practice of parsimony analysis*, 2nd edn, Oxford: Oxford University Press.

Lynch, J. M., Wood, C. G., and Luboga, S. A. (1996) 'Geometric morphometrics in primatology: craniofacial variation in *Homo sapiens* and *Pan troglodytes*', *Folia Primatologia*, 67, 15–39.

MacLeod, N. (1999) 'Generalizing and extending the eigenshape method of shape visualization and analysis', *Paleobiology*, 25, 107–138.

MacLeod, N. (2001) 'The role of phylogeny in quantitative paleobiological analysis', *Paleobiology*, 27, 226–241.

MacLeod, N. 'Landmarks, localization, and the use of morphometrics in phylogenetic analysis', in Edgecombe, G., Adrain, J. and Lieberman, B. (eds) *Fossils, phylogeny, and form: an analytical approach*, New York: Plenum (in press).

MacLeod, N. and Rose, K. D. (1993) 'Inferring locomotor behavior in Paleogene mammals via eigenshape analysis', *American Journal of Science*, 293-A, 300–355.

Mayr, E. (1982) *The growth of biological thought: diversity, evolution, and inheritance*, Cambridge: Harvard University Press.

Michevich, M. F. and Johnson, M. F. (1976) 'Congruence between morphological and allozyme data', *Systematic Zoology*, 25, 260–270.

Monterio, L. R. (2000) 'Why morphometrics is special: the problem with using partial warps as characters for phylogenetic inference', *Systematic Biology*, 49, 796–800.

Naylor, G. J. P. (1996) 'Can partial warps scores be used as cladistic characters?', in Marcus, L. F., Corti, M., Loy, A., Naylor, G. J. P. and Slice, D. E. (eds) *Advances in morphometrics*, New York: Plenum Press, pp. 519–530.

Nelson, G. (1989) 'Species and taxa: systematics and evolution', in Otte, D. and Endler, J. A. (eds) *Speciation and its consequences*, New York: Plenum, pp. 63–81.

Nelson, G. and Platnick, N. (1981) *Systematics and biogeography: cladistics and vicariance*, New York: Columbia University Press.

Otte, D. and Endler, J. A. (1989) *Speciation and its consequences*, Sunderland, Massachusetts: Sinauer Associates.

Owen, R. (1843) *Lectures on comparative anatomy of the invertebrate animals, delivered at the Royal College of Surgeons in 1843*, London: Longman, Brown, Green, and Longman.

Patterson, C. (1982) 'Morphological characters and homology', in Joysey, K. A. and Friday, A. E. (eds) *Problems of phylogenetic reconstruction*, London and New York: Academic Press, pp. 21–74.

Pimentel, R. A. and Riggins, R. (1987) 'The nature of cladistic data', *Cladistics*, 3, 201–209.

Rae, T. (1995) 'Continuous characters and fossil taxa in phylogenetic reconstruction', *American Journal of Physical Anthropology*, **Supplement 20**, 176–177.

Rae, T. C. (1998) 'The logical basis for the use of continuous characters in phylogenetic systematics', *Cladistics*, 14, 221–228.

Rieppel, O. (1980) 'Homology, a deductive concept', *Zeitschrift für Zoolgische, Systematik und Evolutionforschung*, 18, 315–319.

Rieppel, O. (1994) 'Homology, topology, and typology: the history of modern debates', in Hall, B. K. (ed.) *Homology: the hierarchical basis of comparative biology*, London: Academic Press, pp. 63–100.

Reyment, R. A. (1991) *Multidimensional paleobiology*, Oxford: Pergamon Press.

Reymont, R. A., Blackith, R. E. and Campbell, N. A. (1984) *Multivariate morphometrics* 2nd edn, London: Academic Press.

Rohlf, F. J. (1990) 'Rotational fit (Procrustes) methods', in Rohlf, F. J. and Bookstein, F. L. (eds) *Proceedings of the Michigan Morphometrics Workshop*, Ann Arbor: The University of Michigan Museum of Zoology, Special Publication No. 2, pp. 227–236.

Rohlf, F. J. (1998) 'On applications of geometric morphometrics to studies of ontogeny and phylogeny', *Systematic Biology*, 47, 147–158.

Simon, C. (1983) 'A new coding procedure for morphometric data with an example from periodical cicada wing veins', in Felsenstein, J. (ed.) *Numerical taxonomy*, Berlin: Springer-Verlag, pp. 378–382.

Smith, A. B. (1994) *Systematics and the fossil record: documenting evolutionary patterns*. London: Blackwell.

Sokal, R. R. and Rohlf, F. J. (1981) *Biometry: the principles and practice of statistics in biological research*, 2nd edn, San Francisco: W. H. Freeman.

Stevens, P. F. (1991) 'Character states, morphological variation, and phylogenetic analysis: a review', *Systematic Botany*, 16, 553–583.

Strait, D., Moniz, M. and Strait, P. (1996) 'Finite mixture coding: a new approach to coding continuous characters', *Systematic Biology*, 45, 67–78.

Strauss, R. E. and Bookstein, F. L. (1982) 'The truss: body form reconstruction in morphometrics', *Systematic Zoology*, 31, 113–135.

Stuessy, T. F. (1990) *Plant taxonomy: the systematic evaluation of comparative data*, New York: Columbia University Press.

Swiderski, D. L., Zelditch, M. L. and Fink, W. L. (1998) 'Why morphometrics is not special: coding quantitative data for phylogenetic analysis', *Systematic Biology*, 47, 508–519.

Thiele, K. (1993) 'The holy grail of the perfect character: the cladistic treatment of morphometric data', *Cladistics*, 9, 275–304.

Thiele, K. and Ladiges, P. Y. (1988) 'A cladistic analysis of *Angophora* Cav. (Myrtaccae)', *Cladistics*, 4, 23–42.

Thompson, D. W. (1917) *On growth and form*, Cambridge: Cambridge University Press.

Thorpe, R. S. (1984) 'Coding morphometric characters for constructing distance Wagner networks', *Evolution*, 38, 244–355.

Wagner, G. P. (1994) 'Homology and the mechanisms of development', in Hall, B. K. (ed.) *Homology: the hierarchical basis of comparative biology*, San Diego: Academic Press, pp. 273–299.

Wiley, E. O. (1981) *Phylogenetics: the theory and practice of phylogenetic systematics*, New York: Wiley.

Zahn, C. T. and Roskies, R. Z. (1972) 'Fourier descriptors for plane closed curves', *IEEE Transactions, Computers*, C-21, 269–281.

Zar, J. H. (1974) *Biostatistical analysis*, Englewood Cliffs: Prentice Hall.

Zelditch, M. L., Bookstein, F. L. and Lundrigan, B. L. (1992) 'Ontogeny of integrated skull growth in the cotton rat *Sigmodon fulviventer*', *Evolution*, 46, 1164–1180.

Zelditch, M. L. and Fink, W. L. (1995) 'Allometry and developmental integration of body growth in a piranha *Pygocentrus nattereri* (Teleosti: Ostariophysi)', *Journal of Morphology*, 223, 341–355.

Zelditch, M. L. and Fink, W. L. (1998) 'Partial warps, ontogeny, and phylogeny: a reply to Adams and Rosenberg', *Systematic Biology*, 47, 345–348.

Zelditch, M. L., Fink, W. L. and Swiderski, D. L. (1995) 'Morphometrics, homology, and phylogenetics: quantified characters as synapomorphies', *Systematic Biology*, 44, 179–189.

Zelditch, M. L., Fink, W. L., Swiderski, D. L. and Lundrigan, B. L. (1998) 'On applications of geometric morphometrics to studies of ontogeny and phylogeny: a reply to Rohlf', *Systematic Biology*, 47, 159–167.

Chapter 8

Creases as morphometric characters

Fred L. Bookstein

ABSTRACT

The real number line, with its familiar ordering and its geometry of gaps and averages, misleads the systematist's intuition in predictable ways when imagined in the broader context of morphospaces for organismal shape variation. This chapter illuminates one frequently encountered misconception arising from the distinction between continuity of natural phenomena and discontinuity of descriptions of those phenomena. For the specific problem of shape description in systematics, a new semantics of discontinuous descriptors is sketched, the method of creases, using a calculus borrowed from classical catastrophe theory via the thin-plate spline. Creases support claims about homology considerably stronger than other, earlier translations of landmark-location data into the language of systematics. They may supply the long-sought after passage from continuous, descriptive morphospaces to discrete characters.

Introduction: on themes visual and methodological

Take a tetrahedron. Or, better, *make* a tetrahedron, by unfolding a wire paperclip into the shape of two equilateral triangles sharing an edge at which they meet at 60°. Number the corners or thread colored objects through the wire in their vicinity, so you can tell them apart: 1, 2, 3, 4.

The vertices of the tetrahedron are meant as an ordination of the shapes of four biological taxa arranged in a three-dimensional space at their 'shape coordinate' locations. The concern of this chapter is the range of potential measurements by which they can be ordered, individually or in combinations. This chapter will use the word 'ordination' to mean a representation of taxa by positions of points in some Euclidean space, such as the vertices of the tetrahedron here, whereas the words 'ordering' or 'sorting' will apply to one-dimensional representations.

The easiest way to investigate this measurement space is via the ways the tetrahedron can lie in your hand: the group of rigid rotations of the tetrahedron, 'represented' (as the mathematicians like to say) by its action on the taxa at the corners. By tumbling the tetrahedron in your hand you begin to understand the domain of shape variables that characterize the relationships among its vertices.

So tumble it, then, until you understand how to produce all possible orderings of these taxa as the top-to-bottom ranking of the vertices after a variety of rotations.

140   Fred L. Bookstein

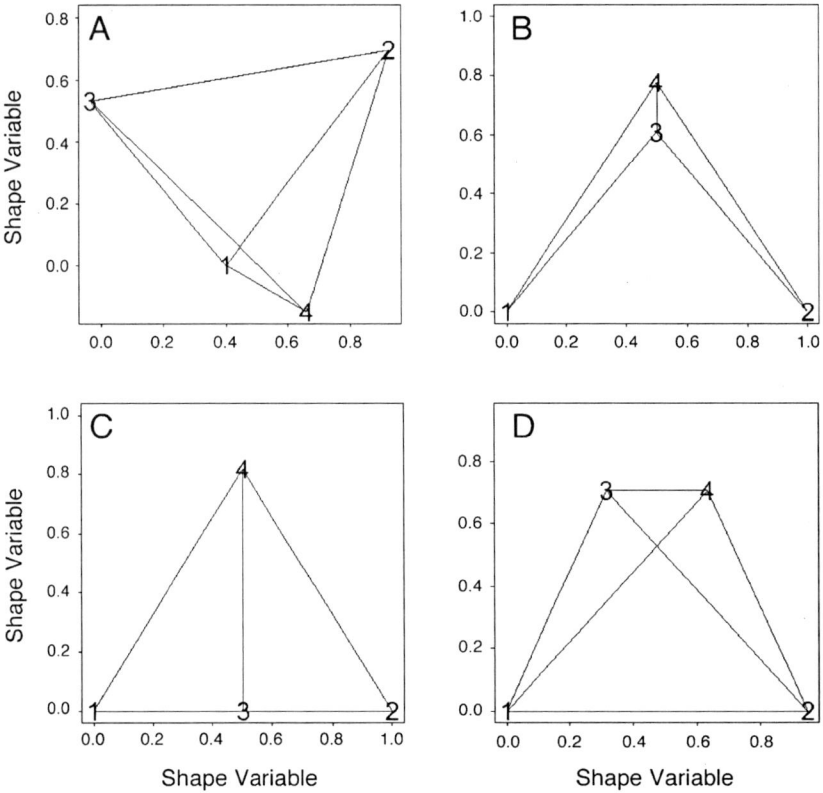

Figure 8.1 The 74 shape orderings of four specimens: distinct orderings of the vertices of a tetrahedron by projection onto an axis. (A) Four distinct values; this can be obtained in 24 different vertical orderings. (B) Three distinct values, 36 versions. (C) Three-way tie, eight versions. (D) Two two-way ties, six versions. The shape variable embodying the ordering is vertical in all panels.

There are 74 of these orderings, as follows:

- Four distinct quantities (Figure 8.1A), in any of the 24 possible rankings of the vertices (e.g., 1 > 2 > 3 > 4, 4 > 3 > 2 > 1, 2 > 3 > 1 > 4). This can be done in 24 different ways; note that turning the tetrahedron upside-down yields a different ordering.
- Three distinct quantities (Figure 8.1B), whenever the tetrahedron is posed with some edge horizontal (e.g., 1 > 3 = 4 > 2, 4 > 3 > 1 = 2). This can be done in 36 different ways: six choices of edge, times six orderings each of the resulting three values.
- Two distinct quantities involving a three-way tie (Figure 8.1C), which obtains when viewing the tetrahedron with one face horizontal (e.g., 4 > 1 = 2 = 3). This can be achieved in eight different ways.
- Two distinct quantities each encountered twice (Figure 8.1D), which obtains when viewing the tetrahedron with pairs of opposite edges both horizontal (e.g.,

1 = 2 < 3 = 4). This can be achieved in six different ways. In each such configuration, the plane of your hand is parallel to both of the edges spanned by the tied taxa, and the vertical is their common perpendicular. For this particular tetrahedron, that vector also connects the centroids of the two edges in question, but that is not the case in general.

Thus, for every ordering of the four vertices, except the completely unresolved report 1 = 2 = 3 = 4, there is some range of tetrahedron orientations that can be represented by a vertical coordinate (the distance from your hand). If the original ordination is by Procrustes distances among some set of shape coordinates, then these 'possible verticals' are all themselves shape variables (Bookstein 1991). Thus, every ordering, except the uninformative 1 = 2 = 3 = 4, is accessible through some shape variable that is a linear projection of those shape coordinates.

This absolute ambiguity is neither an illusion nor a contradiction. Instead, it illustrates the limits of the naïve systematist's intuition in matters of quantification (see Bookstein 1991, 1994). Just as the methodology of physics is a terribly misleading model for quantitative praxis across the sciences in general, so the real number line – with its familiar ordering, gaps, and arithmetic – is an inappropriate guide to quantitative methodology in comparative biology. In particular, the familiar number line is of no use in describing shape. The geometrical grammar of quantitative morphological descriptions and their comparisons arises from notions akin to rotating our tetrahedron, not sorting the number line. The role of morphometrics within systematics is to sort through all the possible orderings of shape-based ordinations – all the possible vertical coordinates that could be generated out of the original shape measurements – in order to find some that are useful for testing hypotheses about the shape data in a specific systematic context. Geometrically, all shape variables are equally valid descriptors, and it is part of the morphometrician's task to preserve access to them all.

Through practice, this morphospace for your tetrahedron can become nearly as familiar as that of the real number line. Yet, as soon as there are more than four forms in a data set, the geometry of alternative linear descriptors is no longer that of familiar three-dimensional Euclidean space. In these cases we will need other visual aids if we are to understand the range of possible form descriptors. Consider, for instance, the simplest interesting morphospace for the shape of configurations of landmarks: the four-dimensional space that arises from sets of four labeled points of the plane when one ignores size, orientation, and position. The four points are now not the vertices of a tetrahedron, each standing for one whole shape, but rather, each set of four points makes up a shape of its own. For a review of this version of shape space, see Bookstein (1997a, 1998a) or Rohlf (this volume).

Figure 8.2 shows, at upper left, a random sample of five four-landmark shapes from the Dryden–Mardia distribution (Dryden and Mardia 1998) arising from circular Gaussian variation of landmarks independently around a fixed mean form. The forms are superimposed in the so-called Procrustes pose – with position, orientation, and scale all constrained by a single least-squares criterion – and then flattened so that the data lie in a four-dimensional subspace of the eight dimensions (four sets of two Cartesian coordinates) in which they were originally digitized. The geometry of possible shape measurements of these five forms is the four-dimensional equivalent of

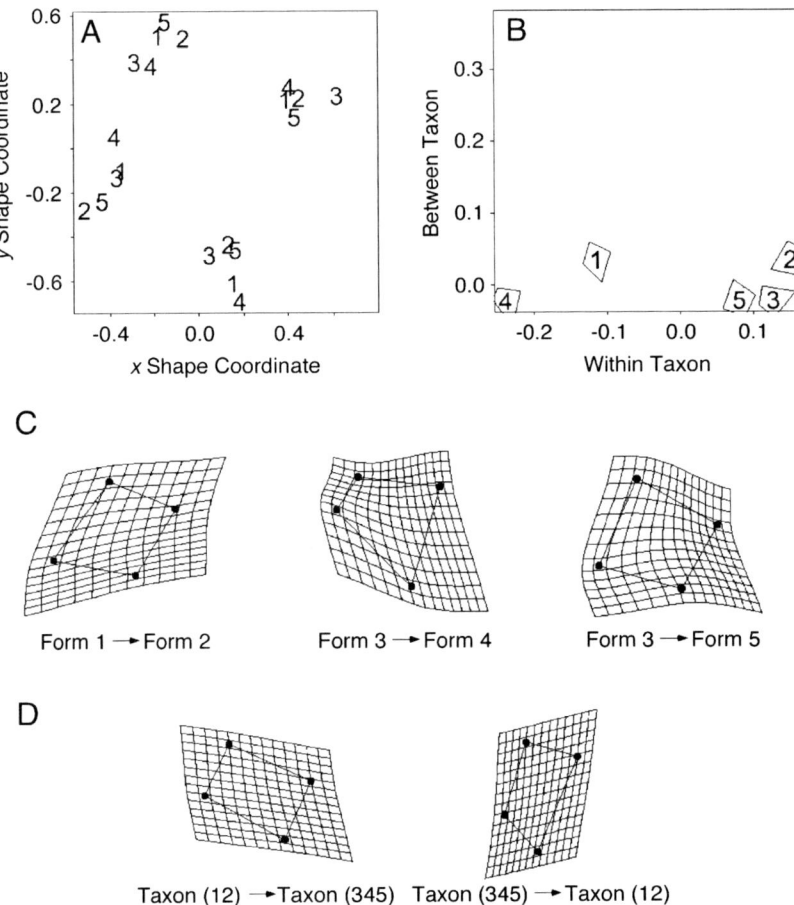

Figure 8.2 Morphospace for an arbitrary dichotomy of five randomly generated quadrilateral shapes (A). The axis labeled 'between-taxon' (B) is the shape variable that perfectly confirms a discrimination of shapes 1 and 2 from shapes 3, 4, and 5. The other splines (C) represent 'within-taxon' contrasts in perpendicular directions of shape space. At each taxon is the icon of its quadrilateral of landmarks. (D) This same variable visualized as a spline of Procrustes length 0.25 in either direction; it is mainly a uniform transformation.

the tetrahedron. Mathematicians call this configuration a 4-simplex, where a line segment is a 1-simplex, a triangle a 2-simplex, and your familiar tetrahedron a 3-simplex. As there were 74 different postures of the tetrahedron – 74 different rank-orders of the shapes of four taxa – there are 540 for the 4-simplex. Of these, the most relevant for systematics are those that reduce to binary variables: contrasts such as $2 < 1 = 3 = 4 = 5$ of one taxon against the other four, and contrasts such as $3 = 4 < 1 = 2 = 5$ of two taxa against the other three. Notice that there is no concern for gap coding in morphospaces like this one, inasmuch as by altering the measurement formula the coding of a variable can be simplified to the irreducible minimum of two values separated by one gap.

Figure 8.2 goes on to show the complete geometry of the four-landmark shape morphospace for one arbitrarily chosen exemplar of this binary class: the contrast of forms 1 and 2 against the others. The ordination is as shown in the upper right panel, in which the contrast of interest is the 'between-taxon' axis running vertically ('upward' from the hand holding the hypertetrahedron). All the other dimensions of shape space are concealed in the two substructures viewed on edge along the horizontal. The upper is a simple line (the join between shape 1 and shape 2), the lower the plane through shapes 3, 4, and 5 viewed in some arbitrary foreshortening of its own.

Looking at the tetrahedron under the two-pairs case (Figure 8.1D), the direction of the vertical was fixed by the statement of the problem (to orient the tetrahedron with the edges spanning both tied pairs parallel to your hand). The same constraint pertains to this five-taxon version: only one vector is perpendicular to both domains of within-taxon variation, the line between taxon 1 and taxon 2 and the plane through taxa 3, 4, and 5. Here, perpendicularity is in terms of the underlying Procrustes geometry of shape distance (see Bookstein 1997a, 1998a).

Directions like these can be diagrammed helpfully by moving each landmark of the mean shape to the location it would have if only the vertical coordinate in this plot were changing (not any of the other shape dimensions) and then relating the final form to the mean form by a thin-plate spline. In Figure 8.2C are the splines that result from going 'down' (in the direction from taxon 12 to taxon 345) and then from going 'up' (the direction from taxon 345 to taxon 12). In Figure 8.2D are representatives of the remaining directions in morphospace perpendicular to this particular contrast: the contrasts between forms 1 and 2, between forms 3 and 4, and forms 3 and 5. These three shape contrasts, together with either of the two at far right, span the complete space of all possible contrasts of shape for this quadrilateral. In other words, combinations of these deformation patterns are the equivalent of the three-dimensional set of re-orientations of the tetrahedron that applied in the four-form (paperclip) example. To name this contrast, we read the grid (or the equivalent vector of partial warp loadings). The feature that is 'vertical' here, the contrast between (12) and (345), is mainly a uniform transformation, relative extension or compression of the form in the direction of the grid's diagonals, with hardly any bending. Conversely, the transformations that are 'horizontal' – that are ignored by this character – involve a great deal of bending (as in the grids for shape 3 to shape 4 or shape 5) or the orthogonal uniform transformation that changes the aspect ratio, vertical : horizontal. For alternative descriptions of transformations of quadrilaterals, see the discussion in Bookstein (1991).

Now just as the tetrahedron could be rotated to generate any ordering of the four vertices that we wished along its vertical, so this 4-simplex can be rotated to any other vertical ordering. Figure 8.3 shows the same set of panels for the 'taxa' that arise when 2 and 5 are reassigned, so that now shape of taxon 5 goes with that of taxon 1 and shape of taxon 2 with taxa 3 and 4. The vertical coordinate, the perpendicular to both within-taxon morphospaces, is now a different sort of shape variable entirely, as can be seen in the geometry of its spline (Figure 8.3C): it combines a mainly vertical/horizontal uniform term with a substantial amount of bending. The between-taxon axes make an angle of about 73° between Figure 8.2 and Figure 8.3.

We could continue in this way for the remaining eight possible two-*versus*-three bipolar contrasts of this (or any other) set of five four-landmark forms. For each,

144 Fred L. Bookstein

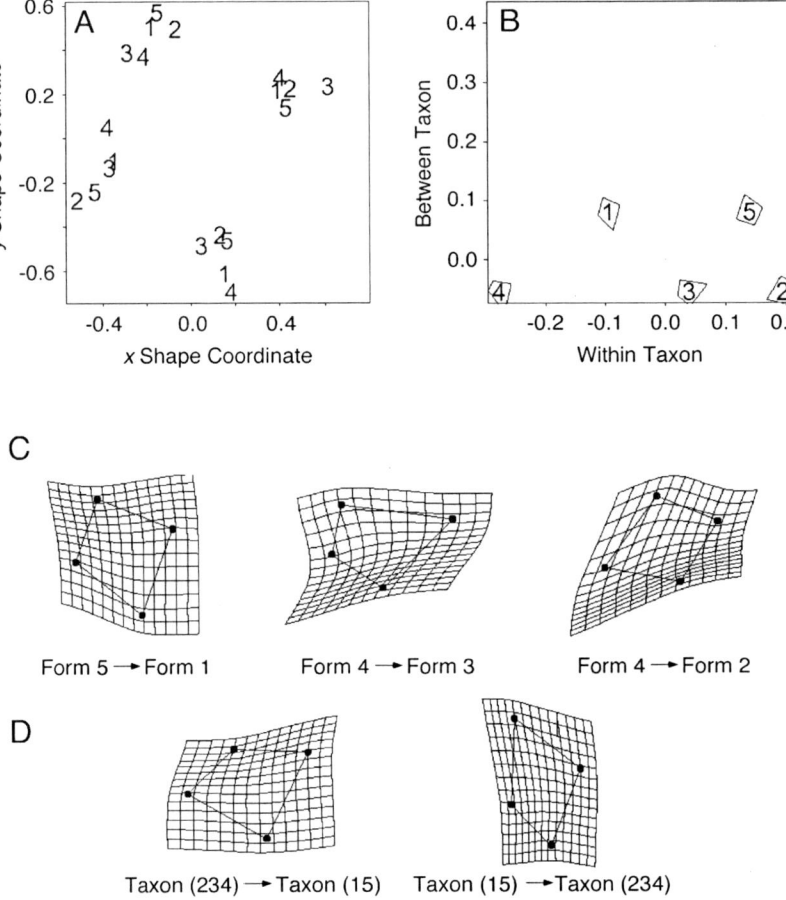

Figure 8.3 A different dichotomy of the same five quadrilaterals. See Figure 8.2 caption for labeling conventions. Again some shape variable confirms the perfect bifurcation between these 'taxa'.

there is a shape variable aligned with that contrast and perpendicular to both of the within-'taxon' variance structures that the geometry is arranged to ignore. For five landmarks, the statement would be true of any bipolar contrast among seven forms; for six landmarks, among nine forms; ..., for k landmarks, $2k-3$ forms, corresponding to the $2k-4$ dimensions of the full Procrustes shape space. Since, in practice, landmarks are not very difficult to come by – in general much easier to produce than new taxa – there results a praxis for producing unique morphometric variables congruent with any imaginable node of a cladogram.

To readers who have not had this situation explained to them before, the proposition that given enough landmarks we can extract a variable confirming any bifurcation whatsoever in a data set of three or more taxa, regardless of its actual systematics, seems absurd. Yet, that is the way rotations of sufficiently high-dimensional descriptor spaces will be exploited below. The manner in which these shape spaces yield descriptor variables surely seems not to map onto phylogenetics the way we expect variables ought to: onto a real line or a circle, some ratio or some angle, capable of stable arithmetic summaries in terms of order and gaps between groups.

The problem is not with the geometry, but with our systematic expectations. Quantification in biometry is not like the number line, but rather like the tetrahedron. There is a space of possible measures a great deal richer than the actual data, a space not only continuous but multidimensional and non-Euclidean. The task of describing a contrast of shape lies primarily in the formulation of variables in some manner that is sensible for an ultimate systematic purpose, whereas the issue of collecting and sorting the variables' values is only secondary. The geometric simplicity of the number line, with its too-familiar relationships of greater-than and less-than arising out of human social history, conceals the myriad of ways in which ordering principles themselves arise flexibly before being constrained optimally to illuminate a systematic question. Before taxa can be contrasted, the space of variables spanning all the possible descriptions must *itself* be ordered. And this ordering – the selection of some single variable from the complete hypersphere of possibilities – needs to be logically coherent, just as the ordering afforded by the tetrahedron was equivalent to the task of deciding 'which way is up' from the full spherical space of possibilities. This decision actually entrains all of the final 'systematic findings' – they are begged at the time the variables are selected.

Consequently, it will not do to guess at the identity of useful variables a priori. There are far too many candidates available. This sort of guesswork cannot result in outcomes having any validity as guides to morphospace, whether angles, ratios, or partial warp scores. To select a descriptor of shape variation in advance of inspecting the complete morphospace of that shape representation is to impose an arbitrary discontinuity at will: a stance unlikely to lead to biological inferences having any authority.

There are only two sensible ways to break the symmetry of description in the space of 'all possible contrasts' (the space of simplexes and their rotations). Either one must have directions of contrast specified a priori as covariates of exogenous specimen properties (e.g., membership in particular taxon groups predicated in advance), or else these directions must arise as functions of sample variation likewise specified in advance (e.g., relative warps [= principal components of empirical shape covariance matrices in the Procrustes metric]). In this way, there is no guesswork or mysticism involved in the production of interesting directions in shape space. The directions are embodied in the systematist's a priori choice or definition of 'interesting contrast' or 'figure of merit'.

Localization in biometry

The task of selecting an interesting direction in organismal shape space is not statistical, but systematic: the visualization of a contrast supplied by the data in some automatic fashion such as a planned mean difference. Any such direction is a topic of potentially informative discussion. But to be useful in a systematic context, the discussion cannot be in the language of 'directions in shape space' (the geometry of Figures 8.2 and 8.3). It must be phrased instead in the ordinary, extensive language of organismal form, exploiting the classic semantics of references to parts and their

146 Fred L. Bookstein

proportions, direction and magnitude of gradients, and the like. In the contemporary literature of morphometrics (which has been couched more often in the language of statistical signal analysis than that of biological interpretation) this entire concern had been 'put off until later', until appropriate tools had been invented that might formalize the systematist's intuition. One such tool is now available, the method of creases (Bookstein 1998b,c,d, 2000). The remainder of this chapter deals with the geometric and biometric language of creases.

Any discussion of localization in morphometrics must begin with an exploration of that other system of localization we intuit without any special education: localization within a physical landscape. For instance, this analogy is built into the grid visualization scheme for the thin-plate splines we are using. The typical viewer of Figure 8.4 does not see a landscape in the landmark displacement plot at upper left, but it is very difficult to nullify the perception of the same displacements as a landscape once the 'contours' of the landscape have been indicated (Figure 8.4, upper right).

Figure 8.4 goes on to present a random assortment of 'points of view' (elevations) of this surface. Within a broad range of angles we see the same localization regardless

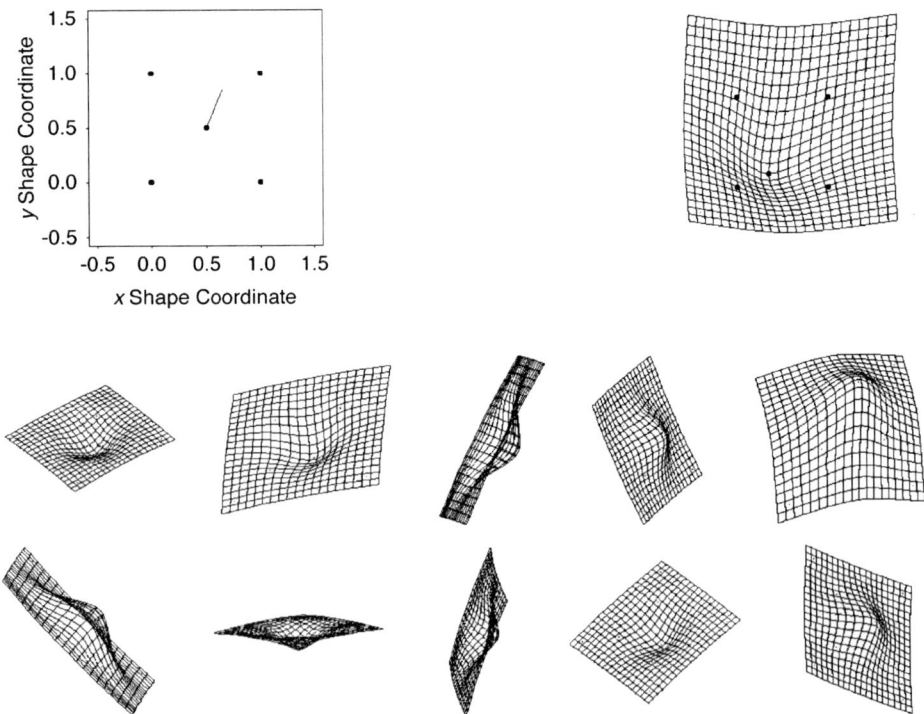

Figure 8.4 Shape changes as landscapes. A shift of the midpoint of a quincunx (upper left), visualized as a splined grid (upper right), is immediately decoded as a surface shape (a landscape). The central landmark on the grid is in the same relative position as the undotted end of the segment in the panel to its left. As this shape is tumbled (lower 10 images), one proves particularly informative about the shape change: the panel at center right, in which grid lines just touch without crossing.

of optical elevation or foreshortening (but turn the page upside-down, and those that looked like peaks now look like pits – a classic optical illusion). Linear transformations such as these orthogonal projections are uniform throughout the image and thus cannot add any local information of the sort systematists seek. Hence, all of these new perspectives convey the same information about non-uniformities of the grid. Nevertheless, they do not all appear to be equally informative. Our eyes find some of them quite a bit more striking than others. Careful inspection of figures like these suggests that the most interesting or intuitively accessible grids are those resembling the exemplar at the center right: grids for which one of the warped lines conveying the deformation comes to just touch another without quite crossing. This is an instance of the general tendency of image semantics to emphasize points at which atypical coincidences occur (Koenderink 1990).

There are two generic (= mathematically typical) ways that this can happen, each demonstrated several times in the grids of Figure 8.4. Projections of curving surfaces onto specific planes can fold (Figure 8.5A), or they can have cusps (Figure 8.5B). At a fold, nearby parts of one surface are mapped to the same region of the image surface after one side is mirrored away from the fold line. At a cusp, one part of the surface rises up to occlude another part. To say these are generic means that small changes in the mapping perspective do not make them vanish, only move a bit. That is why they occur more than once in Figure 8.4.

The specific perspective at center right in Figure 8.4, where nearly parallel grid lines touch but do not cross, is not generic in this sense. Small changes in the point of view turn the touching locus into a combination of cusps and folds, or else make it go away entirely (as shown in Figure 8.6). For any smooth landscape that is nearly flat, but not exactly flat, there is a direction of view closest to the zenith for which some grid on the surface appears to touch without crossing, at which both that point and that view are isolated (discrete, distinctive). But whereas contour maps of landscapes are maps from the underlying plane to one real value (height), creases represent this

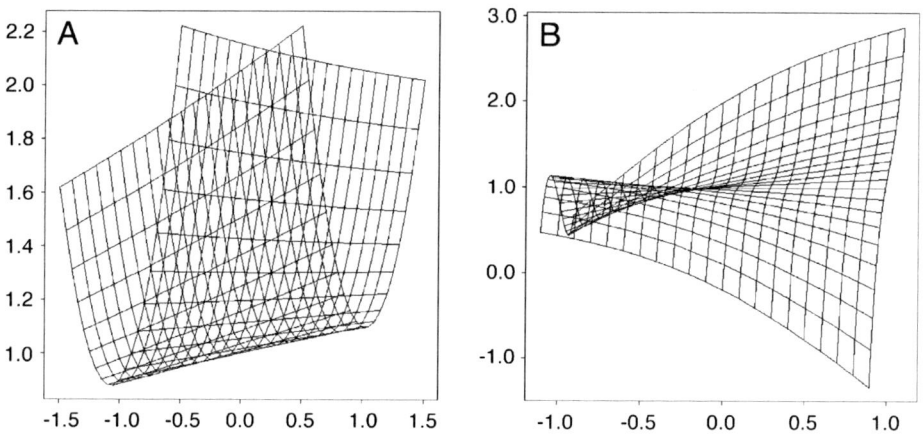

Figure 8.5 The two generic (typical) ways in which grid diagrams fail to be one-to-one: (A) folds and (B) cusps. These are both exemplified in many of the panels of Figure 8.4.

148 Fred L. Bookstein

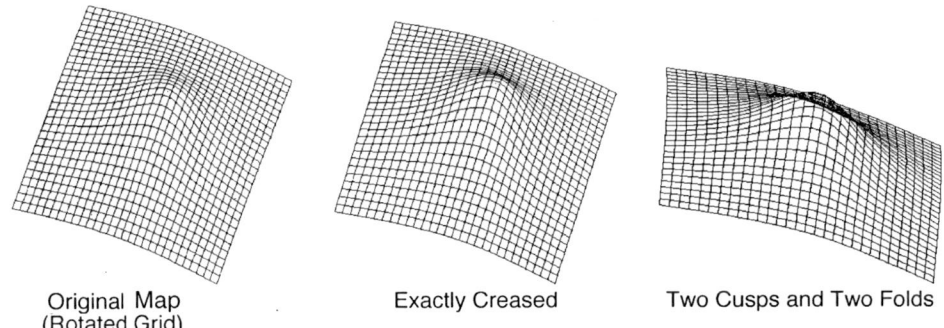

Original Map   Exactly Creased   Two Cusps and Two Folds
(Rotated Grid)

Figure 8.6 The crease (center) is the transitional case between the one-to-one map (left) and the map with interrelated folds and cusps (right).

same just-touching construction for maps into a space of two real values, that is, the deformation of one image into another.

Geometry of creases in one and two dimensions

Thus far this chapter has emphasized the intuition of possibilities at the expense of mathematical rigor. Now it is time to replace talk of tumbling and touching by descriptions of actual algorithmic approaches. Because the subject of study pertains to a four-dimensional geometry (the two dimensions of the original square grid, mapped into the two others of the deformed version), we can use the underlying mathematics of catastrophe theory (Poston 1978) – the calculus of ways in which the Cartesian quality of a projected grid can break down. The following discussion is based mainly on Bookstein (2000).

Consider the function $f : x \to y = (x^4 + x^3 - 3x^2 + 4x)/3, [0, 1] \to [0, 1]$, construed as a map between two one-dimensional continua (intervals) x and y registered at their endpoints (Figure 8.7A). In the more usual presentation as a Cartesian graph (Figure 8.7B) an inflection is apparent where the tangent crosses the curve near its midpoint. One can locate this inflection by a dynamic graphical maneuver (Figure 8.7C), shearing the curve steadily more and more downward into curves $x \to y - \varepsilon x$ for increasing ε, in search of the first value of ε for which the corresponding curve has an isolated zero slope. This happens for $\varepsilon = \frac{3}{4}$, the particular sheared graph indicated with an arrow and drawn by itself in Figure 8.7D. The corresponding one-dimensional continuum simply projects this curve onto a vertical axis, where the inflection – now horizontal – is visualized as the pile-up of dots near the middle of the range (on the right).

For functions near the identity (such as the example here) ε is near 1 and the range of the sheared vertical graph is the small quantity $(1 - \varepsilon)$. Then the singularity of the dot map is hard to read. If we rewrite $x \to y - \varepsilon x$ as $x \to (1 - \varepsilon)(x + (y - x)/(1 - \varepsilon))$, dropping the prefactor, we arrive at a family of maps $x \to x + \alpha(y - x)$, where α is $(1 - \varepsilon)^{-1}$. These are linear extrapolates of $x \to y$ from the identity as shown in Figure 8.7E. The corresponding one-dimensional versions are ordinary magnifications

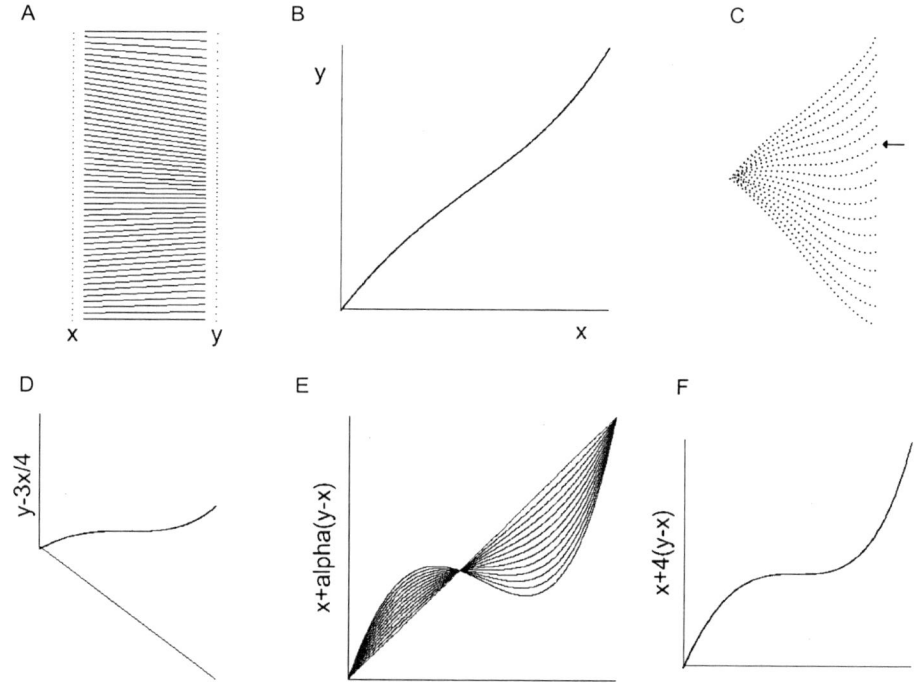

Figure 8.7 The one-dimensional equivalent of the crease is the graphical location and enhancement of an inflection point by a combination of shearing and vertical expansion. See text for discussion.

of the dot patterns generated by $y - \varepsilon x$. As before, there is one member of this family having isolated derivative zero somewhere, the function $x \to x + 4(y - x)$ shown by itself in Figure 8.7F. This curve is just a fourfold vertical distortion of the curve in Figure 8.7D. Because the corresponding one-dimensional continuum is now full-scale within its copy of the real line R^1, the viewing eye can more precisely discern the locus at which the dots pile up. The search for inflections at slope ε has become a search for values of $\alpha = 1/(1 - \varepsilon)$ for which the map $x + \alpha(y - x)$ first 'overruns itself': first has a zero of its derivative just before incorporating a domain of going backward.

For applications to organismal form, our task is to construct a graphical manipulation for maps $R^k \to R^k, k > 1$, as close as possible in spirit to what was shown in Figure 8.7. The discussion here takes $k = 2$: a matched pair of labeled point sets **S** and **T** (Starting and Target forms) in the Cartesian plane. Consider, for example, the two point sets in Figure 8.8: a starting form **S** (Figure 8.8B), a quincunx (pattern of spots in an ×, like the 5-side on the dice used for gambling), and a target form **T** (Figure 8.8A) in which the central spot has been displaced by a modest amount (in this instance, 15 percent of the diameter) along its diagonal. We have already seen maps like this in Figure 8.4. Now we are ready to construct the algorithm that automatically produces the special points of view suited to visualizing the crease.

At any point, the derivative of such a map is a 2×2 tensor $\mathbf{O}_{\theta_T} \mathbf{D} \mathbf{O}_{-\theta_S}$, where the **O**'s are orthogonal matrices (rotations) and **D** a diagonal matrix of principal strains

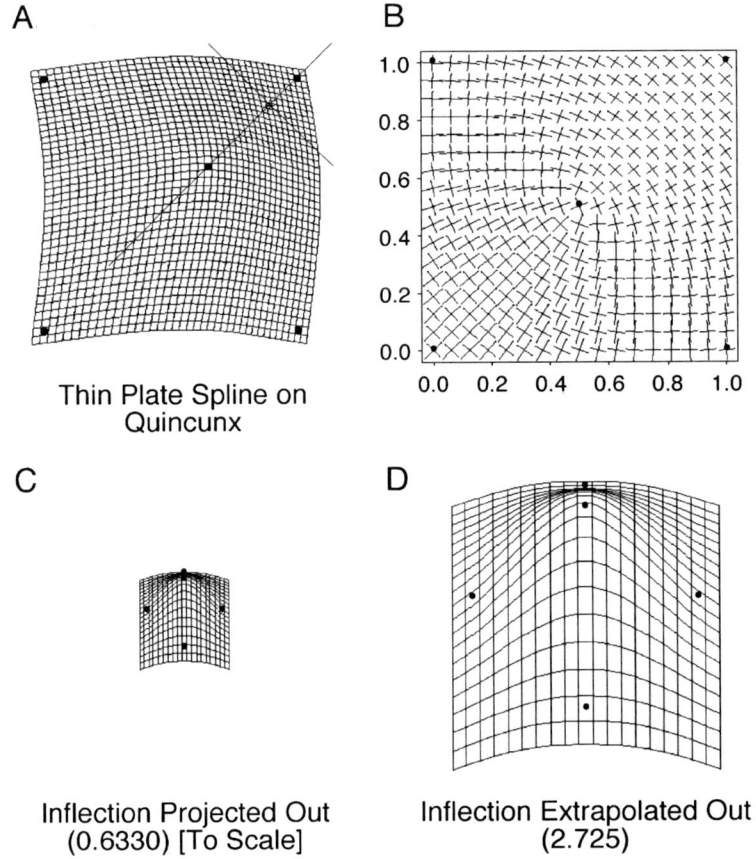

Figure 8.8 Computation of the crease for the transformation of a quincunx **S** into a target form **T**. (A) Original spline, with dots indicating the target form **T** and crosshairs locating the maximum strain. (B) Derivative of the spline as a 2 × 2 symmetric tensor on the starting form **S**. (C) Spline after a suitable multiple of the identity transformation is rotated out. (D) The same as an extrapolation instead, so as to preserve graphical scale. See text for discussion.

(Bookstein 1991). The symmetric part is $O_{\theta_S} D O_{-\theta_S}$, the rotational part $O_{\theta_T - \theta_S}$. The symmetric part can be drawn on the S form as a symmetric tensor field (Figure 8.8B) of axes of the ellipses into which little circles in the starting form are taken by the interpolation at hand. One equivalent in two or more dimensions for the inflection produced in Figure 8.7 is the locus at which some principal strain is an extremum σ_{ext} as a function of position. For an interpolant of closed algebraic form, like the thin-plate spline here, these are easily located by mesh refinement. In this example, there is one single extremum of compression, toward the top of the major diagonal, indicated by its principal directions atop the original spline diagram. The extremal strain happens to be a compression to 0.6330 of the original length; the thin plate, because it smoothes, has reduced this ratio below the 30 percent shrinkage of the interlandmark distance spanning the locus. In both forms, the principal axes of the interpolation at this point lie at ±45° to the horizontal.

While the location of these crosshairs is plausible by eye, it is no more verifiable than the equivalent problem of localization in Figure 8.7A. By analogy with that $R^1 \to R^1$ case, we first subtract a multiple of the identity map corresponding to the 'slope' (= the isotropic map having the same derivative as that extremal strain in its direction) to arrive at the mapping $O_{-\theta_S}S \to O_{-\theta_T}T - 0.6330 O_{-\theta_S}S$, where both θs are taken at the locus of extremal strain. Dropping the θs from the notation in this case of $\theta_S = \theta_T$, this is the map $S \to T - 0.6330S$. Arithmetic here is carried out vectorwise on the points of S and T separately, and the interpolation engine applied to the result: by linearity of the thin plate, this is the same as the difference of splines $S \to T$ and $S \to 0.6330S$. Its grid (Figure 8.8C) is informative, but there remains an undesirable shrinkage of scale exactly like that shown at lower left in Figure 8.7. Just as in the one-dimensional case, this shrinkage is circumvented by converting the problem to extrapolation from the identity. Because the thin-plate spline is linear in its right-hand side, the grid for the map $S \to T - 0.6330S$ differs only in scale from the grid for the map

$$S \to (1 - 0.6330)^{-1}(T - 0.6330S) = 2.725T - 1.725S = S + 2.725(T - S) \tag{8.1}$$

Again arithmetic is done coordinatewise by landmark, and again, by linearity, the resulting map is a sum of its components: the identity plus 2.725 times the spline $S \to T$. This is the spline that moves every point of S by a vector just 2.725 times as long as the vector by which it was displaced to get to its position in T.

In the general case, with $\theta_S \neq \theta_T$, the projection $S \to O_{\theta_S - \theta_T}T - \sigma_{\text{ext}}S$ becomes the extrapolation

$$S \to S + \frac{e^{i(\theta_S - \theta_T)}}{e^{i(\theta_S - \theta_T)} - \sigma_{\text{ext}}}(T - S) \tag{8.2}$$

up to an arbitrary rotation. The figures that follow take this rotation as $O_{(\pi/2) - \theta_S}$, so that in all cases the extrapolated map has principal strains horizontal and vertical with vertical principal strain zero.

In the grid corresponding to the extremal extrapolation (Figure 8.8D), the structure of the original $R^2 \to R^2$ map (Figure 8.8A) has been enhanced until it is legible. The extrapolation has pushed the deformation of the original quincunx along the diagonal (the locus of extremal strain) until it 'overruns itself', just as in Figure 8.7. Figure 8.9A is a graphical enlargement of the interesting part of this grid, the locus at which the map has directional derivative precisely zero.

In that both of the derivatives of y' at $(0, 0)$ are zero, this map has a singularity there. The generic singularities of maps $R^2 \to R^2$ are folds and cusps (Whitney's theorem; see Bruce and Giblin 1992), having canonical forms $(x, y) \to (x, y^2)$ and $(x, y) \to (x, xy + y^3)$, respectively. The singularity here is neither of these, taking the form $y' = x^2 y + y^3$ up to a function of x: the 'ideal type' shown in Figure 8.9B. As a map $R^2 \to R^2$ it seems not to have a conventional name, at least, not in English. In earlier publications (Bookstein 2000) I have named it the *crease*. The vertical line $x = 0$ is mapped into itself by this transformation, but the derivative of that map is zero at the singularity. Lines $y = sx$ of non-zero slope through the singularity are mapped

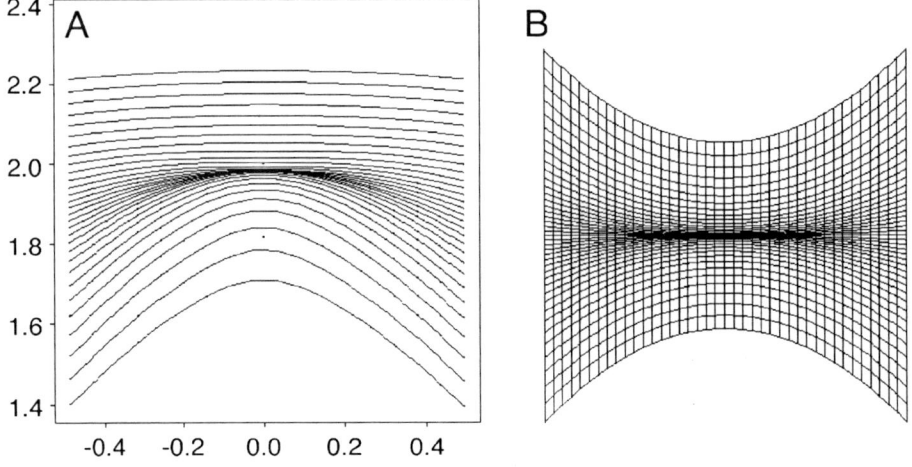

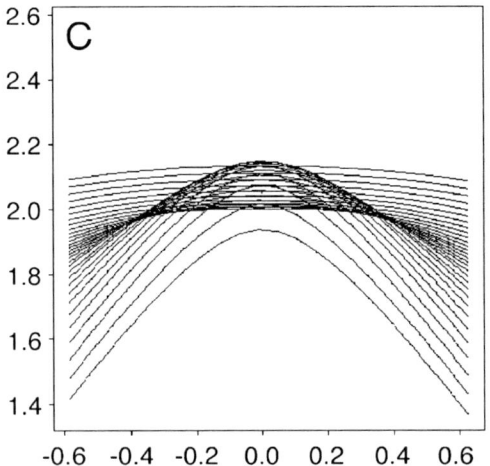

Figure 8.9 The form of the grid at the crease (A) for **S** → **S** + 2.725(**T** − **S**) is a bent version of the polynomial $x^2y + y^3$ (B). Overextrapolations, such as **S** → **S** + 3.5(**T** − **S**) (C), generate the familiar configuration of two folds joined at two cusps.

into cubics $y' = (s + s^3)x^3$ all having slope zero at the singularity. That is, only one line through the origin of the starting grid S has an image that is a proper geometric traversal of the x-axis; all others cross at an angle of $0°$. If the value of α is too large, the resulting mapping extends 'past the singularity' into a region that reverses the sign of area, and thus no longer has an inverse. These regions of overextrapolation are bounded by a pair of cusps joined by two folds (Figure 8.9C) between which the map turns areas upside-down.

Example: effect of schizophrenia on a midsagittal brain polygon

The data set for our first example (DeQuardo et al. 1996) was acquired and analyzed in 1994 in the course of research into the neuroanatomy of schizophrenia. These data are the locations of 13 landmarks in a sample of 28 approximately midsagittal sections of three-dimensional magnetic resonance (MR) scans of human brains, 14 from normal adults and 14 from persons diagnosed with schizophrenia. The landmarks concentrate in the midbrain, where fluid volumes jostle with many different clearly delineated nuclei carrying out many different functions. Actually, only eight of these are proper Type-1 landmarks (Bookstein 1991); the rest sample extended curves. We ignored this distinction in the originally published analysis, and continue to ignore it here (but see Bookstein 1997b, 1998a).

For multivariate analysis it is convenient again to use the Procrustes shape coordinates (Figure 8.10A), locations of the landmarks of all the original specimens of the data set after each configuration is fitted to their average by minimizing summed squared discrepancy over the similarity group. For the two-group comparison at hand, one averages each fitted coordinate pair in this same coordinate system (Figure 8.10B). The landmarks of the mean shape for the normal adults are indicated by +; for the schizophrenics, by ×. There are a total of $2 \times 3 - 4 = 22$ degrees of freedom in this multivariate space, but we will always draw its points as vectors of multiple landmarks in the original Cartesian plane. One pair of landmarks that will concern us presently is highlighted by a bracket.

Contrasts like this one can be tested for statistical significance by one general-purpose procedure, a permutation test (Good 2000) of Procrustes distance. Here, the squared Procrustes distance between the mean shapes is 0.001444. Without making any probabilistic assumptions that require verification (e.g., Gaussian distribution of landmark variability jointly or separately, see Goodall 1991), we can compare this quantity to the distribution of distances between averages of 14 of these 28 forms and averages of the other 14 when these subsets of 14 are selected at random from the full set of 40,116,600 possibilities. In 5,000 such random subsettings of the 28 cases, the

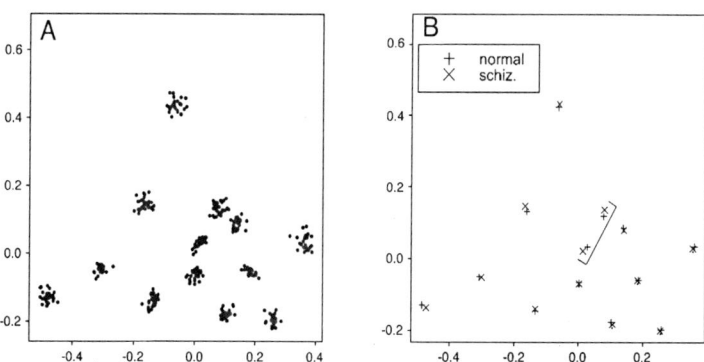

Figure 8.10 A data set of 13 midplane brain landmarks. (A) Procrustes shape coordinates for all 28 subjects. (B) Averages by group (+, 14 normal subjects; ×, 14 adult male schizophrenics). The mean difference is significant at 0.036 by omnibus permutation test (see text). The bracketed landmarks underlie the crease that will be found in the next figure.

154   Fred L. Bookstein

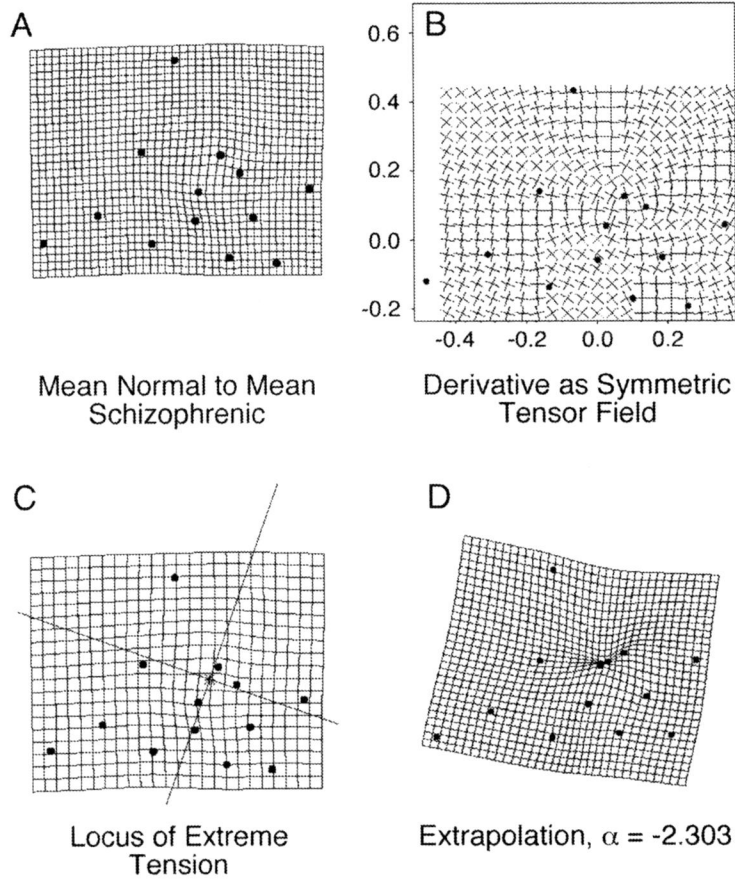

Figure 8.11 Crease analysis of the schizophrenia data set: the spline $N \to S$ from normal mean to schizophrenic mean. Panels are as in Figure 8.8. At extrapolation ($\alpha = -2.303$) a crease appears between the landmarks bracketed in Figure 8.10. It is evidently the only localized characteristic of this shape comparison, and tested as a crease, it is significant beyond the 0.001 level.

observed Procrustes distance is exceeded 183 times, for an empirical significance level of 0.0366. It is this quantity that has been declared 'the significance level of the group difference' in earlier publications.

As 0.0366 is less than the conventional 0.05, we are authorized to talk about the features of this shape difference. But what shall we say about it? The thin-plate spline interpolation grid (Figure 8.11A) shows some sort of bulge near the center. Perhaps it is this region that is responsible for the significance of the permutation result. The method of creases introduced in the previous section can be used to help make sense of the geometrical signal in cases like these. The derivative of the interpolation between these average shapes, diagrammed as a symmetric tensor field (Figure 8.11B), has only one region of high strain, located in the vicinity of the visible bulge. The extreme principal strain there is 1.434; its orientation (Figure 8.11C) suits the analysis of the bulge as well. Extrapolation to the crease singularity (Figure 8.11D) is for $\alpha = 1/(1 - 1.434) = -2.303$. In this panel, the starting grid has been rotated so that its vertical is aligned with this same principal strain. Visually, the structure of this singularity is just what we were led to expect from Figure 8.9. More to the point, there seems to be no other signal anywhere in the diagram. The crease is all there is here – a monolithic organizing icon for reporting the deformation 'as a whole' with great graphical force.

By virtue of this degree of localized organization, the evidence of shape change in this example is considerably more persuasive than the omnibus permutation test implied. Customized for localized phenomena such as this, a better permutation test would estimate the frequency with which extrapolations to the same factor of -2.303 result in grids that include or over-shoot a singularity when one compares the mean of one randomly selected half of the data base to the mean for the other half. It is sufficient to examine the permutation distribution of this largest principal strain to see how frequently it exceeds 1.434 in magnitude. In 1,000 random subsettings of these 28 cases into halves, no principal strain anywhere larger than 1.434 was found.

Back at the right in Figure 8.10, it is clear that the two landmarks indicated by the bracket have moved more than the other 11. Their motions, which are in opposite directions, are collinear with the segment between their mean positions and, nearly enough, with that principal strain of 1.434. We know now that these landmarks straddle the crease just estimated. Hence, the distance between these two landmarks, relative to the overall central moment of the configurations (the scale normalized out in the course of the Procrustes registration), is likely to be a very sensitive indication of this shape change. Indeed this separation averages about one-third more in the syndromal subgroup than in the normal adults, a comparison ostensibly 'significant' by separate t-test at $\rho \sim 0.0013$. As there are 78 such separations, a Bonferroni-corrected probability would be $78 \times 0.0013 \sim 0.10$. But we did not come to be thinking about this by selecting it as the best of the interlandmark separations. Rather, over this featureless background, the only possible signal appears aligned with this particular pair of landmarks, falling off toward something uninteresting in all directions. The crease is a convenient way of wielding this evidence of spatial coherence in the course of reasoning about the reality and implications of this shape difference, and the relative separation of this pair of landmarks (relative, that is, to the overall scaling of the configuration, which is by Centroid Size) seems like the most useful possible quantitative character.

What is the meaning of such a finding? In the context of geometric information, the analysis in Figure 8.11D seems to have uncovered a *focal* feature. This algorithmic approach has called attention automatically to the same bulge on which the viewing eye fastened in Figure 8.11A. Far richer than any alternative style of reporting, Figure 8.11D organizes crucial information about change of length and its gradients throughout the image in a way that shows us how dominant this particular singular feature is. Quite literally, there is nothing else to talk about in this geometric diagram. In this sense, what we have accomplished is an automated caricature that takes the atypical features of a scene and exaggerates them for ease of communication.

The same finding can be rephrased in the language of anatomy. We have located the center of directional compression of a structured dysmorphogenic field that, in this example, seems to summarize all that might be interesting about the schizophrenia-associated deformity as a whole. The expansion (actually, an erosion of neural matter) that we are examining is centered directly over the quadrigeminal cistern, an extension of the third ventricle under the splenium of corpus callosum, so that the discriminator we have uncovered here has the semantics of a good morphometric character – the relative extension of the quadrigeminal cistern along the direction toward colliculus. This is certainly not any sort of standard measurement in the geometry of schizophrenia (indeed there are no such standard geometric measurements in the biomedical literature). But it is entirely in keeping with the grammar of quantitative keys, which often deal with the relative sizes of parts. In this case, the part that we discerned algorithmically was already there to be measured, even having a Latin name. The shape variable we have produced is not, in the language of Figure 8.1, an 'edge' of the ordination. It is not the mean difference between the clinical groups, but a picture of the most local aspect only. If these two groups were proper taxa, this particular descriptor would be a fine candidate for a character. Of this interpretation there will be more below.

To complete the analysis of this data set, we need to look at extrema of the smaller principal strain, the one corresponding to peaks of compression. The finding in Figure 8.12 will come as no surprise: the greatest compression is located approximately parallel to the expansion of Figure 8.11 where it springs from the landmarks just beneath. The extremal strain is 0.79, and so yields a crease at extrapolation factor

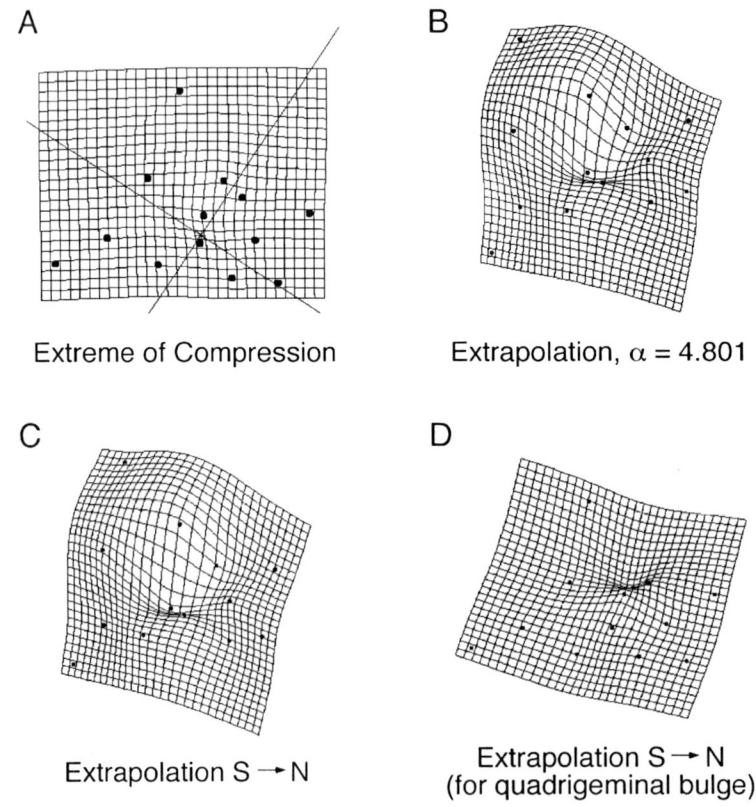

A Extreme of Compression

B Extrapolation, α = 4.801

C Extrapolation S → N

D Extrapolation S → N (for quadrigeminal bulge)

Figure 8.12 (A, B) Analysis of the minimum strain for N → S. (C, D) Analyses of the reverse spline S → N result in the same features.

$\alpha = 4.8$. Because this strain is closer to unity than the larger principal strain of 1.43 in the same vicinity, it is inappropriate to test it for significance separately.

Readers often ask about the dependence of this sort of analysis on the choice of starting form – in these figures, it is the normal mean form on which the grid is squared, the syndromal that is shown as deformed. When the forms N (normals) and S (schizophrenics) are similar, as is the case here, the map $S \to N$ (reverse spline) closely resembles the map $N \to N + \alpha(S - N)$ with $\alpha = -1$, which is to say, the map $N \to 2N - S$. To the accuracy of this approximation, the crease diagram will look the same whichever form, N or S, is taken as the 'starting' form. Compare the two versions of the 'normalization of schizophrenia' in Figure 8.12 or the alternative analysis of the quadrigeminal bulge in Figure 8.12D. Crease analysis in general is not polarized: no direction need have been specified in advance.

Example: sexual dimorphism in a curving form without landmarks

The data for the second example come courtesy of Christos Davatzikos of Johns Hopkins University. As explained in Davatzikos *et al.* (1996), these are outlines tracing around the human corpus callosum, the thick bundle of neural matter connecting the two cerebral hemispheres. The tracing is of the intersection of this bundle by the midplane of the head. The sample consists of 16 three-dimensional MR brain images: eight from elderly males and eight from elderly females. The outlines were traced automatically by an active contour method that searches the image for a locus of steep grayscale gradient without distorting a prototype too far (a standard boundary-tracking method, see citations in Xu *et al.* 2000). On one subject, 100 points were evenly spaced around the outline. Then sets of 100 points for each of the other 15 forms were selected on the polygonal outlines by an elastic method of 'slipping' with respect to the evenly-spaced points on the first form. When points are attached to outlines in this way, they bear information along only one Cartesian coordinate, the direction normal to the outline. Such points are called semilandmarks. The full data set, 100 points for 16 cases, is shown in Figure 8.13. These configurations were Procrustes-averaged and converted to shape coordinates by superposition over that average just as for landmarks. The full sample of shape coordinates and their averages by sex are shown in Figure 8.14.

The mean shapes (Figure 8.14B) look considerably more divergent in shape than the group averages of the schizophrenia example (Figure 8.10B). On the other hand, these sample sizes are smaller (only eight per group) and the points more numerous. In the Procrustes formulation for landmarks, distances between matched points are computed using all the information available, two coordinates per point. Semilandmarks are originally acquired and processed in Procrustes scatters as coordinate pairs, just like landmarks. But for statistical analysis they are better considered in a different Cartesian coordinate system that varies its orientation from point to point: the system of tangent and normal to the mean outline curve. The real import of data sets like these characterizes outlines only in the direction perpendicular to the (typical) curve. The usual Procrustes formula, by attending to shifts in both directions, is incorporating too great a degree of noise into the computation for the signal here to emerge clearly. We can correct this problem by reducing the 'distance' squared and summed

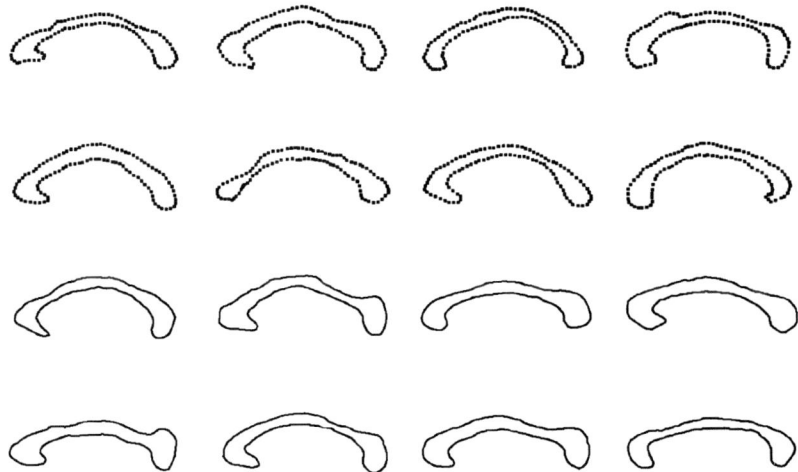

Figure 8.13 Hundred-semilandmark outline shapes of the midline corpus callosum for eight elderly males (above) and eight elderly females (below).

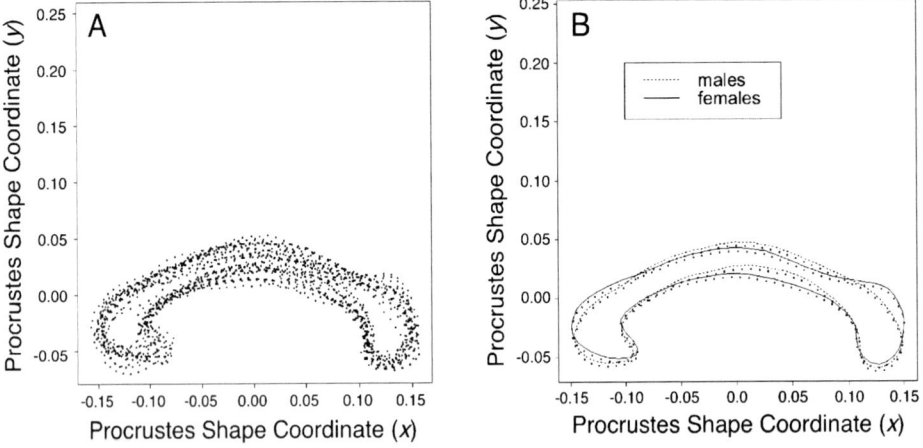

Figure 8.14 Procrustes analysis of this data set. The mean difference is significant at 0.005 by a suitably modified permutation test. See text for discussion.

in the Procrustes formula to only one of its coordinates, the distance normal to the average curve.

Following this modification, the permutation test goes forward exactly as before. The true summed squared normal difference between the group averages is calibrated against the distribution of that same sum of squares, all the way around the outline, when group is randomized over the 16 cases. There are 12,869 other groupings of this data set into 8 *versus* 8; a thousand of these were taken at random. Out of the thousand, only five permutations generated pseudogroups having a larger summed squared distance taken in this locally directional sense. In other words, the two group mean shapes in Figure 8.14 are significantly different at about the 0.5 percent level. It is telling that this same contrast could not be shown to be statistically significant by Davatzikos *et al*. (1996) using analysis of areas.

As in the first example, it remains to extract the features of this difference – to put the contrast into words. Again, the thin-plate spline is the best available tool. In Figure 8.15A is the spline, extrapolated twofold, from the average of the male forms to the average for the females. There is clearly a reshaping of the region of splenium involving some directional size change and some adjustment of its connection with the isthmus. In a somewhat focused-down version of the entire Procrustes analysis, for

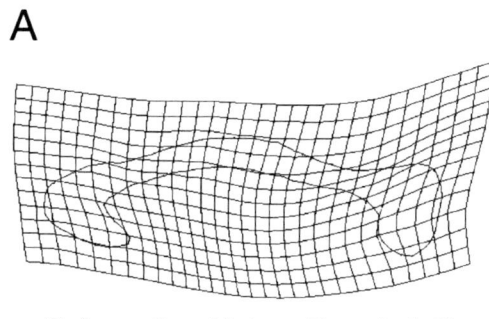

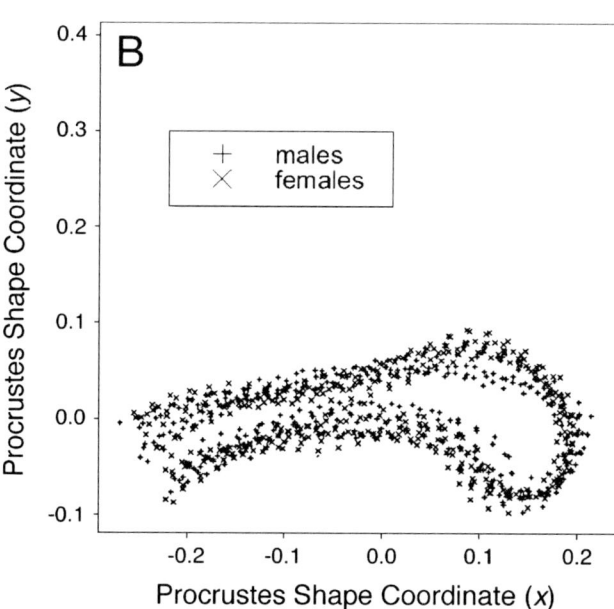

Figure 8.15 Inspection of the thin-plate spline between the means suggests a superposition on posterior structures only. There is now a strikingly localized divergence between the samples along the upper curve of splenium.

the posterior half of the form, we can see some of this difference in the nearly perfect separation of the outlines by sex in the vicinity of the 'bump' at the top of splenium (Figure 8.15B).

When I first inspected these data in 1996, the decision to attend to the vicinity of splenium was just an intuition. Today, however, the method of creases produces it directly and automatically as an explicitly quantitative descriptor. In other words, the shape contrast is 'characterized' – turned into an objective character. Figure 8.16 exploits the crease in this second data set (which, you will recall, has no landmark points at all, only semilandmarks). Figure 8.16A is the analytic computation corresponding to our informal report of Figure 8.15A, the precise location and orientation of the global maximum of strain from the male to the female mean form. The maximum strain, computed analytically from the spline's explicit formula, is 1.413. If we were to replace the female mean form, then, by the form that deviated from the male form by a multiple of $-0.413^{-1} = -2.42$ of the actual transformation here – the form for which landmarks shifted 2.42-fold as far in the opposite direction – then this optimal derivative would be precisely zero, giving the creased grid in Figure 8.16B. Algebraically, if the grid in Figure 8.16A corresponds to the spline $M \to F$, where M and F are the male and female mean forms in the Procrustes geometry, then Figure 8.16B is the result of applying the standard splining technique to the map $M \to (M - 2.42(F - M))$. Also, the original Cartesian system has been rotated so that the direction in which strain was artificially sent to zero is now the y-axis of the original coordinate system. (The grid in Figure 8.15 is $M \to M + 2(F - M) = 2F - M$.) There results a great clarification of the analysis.

However distorted the form of the 'callosum' here, its message about the geometry of the original difference of means by gender is astonishingly clear. We knew this strain had to be aligned with the boundaries of the callosal outline in its vicinity, as there was no possibility of shear along that boundary once the data were 'slipped' along that direction top and bottom. But one could not anticipate how spatially focused that

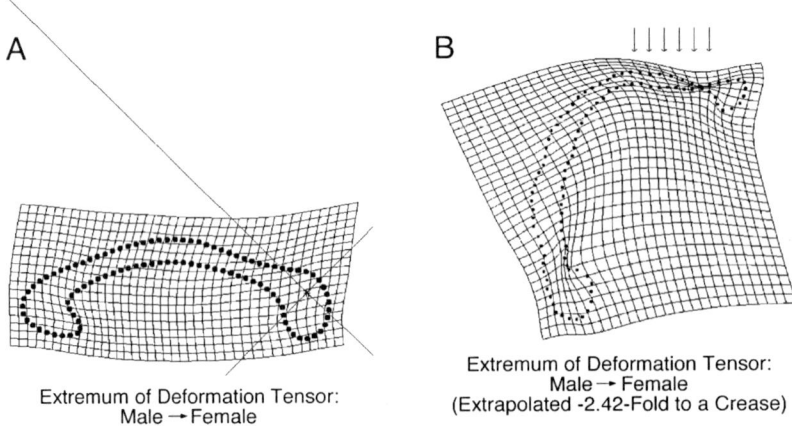

Figure 8.16 Automatic production of this same feature by the method of creases. (A) locus of extreme extension (1.413); (B) the corresponding extrapolation to the crease. At somewhat higher extrapolation a second crease will appear at the other end of the arch.

expansion of thickness would be found to be. The crease organization that ordinarily (Figure 8.9) looks like a 90° rotation of a pair of abutting parentheses,)(, here takes on an appearance like an hourglass instead – a highly local squeezing-inward (in the real contrast $M \to F$, extension outward) at the posterior end of an arch that otherwise manifests a more-or-less constant strain in this direction (arrows in the figure). The width of this 'waist' is roughly the same as that of the 'bump' in Figure 8.15B. There is an additional center of compression in the crease picture, at the crook of genu, showing a similarly localized aspect of relative female hypertrophy at a slightly lower value of maximum strain. It was presaged at far left in the grid of Figure 8.15B, just as the crease here was already visualized at far right. In between these two foci of compression, the extrapolation in Figure 8.16 is quite smoothly graded. We might have anticipated this in the smooth progression of grid cell shape within the arch in Figure 8.15A.

In a comparison of a mere eight forms of each gender, it would be unwise to speculate further on this localization. But the technique of creases is likely to be a great aid to hypothesis generation in larger samples, for which the underlying uncertainty of these foci is often less. In the systematic context, the bulge in the female splenium (specifically, the relative height of the arch at that locus) would serve as a perfectly satisfactory quantitative morphometric character if the two groups here were different taxa; and we have evidently produced it by an entirely automatic procedure. Thus the method of creases is not restricted to landmark point data, but extends to representation of curves without any formal change. We will exploit this extension in the phylogenetic example to come.

Lab rats

A third preliminary example treats an octagon of landmarks from 20 male laboratory rat skulls observed over 83 days of growth. The original Cartesian coordinates are set out in full in Bookstein (1991) for these 20 rats and one other with incomplete data. In Figure 8.17, changes of sample mean shape between contiguous age classes, extrapolated eightfold, are displayed by thin-plate spline. These splines draw our attention to a stable focus of interest at upper left, but it is quite difficult to put into words just what it is that we are looking at, or what other signals there may be elsewhere in this sequence of diagrams. From the crease representation, Figure 8.18, it is startling to see that the maximum strain near spheno-occipital synchondrosis (SOS, Figure 8.17), ranging over 1.05–1.07 per time interval, is, from age 14 days on, approximately stable in position, and that its gradient varies only slowly in orientation. This phenomenon was far from apparent in the earlier figure, as our eye was drawn away from it by the less stable (but more dramatic) features toward the back of the skull.

The corresponding analysis of the minimum directional derivative (Figure 8.19) shows even greater stability over the second through fourth time intervals, from age 14 days to 40 days. (For legibility, changes within the subspace of uniform variation have been suppressed.) The minimum strains for the three temporal intervals in question are 0.927, 0.943, 0.928. The directions of their creasing are nearly the same, and their product is 0.811. Then extrapolation by a factor of $(1 - 0.811)^{-1} = -5.29$ can be expected to drive their composite to minimum directional derivative zero. Figure 8.20 shows this extrapolation of 14-to-40-day change for each of the 20 animals having

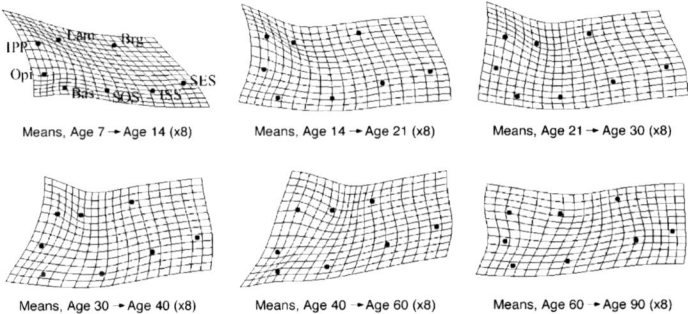

Figure 8.17 The Vilmann rat neurocranial data set (Bookstein 1996), revisited: mean growth changes for consecutively observed shape changes of a configuration of eight landmarks, magnified eightfold. The task is to describe what is systematic in this series. Landmarks: Bas, basion; Opi, opisthion; IPS, interparietal suture; Lam, lambda; Brg, bregma; SES, spheno-ethmoid synchondrosis; ISS, intersphenoidal synchondrosis; SOS, spheno-occipital synchondrosis.

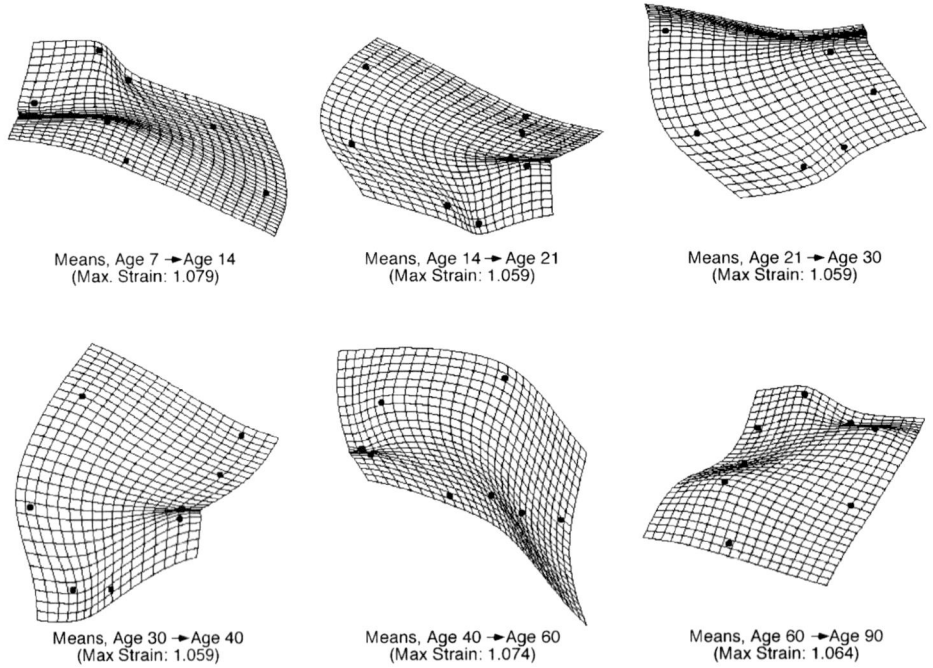

Figure 8.18 Crease analysis of Figure 8.17, uniform component suppressed, for maximum extensions. These are usually located in the cranial base. See text for discussion.

data at both ages. The scenes are evidently similar, and indeed the minimum strain is always near zero.

We do better to analyze these same comparisons by the crease method, so as to explicitly parameterize that local minimum along the calva. Figure 8.21 analyzes the 20 animals separately, one at a time. The result is quite unanticipated, and clearly

Creases as morphometric characters 163

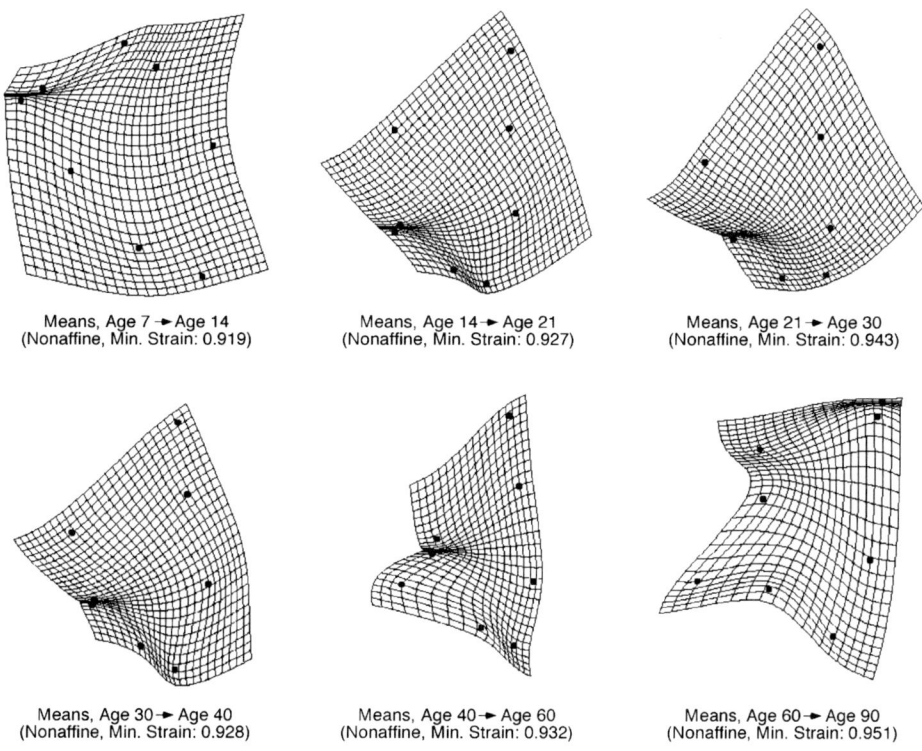

Figure 8.19 Crease analysis of Figure 8.17, uniform component suppressed, for minimum extension. The grids for growth 14–21, 21–30, and 31–40 are startlingly similar.

invalidates the implication in Bookstein (1996) that this data set could be analyzed more or less in its entirety by the method of relative warps. While the minimum σ_{ext} of relative growth is highly variable among these animals (from 0.878 to 0.678 of the change in overall Centroid Size), both the locus of the minimum and its orientation are remarkably invariable across this small sample. Here is startlingly strong evidence for a hypothesis of canalization (stable regulation of certain growth parameters individual by individual).

A phylogenetic example

I combine the logic of the tetrahedron and the method of creases in revisiting a data set from extant and fossil hominoids, along with an outgroup, originally published to illuminate another issue entirely (Bookstein *et al.* 1999). The data set comprises 11 calvarial landmark points and 20 semilandmarks on the midsagittal planes of 16 modern humans, 5 archaic *Homo* (*H. heidelbergensis* Bodo I, Kabwe I, and Petralona, Atapuerca SH5, and the *H. neanderthalensis* Guattari I), the australopith STS5, and 2 modern *Pan troglodytes*. (The original publication referred as well to external lambda and inion and three landmarks on the maxilla, all omitted here.) Landmarks were

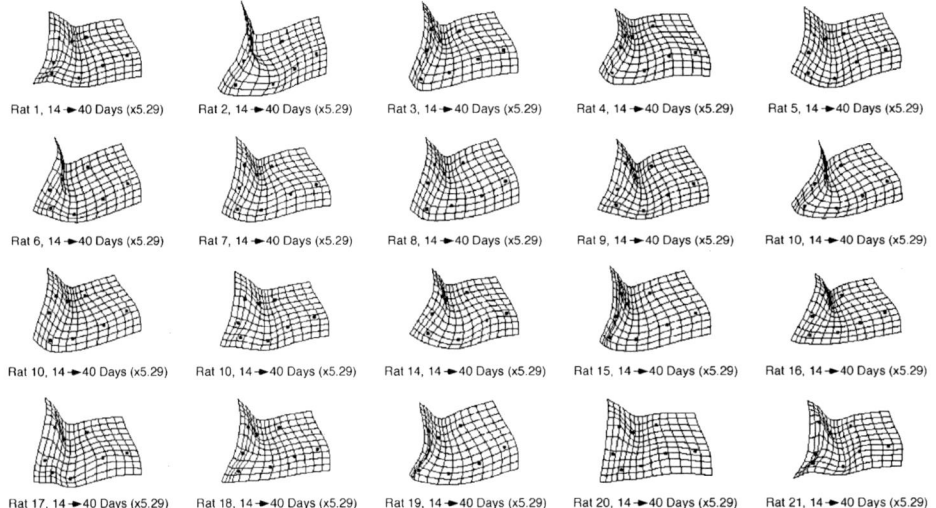

Figure 8.20 Analysis of animals individually using an extrapolation factor of $(1 - 0.811)^{-1} = 5.29$, where 0.811 is the product of the minimal strains on three of the creases of the previous figure.

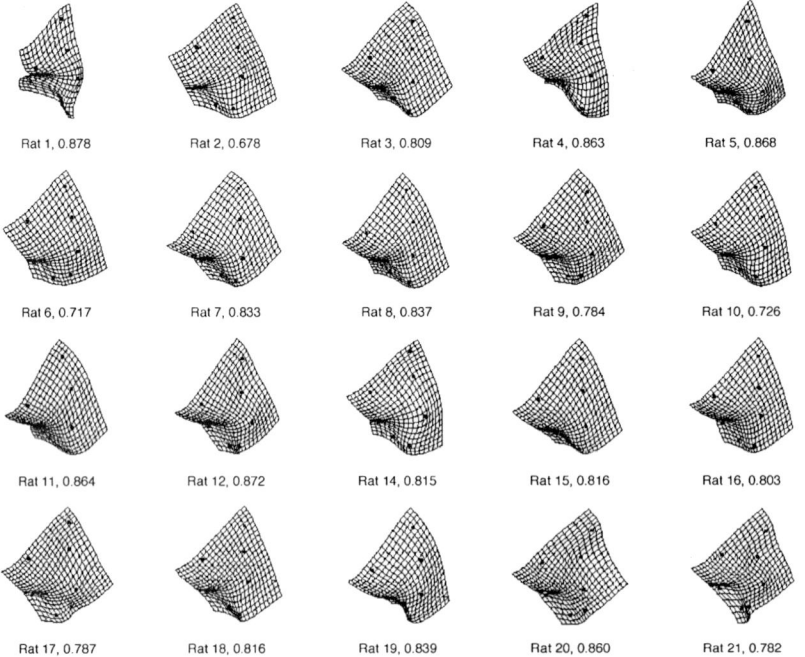

Figure 8.21 Crease analysis of the animals individually. All have a crease in very nearly the same position, but the extremal strain varies widely, from 0.678 to 0.878, over this fixed age range. The location of this apparent invariant of ontogenetic shape change is suggested as a good morphometric character that would be extremely difficult to construct by manipulation of conventional shape variables.

Creases as morphometric characters 165

located by hand upon actual stereolithographs of CT scans of the skulls, were transferred to a computed midsagittal section according to textural cues, and were then augmented by 10 semilandmarks each on the inner and outer tables of the frontal bone in this same plane section. Semilandmarks were extracted in Edgewarp (Bookstein and Green 1994) using the principle of jointly sliding points to a position of minimum bending energy (Bookstein 1997a), and all configurations of 31 points were reduced to Procrustes shape coordinates in the usual way.

For this demonstration, the 16 *H. sapiens* are averaged, likewise the 5 archaic *Homo* and the 2 *Pan*, and the single australopith comprises a group of one. The resulting four shapes can be placed at the vertices of a tetrahedron according to Procrustes distance. Figure 8.22 shows this tetrahedron as projected onto its principal coordinate planes. Each edge of this tetrahedron can be visualized as a thin-plate spline deformation of one vertex (one taxon mean) onto another. Figure 8.23 presents one of the two ways of showing these, as deformations pointing backwards in grade. With

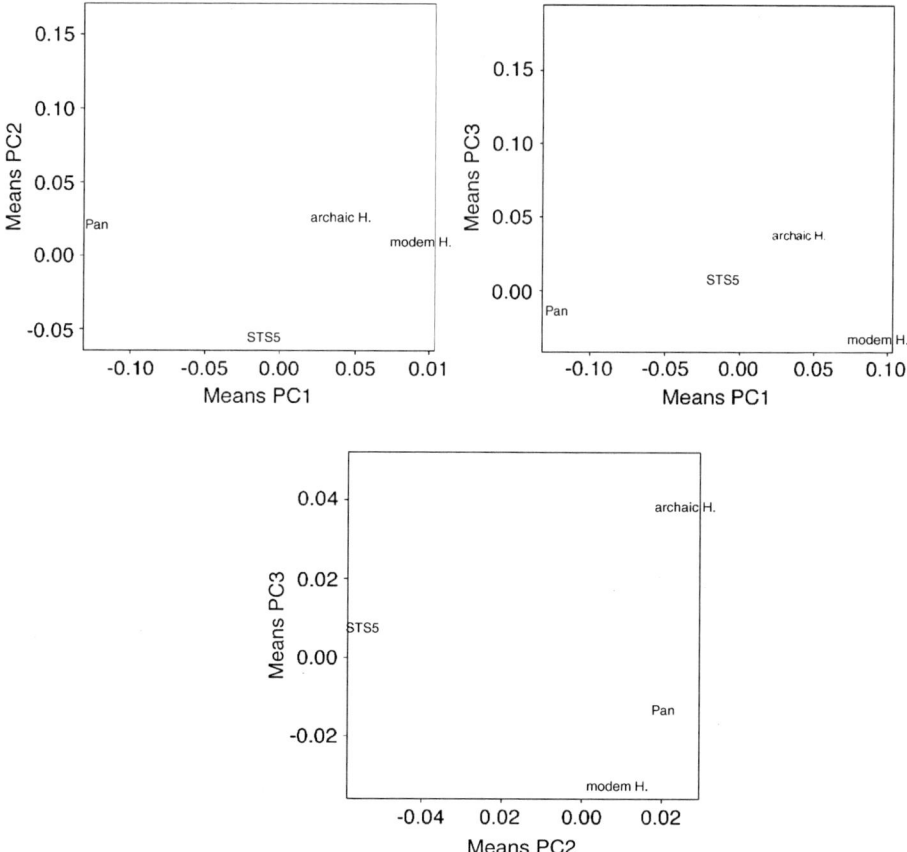

Figure 8.22 Principal coordinate projections for four taxa, ordinated by the Procrustes distances among their Procrustes mean shapes for configurations of 11 landmarks and 20 semilandmarks in the midsagittal calvarium. H. *Homo*.

166   Fred L. Bookstein

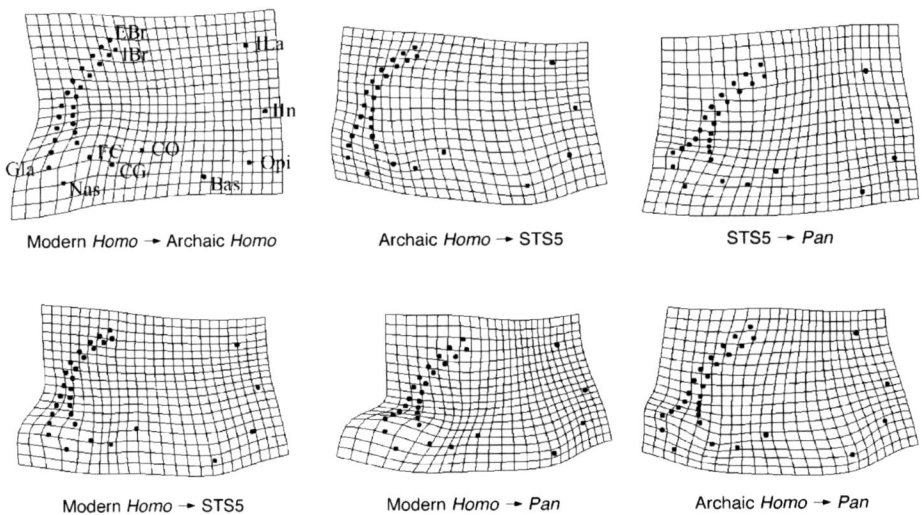

Figure 8.23 Thin-plate splines for all pairs of taxa in Figure 8.22, each taken in one of the possible two directions. Landmarks: Gla, glabella; Nas, nasion; FC, foramen cecum; CG, crista galli; CO, canalis opticus; EBr, external bregma; IBr, internal bregma; ILa, internal lambda; IIn, internal inion; Opi, opisthion; Bas, basion.

this polarity, all deformations involve thickening of the frontal bone in the vicinity of glabella (Bookstein *et al.* 1999), but there seem to be other features varying in a manner that is difficult to verbalize.

The descriptive task becomes easier if we convert from these grids, which are composites of many features, to the corresponding crease images, Figure 8.24, which are designed to present one feature at a time. To save space, the figure restricts its attention to three of the six pairwise comparisons, and please recall that all creases are rotated to lie horizontally in these displays, so that individual frames are not in any consistent orientation. In the left column is the analysis of archaic *Homo* as a deformation of modern. Above, the only focus of specific expansion over this comparison is the frontal bone near glabella: its thickness more than doubles. Put another way, the frontal bone in this vicinity has thinned by more than half from the archaic to the modern *Homo* average, and that is the greatest reduction of relative extent anywhere in the thin-plate spline interpolating this pair of configurations of 31 points. At lower left is the greatest relative compression from modern to archaic, which is to say, the greatest relative expansion of the modern average with respect to the archaic average. There is a clear crease here also, one that nearly fills the front half of the cranial cavity with a compression by up to 30 percent (relative to Centroid Size, the normalizing factor for Procrustes shape coordinates). The direction of compression is predominantly vertical, but the arrowhead shape of the grid image conveys some horizontal compression as well.

The greatest relative expansion from STS5 to *Pan*, upper right panel, is seen here again as a graphical compression of the frontal bone. This effect wraps around most

Creases as morphometric characters   167

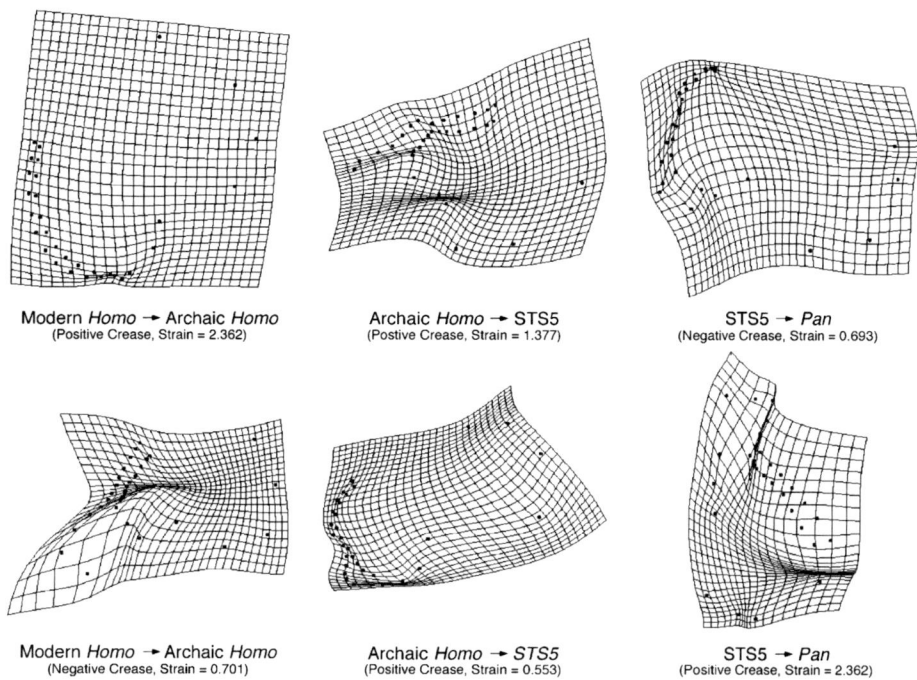

Figure 8.24 Crease analyses for the upper three splines in Figure 8.23. Upper row, creases for expansion for the comparison indicated (expansions backwards in time). Lower row, creases for compression for the comparison indicated (expansion forward in time).

of the arc from glabella to bregma. The frontal is thicker in *Pan* along most of its length; while there is indeed some compression at glabella, it is not focused there. In the lower right panel is shown a striking relative compression of *Pan* not too dissimilar to that between the grades of *Homo*. One of its features, at the top in the diagram, is a compression by the same 30 percent, with the same anatomically vertical orientation, near the vicinity of the midline sinus. But another focus emerges at nearly the same degree of compression that is horizontally aligned at lambda. This second feature is hinted at in the *Homo–Homo* comparison as well, lower left, but there it seems less directional, and also of lower magnitude.

The deformation of archaic *Homo* into STS5 (middle column) has a focus of vertical compression below glabella, where the other two 'chronological edges' of the tetrahedron had local expansions instead; and the focus of greatest expansion is along the anterior cranial base, between crista galli and canalis opticus, not in the vicinity of glabella at all. This contrast seems unrelated to the outer two.

Recall from the Introduction (Figures 8.2 and 8.3) that to generate binary contrasts, instead of gap-coding quantitative features we would rotate the tetrahedron until specific pairs of edges lay horizontally. In these configurations of the tetrahedron, the vertical is indifferent to the contrasts within the pairs of grouped taxa separately, and hence net Procrustes distance between the edges is minimized when it is taken straight

168   Fred L. Bookstein

up or down. In the context of a search for creases, one does the same after substituting bending energy for Procrustes distance in the definition of what counts as 'straight up'. For any pair of groups of taxa in this shape space, there is one unique deformation from some weighted average of the taxa of one grouping to some weighted average of the taxa of the other grouping that minimizes the overall bending energy. The algebra by which these particular composite contrasts are produced is set out in Appendix 1. From the four taxa of this example, three contrasts are particularly meaningful: the outgroup *Pan* against the hominins, *Homo* against the pool of australopith and *Pan*, and modern *Homo* against all of its ancestors. Crease analyses of these three 'orientations of the tetrahedron' are collected in Figure 8.25. In all frames, the starting form is the sample grand mean.

At left in Figure 8.25 is the analysis for the deformation from average modern *Homo* to the 'face of the tetrahedron' facing it, the space of shapes spanned by the other three taxa. In effect, the computation compares modern *Homo* to that 'hypothetical common ancestor', that combination of the other three taxa, to which it relates by the least bent (least focal) transformation possible: the estimated common ancestor requiring the least local changes to pass from ancestral state to modern. Thus, in this approach bending energy serves as the precise morphometric equivalent of parsimony in more conventional methods.

This least-bent contrast plainly resembles the specific contrast of modern with archaic *Homo*, Figure 8.24, in both its positive and its negative focal features. The most

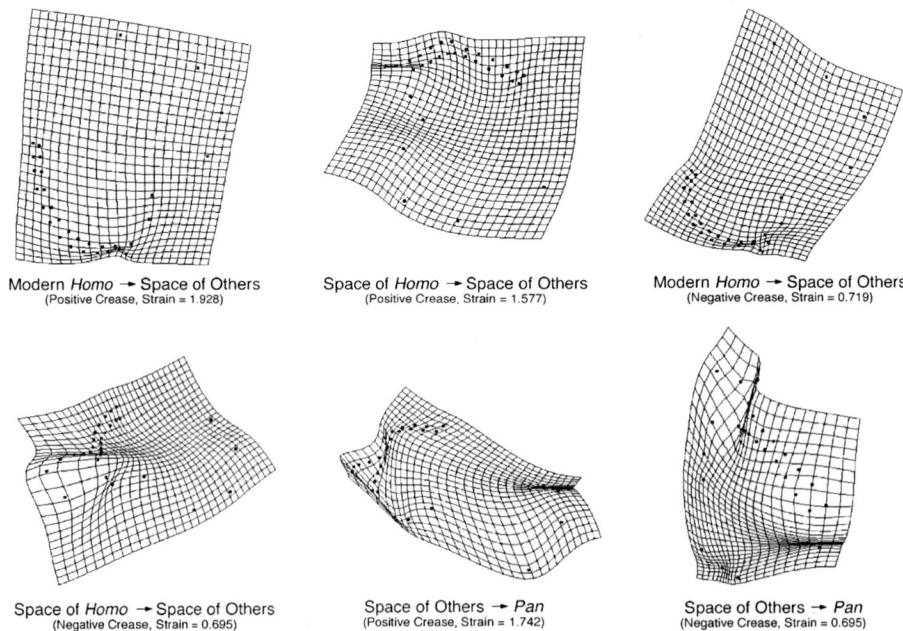

Figure 8.25 Crease analyses analogous to the geometry of Figure 8.1C (one taxon *versus* the span of the other three), left and right columns, or Figure 8.1D, one edge *versus* the other edge, middle column. The strong resemblances between the panels of this figure and the preceding convey considerable phylogenetic information (see text).

extensive (i.e., least bent) feature of decrease in size in modern *Homo* relative to the combination of the three other taxa is the reduction of the frontal bone in a small vicinity of glabella that we have already seen. And the greatest relative enlargement of modern *Homo*, in comparison to this particular combination of the other three taxa, also strikingly resembles that for the direct comparison of modern to archaic *Homo*. In phylogenetic language, this comparison of form looks like a good terminal character, stable over expansion of the range of comparisons against which we are testing it. It is startling that the character produced by this criterion seems to have nothing to do with the shape of the brain in this plane (cf. Bookstein *et al.* 1999).

The contrast of *Pan* to the composite clade of hominins, right-hand column, is also easy to interpret in phylogenetic terms. The greatest relative expansion of *Pan*, upper right, is the same character we have already seen at upper left: specific expansion of the sinus at glabella. The greatest compression of *Pan* with respect to the sister group, however, looks like the comparison to STS5 in particular (preceding figure, same panel): diminutions by almost equivalent rates across the calva at lambda and vertically in the vicinity of the frontal sinus. Thus, we have two more candidate characters here, or, if you prefer, a hint that sinus diminution arose twice, in two different geometric combinations.

The middle column of Figure 8.25 pairs the two *Homo* species against STS5 and *Pan*. If the obvious phylogenetic hypothesis is correct, the comparison of these two groups should strongly resemble the deformation linking the inner pair of sister taxa, archaic *Homo* and STS5, alone. By comparing the middle columns of Figures 8.24 and 8.25 one can see that is indeed the case. Notice in each comparison that what was the focal compression of greatest strain for one figure becomes the focus of second-greatest for the other. For instance, in comparing the upper center panels of Figures 8.24 and 8.25, the region of the crease at the top of Figure 8.25 is indeed compressed at the top of Figure 8.24, but not to zero area; and the crease in the middle of the Figure 8.24 panel is compressed, again not quite to zero, in the corresponding fabric of Figure 8.25. Whether this should count as one character or two cannot be settled without data more detailed about hominid/hominin grades, better anatomical coverage, and information from the third dimension.

In a related demonstration, O'Higgins (2000) visualizes the crease corresponding to the first principal coordinate of a data set comparing modern humans to Gibraltar I, Monte Circeo, and Kabwe I Neanderthals using 10 landmark points in the same general vicinity as those of the example here. His finding confirms the attention directed by the creases of Figures 8.24 and 8.25 at the hafting zone (anterior cranial base to superior maxilla) as a promising domain of derived character states for *H. sapiens*.

Discussion

As the title of this chapter indicates, and as I have hinted a few times earlier in the text, I believe that the crease should serve as prototype for the long-sought phylogenetic operationalization of the morphometric character, the summary of organismal form that is supposed to map onto a cladogram. Creases are, after all, discrete features located on drawings in a sensible way and arising out of image geometry so as to confirm a hypothesis of contrast among operational taxonomic units. Unlike all earlier approaches to this problem (distances, angles, principal components), every crease is already spatially organized, with parameters for position, orientation, and even a magnitude of its own (the parameter $|1 - \varepsilon| = |1/\alpha|$ of Figures 8.9ff).

What of homology for these ostensible characters? Homology in the phylogenetic sense pertains to parts of an organism, not geometric loci like these creases and certainly not algebraic parameters like position, orientation, and α. In the Procrustes framework, homology is declared a priori for the landmark and semilandmark locations themselves, and shape variables in general, which are linear combinations of Procrustes shape coordinates, inherit as much of the classic notion as applies to those linear combinations. Thus, forms having the same ratios of distances among nearby landmarks might be reasonably taken as sharing a homologous character state, but not forms sharing a ratio or angle involving large transects of the form (unless it could be argued to be homogeneously generated during development, or selectable by virtue of some biomechanical function), nor a priori patterns of large-scale bending such as partial warp scores (Rohlf 1998). The homology afforded by creases is more powerful than this, in principle, and potentially far more powerful in practice, as the algebra of the crease enforces the localization in space that 'characterizes' good characters in the systematic context. Creases break the rotational symmetry of Bookstein (1994) or Figure 8.1 here by restoring the criterion of anatomical coherence that the general linear combination of shape coordinates could not enforce (Bookstein 1991). In short, creases *per se* have a better claim on homology in the systematic sense than shape coordinates *per se*; something important has been gained in the passage from algebra to geometric localization.

As the text discussion of Figure 8.16 indicated, creases come in series, ordered by the extrapolation at which they appear. Actually there are two of these sequences, one for the extensions (Figure 8.11, say) and another for the compressions (Figure 8.12). All of them, not just the first and minus-first, can be produced algorithmically. (For instance, the two creases in the lower right panel of Figures 8.24 and 8.25 could be identified separately in this way.) Thus, we have moved some distance down the path to automated description of informative contrasts between samples or holotypes, what Richard Reyment has been calling image-based taxonomy.

Of the findings reviewed in this chapter, the focal quality of the contrast between normal adults and schizophrenics was long-suspected (or, rather, if the difference had been some global feature, it would have been found by now), and the difference between male and female callosa was likewise suspected (or Davatzikos would not have collected his 1996 data set in the first place). But the spatial concentration shown in the crease at lower left in Figure 8.24, regarding hominization of the frontal bone in midsagittal section, came as a surprise when it was published as a tentative character for recent hominid grades (Bookstein *et al.* 1999). Considerably more specific in its import than the 'hafting zone' interpretation, the crease calls attention to one particular sinus, over which an interesting argument can be erected (Prossinger *et al.* 2000). The 1999 publication identified this crease, and declared it systematically important, but the present visualization is better at proving localization than the series of permutation tests for difference as a function of position reported there.

The regularity of the growth crease for the rat skulls (Figure 8.21) is even more surprising. The invariance of the location of the crease here is exactly the kind of discontinuous descriptor that a systematist would wish to use for contrasting this particular species with others at lesser or greater remove. It would be interesting to apply the same technique to the extant anthropoid apes, inasmuch as ontogenetic series of our more immediate ancestors are unlikely ever to be obtained. The parameters of 'growth creases' are, by their very algebra, not equivalent to any of the obvious extensive measures of these skulls, nor to the changes of their ratios, but instead encapsulate patterns of the geometric distribution of growth rates. In these rats they proved both geometrically stable and strongly independent of just how much shape change there happened to be (the parameter σ_{ext} incorporated in the panel labels of Figure 8.21). Insofar as this crease is present invariably over these 20 specimens, it must serve as the derived state of a character at some taxonomic level. Such ontogenetic creases, when reliably present, would surely represent a more powerful bridge between morphometrics and systematics than earlier approaches using allometric regressions or principal components models.

The rich extension of morphometric description that creases afford was already implicit in the opening theme of this chapter: the enormous flexibility of the space of shape descriptions available for any contrast of images of organisms. The method of creases exploits the deepest of morphometrics' strengths, the possibility of carrying out direct manipulations on the space of possible variables (here, the domain of the tensor field whose extrema we are locating on the organism), while greatly strengthening the tie to the actual geometry of the organism that was previously brought to the eye, but not to the algebra, by the method of thin-plate splines. Clearly there is far more information in organismal form than systematists are currently exploiting. Now is the time for a direct assault on the underlying methodological task, that of producing shape descriptions that correspond to the scientific uses systematists need to make of them. The time is past when systematists should limit themselves to the variables that they, or their graduate assistants, just happen to know how to measure.

Acknowledgements

The work reported here was supported in part by NIH grant GM-37251 to Fred L. Bookstein. I am grateful to my long-time collaborator Bill Green for the exploratory software incorporated in his ewsh package that permits free play with magnification and grid placement in two and three dimensions. The diagrams of this paper were all produced in Splus, a commercial high-end workstation statistics package; codes for the functions invoked here are available from the author. The original Cartesian coordinates for the schizophrenia data set are available from the Morphmet bulletin board, http://life.bio.sunysb.edu/morph/. The coordinates for the growing rats and for the mean hominins and *Pan* are available from the author. Dr. Hermann Prossinger, Institute of Anthropology, University of Vienna, digitized the midsagittal landmarks, projected midsagittal landmarks, and midsagittal semilandmarks of the data set in Figures 8.22–8.25 with excruciating care for the original Bookstein (1999) publication.

Appendix 1: constructing hypothetical forms by minimizing bending energy

Figure 8.25 concerns comparisons of hypothetical forms produced as composites of grouped taxa. Algebraically, the production of these hypothetical ancestral composites is a straightforward modification of the formalism used for sliding landmarks (Bookstein 1997a). Instead of each landmark sliding separately on its own tangent vector, a configuration of landmarks slides as a whole along a weighted combination of the little vectors that connect corresponding landmarks within the groupings that are to be held 'horizontal'. The classical thin-plate spline computes the interpolant of one configuration of k landmarks $X_1, \ldots, X_k$ onto another set $Y_1, \ldots, Y_k$ that minimizes integral quadratic variation. Using the bending-energy matrix $\mathbf{L}_k^{-1}$ derived from the Procrustes grand mean form (the starting grid in Figure 8.25), arbitrarily select one form of the first group of taxa, say, $X_0 = X_{01}, \ldots, X_{0k}$, and another form $Y_0 = Y_{01}, \ldots, Y_{0k}$ from the second group of taxa. If there are n_1 additional forms $X_1, \ldots, X_{n_1}$ in the first group of taxa, and n_2 additional forms $Y_1, \ldots, Y_{n_2}$ in the second (either n_1 or n_2 can be zero), we seek the forms $X_0 + \sum_1^{n_1} \alpha_i(X_i - X_0)$ and $Y_0 + \sum_1^{n_2} \beta_j(Y_j - Y_0)$ the difference of which has the least bending energy. Since the coefficients of each of these combinations sum to 1, you may think of each as a weighted average of the shapes of the taxa in its grouping.

To minimize this bending energy, set up a matrix $\mathbf{U}$ of $2k$ rows by $n_1 + n_2$ columns in which the ith column, $i = 1, \ldots, n_1$, represents $X_i - X_0$, first all its x-coordinates and then all its y-coordinates, and the $(n_1 + j)$th column, $j = 1, \ldots, n_2$, similarly represents $Y_j - Y_0$ written out as $2k$ differences of Procrustes coordinates in the same way. The bending energy to be minimized by suitable α's and β's is the quadratic form

$$\mathbf{Y}^t \begin{pmatrix} L_k^{-1} & 0 \\ 0 & L_k^{-1} \end{pmatrix} \mathbf{Y} \equiv \mathbf{Y}^t \mathbf{L}_k^{-1} \mathbf{Y} \tag{8.3}$$

and the minimum is to be taken over the hyperplane $\mathbf{Y} = (Y_0 - X_0) + UT$ of possible differences of weighted composites, where T is the vector $(\alpha_1, \ldots, \alpha_{n_1}, \beta_1, \ldots, \beta_{n_2})$ governing the weighted averaging.

The solution to this familiar generalized or weighted least-squares problem is achieved for parameter vector $T = -(U^t L_k^{-1} U)^{-1} U^t L_k^{-1}(Y_0 - X_0)$. Up to sign, the solution $(Y_0 - X_0) + UT$ is independent of the choice of base vectors X_0 and Y_0 from the groups and from the decision about which set is to be called the X's and which the Y's. For a different application of generalized least squares to the same sort of problem, now modified to take phylogenetic covariances into account, see Rohlf (2000).

To ease graphical interpretation, the algorithm removes the uniform component (Bookstein 1997a) from these deformations before sending them for crease analysis as explained earlier in this chapter.

Appendix 2: how to approximate your own creases

The commercial program package Splus in which all the crease examples in this paper were produced is not as widely available as Jim Rohlf's very convenient series of programs (http://life.bio.sunysb.edu/morph/) for a variety of morphometric manipulations of landmark data. You can approximate creases using Rohlf's program tpsRegr, even though it was designed for regressions, not graphical extrapolations. Instruct the program to carry out a multivariable (not multiple) regression of Procrustes shape coordinates on a dummy variable for the two sides of the contrast of interest: an 'independent variable' that is -0.5 for one group and 0.5 for the other. The uniform component should be omitted, as it complicates the reading of extrapolated grids while affording no possibility whatever of any localized information.

Under this combination of switch settings, the vector of regression coefficients that tpsRegr produces is exactly equal to the non-uniform part of the group mean difference in Procrustes coordinates. To extrapolate the group difference, use the keyed window option (not the slider bar) to produce 'predicted forms' for predictor values much larger than ± 0.5, for instance, ± 10.0. The corresponding deformation grid is likely to resemble the scene at right in Figure 8.9 in many different places: there will be several of these assemblages of paired cusps and folds. Slowly decrement the 'predictor value' backwards towards zero, until all the creases have evaporated. Then slide it back outward until the first one just appears somewhere, then a second, etc. Reverse analyses (e.g., Figure 8.12) are generated in the same screen when the sign of the 'predictor value' is reversed.

Because regression requires forms to stay in linearized Procrustes coordinates, this approximation has no access to the complex parameter $e^{i(\theta_S - \theta_T)}$ of the complete Equation 8.2, the effect of which is to rotate the forms out of the correct Procrustes superposition in order to see the crease most clearly. Without that rotation, creases will be produced by tpsRegr in approximately correct position, and at the approximately correct value of α, but may have a different graphical appearance than the figures here, with a patch of S-shaped grid lines that slew sideways as they cross the crease. Bookstein (2000) shows how to improve this representation, still in the Procrustes superposition, by rotating the Starting form upon its grid.

References

Bookstein, F. L. (1991) *Morphometric tools for landmark data: geometry and biology*, Cambridge: Cambridge University Press.

Bookstein, F. L. (1994) 'Can biometrical shape be a homologous character?', in Hall, B. K. (ed.) *Homology: the hierarchical basis of comparative biology*, San Diego, CA: Academic Press, pp. 197–227.

Bookstein, F. L. (1996) 'Biometrics, biomathematics, and the morphometric synthesis', *Bulletin of Mathematical Biology*, 58, 313–365.

Bookstein, F. L. (1997a) 'Shape and the information in medical images: a decade of the morphometric synthesis', *Computer Vision and Image Understanding*, 66, 97–118.

Bookstein, F. L. (1997b) 'Landmark methods for forms without landmarks: localizing group differences in outline shape', *Medical Image Analysis*, 1, 225–243.

Bookstein, F. L. (1998a) 'A hundred years of morphometrics', *Acta Zoologica*, 44, 7–59.

Bookstein, F. L. (1998b) 'Features of deformation grids: an approach via singularity theory', *Proceedings of the XIV Annual ACM Symposium on Computational Geometry*, New York: ACM Press, pp. 214–221.

Bookstein, F. L. (1998c) 'Singularities and the features of deformation grids', in Vemuri, B. (ed.) *Proceedings of the Workshop on Biomedical Image Analysis, I.E.E.E.*, Los Alamitos, CA: IEEE Computer Society Press, pp. 46–55.

Bookstein, F. L. (1998d) 'Singularities as features of deformation grids', in Wells, W. M., Colchester, A. and Delp, S. (eds) *Medical image computing and computer-assisted intervention, MICCAI '98, Lecture Notes in Computer Science*, Volume 1496, pp. 788–797.

Bookstein, F. L. (2000) 'Creases as local features of deformation grids', *Medical Image Analysis*, 4, 93–110.

Bookstein, F. L. and Green, W. D. K. (1994) 'Edgewarp: a flexible program package for biometric image warping in two dimensions', in Robb, R. (ed.) *Visualization in biomedical computing, S.P.I.E. Proceedings*, Volume 2359, pp. 135–147.

Bookstein, F. L., Schäfer, K., Prossinger, H., Seidler, H., Fieder, M., Stringer, C., Weber, G., Arsuaga, J., Slice, F. D., Rohlf, F. J., Recheis, W., Mariam, A. and Marcus, L. (1999) 'Comparing frontal cranial profiles in archaic and modern Homo by morphometric analysis', *The Anatomical Record – The New Anatomist*, 257, 217–224.

Bruce, J. W. and Giblin, P. J. (1992) *Curves and singularities*, 2nd edn, Cambridge: Cambridge University Press.

Davatzikos, C., Vaillant, M., Resnick, S. M., Prince, J. L., Letovsky, S. and Bryan, R. N. (1996) 'A computerized approach for morphological analysis of the corpus callosum', *Journal of Computer Assisted Tomography*, 20, 88–97.

DeQuardo, J. R., Bookstein, F. L., Green, J W. D., Brumberg, K. and Tandon, R. (1996) 'Spatial relationships of neuroanatomic landmarks in schizophrenia', *Psychiatry Research: Neuroimaging*, 67, 81–95.

Dryden, I. L. and Mardia, K. V. (1998) *Statistical shape analysis*, New York: Wiley.

Good, P. (2000) *Permutation tests*, 2nd edn, New York: Springer.

Goodall, C. R. (1991) 'Procrustes methods in the statistical analysis of shape', *Journal of the Royal Statistical Society*, **B53**, 285–339.

Koenderink, J. (1990) *Solid shape*, Cambridge, MA: MIT Press.

O'Higgins, P. (2000) 'The study of morphological variation in the hominid fossil record: biology, landmarks and geometry', *Journal of Anatomy*, 197, 103–120.

Poston, T. (1978) *Catastrophe theory and its applications*, San Francisco, CA: Pitman.

Prossinger, H., Schafer, K. and Seidler, H. (2000) 'Reemerging stress: supraorbital torus morphology in the mid-sagittal plane?', *The Anatomical Record – The New Anatomist*, **261**, 170–172.

Rohlf, F. J. (1998) 'On applications of geometric morphometrics to studies of ontogeny and phylogeny', *Systematic Biology*, **47**, 147–158.

Rohlf, F. J. (2002) 'Geometric morphometrics and phylogeny', in MacLeod, N. and Forey, P. (eds) *Morphology, shape, and phylogeny*, London: Taylor and Francis, pp. 175–193.

Xu, C., Pham, D. L. and Prince, J. (2000) 'Image segmentation using deformable models', in Sonka, M. and Fitzpatrick, J. M. (eds) *Medical image processing and analysis. Handbook of medical imaging*, Volume 2, Bellingham, Washington: SPIE Press, pp. 129–174.

Chapter 9

Geometric morphometrics and phylogeny

F. James Rohlf

ABSTRACT

This chapter reviews some of the important properties of geometric morphometric shape variables and discusses the advantages and limitations of the use of such data in studies of phylogeny. A method for fitting morphometric data to a phylogeny (i.e., estimating ancestral states of the shape variables) is presented using the squared-change parsimony estimation criterion. These results are then used to illustrate shape change along a phylogeny as a deformation of the shape of any other node on the tree (e.g., the estimated root of the tree). In addition, a method to estimate the digitized image of an ancestor is given that uses averages of unwarped images. An example dataset with 18 wing landmarks for 11 species of mosquitoes is used to illustrate the methods.

Introduction

The relatively new field of geometric morphometrics represents an important new paradigm for the statistical study of shape variation and its covariation with other variables. Rohlf and Marcus (1993) give a general overview of the field and Bookstein (1991) supplies a more technical account. Marcus *et al.* (1996) include both introductory material and many examples of applications to biology and medicine. Dryden and Mardia (1998) give a comprehensive coverage of shape statistics and Small (1996) covers some of the important properties of shape spaces. Rohlf (1999a) gives an overview of some of the relationships between shape statistics and the shape spaces on which they are based. The fundamental advances of geometric morphometrics over traditional approaches include the way one measures the amount of difference between shapes (using Procrustes distance), the elucidation of the properties of the multidimensional shape space defined by this distance coefficient, the development of specialized statistical methods for the study of shape, and the development of new techniques for graphical representations of the results.

Traditional morphometric approaches are based on the application of standard multivariate analyses of arbitrary collections of distance measures, ratios, and angles. These variables typically represent only part of the information that may be obtained from the relative positions of the landmarks on which these measurements are based. For example, they do not take into account information about the spatial relationships among the measured variables. Intuitively, one expects methods that are able to take such additional information into account to have greater statistical power. Traditional methods also only allow one to visualize statistical relationships either numerically or as scatter plots, not as estimates of the shapes themselves.

Shape is a function of the relative positions of morphological landmarks.[1] Mathematically, shape consists of those properties of landmark coordinates that are invariant to the effects of object size, location, and orientation. If suitable landmarks are available, the simplest method to capture a shape is to record the coordinates of those landmarks and then mathematically remove the effects of variation in size, location, and orientation. Landmarks must be sufficient to capture the shape of the structure of interest (some shapes are much easier to deal with in this way than others). When landmarks are not sufficient, one can also include points around partial or complete outlines (Bookstein 1996c) but their use is beyond the scope of the present paper. The points must, of course, indicate the location of the same anatomical feature on different specimens. Thus, the structures must be homologous in some sense. Landmarks are simply points used to track the changes in shape of some structure of interest. It is not assumed that the partial warps or other mathematical functions of the coordinates of these points are homologous. Bookstein (1994) discusses some of the problems of considering shape variables as being homologous characters.

More than one approach to geometric morphometrics has been proposed. This is perhaps not surprising given the history of the development of *ad hoc* approaches in morphometrics. However, there is growing evidence (Bookstein 1996a; Rohlf 1996a, 2000a,c) that only methods based on Kendall's shape space can be rigorously applied to a broad variety of applications, have the best statistical power, and impose minimal constraints on the patterns of variation that can be detected. Bookstein (1996a) refers to this realization as the 'morphometric consensus'.

There has been considerable interest in the ways in which these new geometric morphometric methods might be used to solve problems in systematics. Applications such as developing more powerful discriminators and visualizing the key differences in shape do not seem controversial. The use of cluster and ordination techniques on shape data in order to search for structure within a collection of specimens is also straightforward if one is careful to avoid methods that distort the shape space (Rohlf 2000a). Avoiding distortion is especially important if morphometric data are to be used in ontogenetic studies and to estimate phylogenies.

The next section provides an overview of some of the methods used in geometric morphometrics. An understanding of the properties of shape variables and the shape spaces they define is needed in order to appreciate how they can be used in practical systematic applications.

Shape variables and multivariate spaces

The data for each specimen consists of a $k \times p$ matrix of coordinates, where p is the number of landmark points and k is the dimensionality of the physical space within which the objects are digitized ($k = 2$ or 3). For simplicity, the account given

---

[1] The analysis of shapes of outlines are also part of geometric morphometrics but will not be covered here.

below considers just the two-dimensional case. It is often convenient to treat the kp coordinates as a single row vector with kp elements. The order of the elements is arbitrary ($x_1, y_1, x_2, y_2, \ldots, x_p, y_p$ will be assumed here). A sample of n specimens may then be represented conveniently as a matrix with n rows and kp columns, that is, as points in a kp-dimensional space.

However, these raw coordinates contain information about the size, location, and orientation of each specimen. This irrelevant information can be eliminated by optimally superimposing the specimens onto a standard reference shape. Specimens are superimposed by first centering them on the origin, scaling them to unit centroid size (square root of the sum of their squared coordinates, Sneath 1967, Gower 1971, and Bookstein 1991: 93–95), and then rotating to align them with the reference shape so that the square root of the sum of squared differences between the corresponding landmarks is as small as possible. The minimized quantity, often called a Procrustes distance d, has often been used to measure the amount of difference between pairs of biological shapes (e.g., Sneath 1967). As discussed below, there is a related quantity, ρ, to which this term is also applied. Note that reflections are not permitted when rotating unless one knows that the coordinates for a particular specimen are reflected relative to the coordinates of the other specimens (e.g., a right wing in a study where most other specimens are represented by left wings). This is because genuine shape differences may appear to be reflections (see Goodall 1991 and Rohlf 1996).

An average can be defined as the shape whose sum of squared Procrustes distances to the other specimens is minimal. It is also the maximum-likelihood estimate for the average shape in certain statistical models (Dryden and Mardia 1993; Kent 1994). This average shape may be computed using an iterative procedure that has been called generalized least-squares (GLS) Procrustes superimposition method, as described by Gower (1975) and Rohlf and Slice (1990). This method is now called generalized Procrustes analysis (GPA), since it is not what is now commonly called a generalized least-squares procedure in the statistical literature (e.g., McCullagh and Nelder 1989). Weighted means can also be used (Goodall 1991, is an example). The average configuration is usually scaled to have unit centroid size and it is convenient to align the average configuration to its principal axes to give it a standard orientation.

The GPA procedure produces a transformed dataset in which each specimen is aligned to the reference shape (usually the average shape). The matrix of aligned specimens has interesting geometric properties (Rohlf 1999a). The Euclidean distance between the aligned specimens and the reference (both with unit centroid size) is a partial Procrustes distance d_P (Dryden and Mardia 1998) where size is not adjusted to minimize the Procrustes distance. Because shapes correspond to points in a curved shape space, it is natural to consider measuring distance as a geodesic or great circle distance, ρ. These Procrustes distances are related as $\rho = 2 \sin^{-1}(d_p/2)$, with $0 \leq \rho \leq \pi/2$ radians.

Kendall (1984) worked out some of the geometric properties of the space implied by the use of this distance as a metric (the space is now called Kendall shape space, Small 1996). While the GPA procedure and the methods discussed below can easily be carried out for three-dimensional coordinates, their geometry is more complicated and will not be discussed here. Dryden and Mardia (1993) and Small (1996) also address some of the properties of shape space for three-dimensional data.

Special statistical methods (rather than the usual linear multivariate methods) are required to take into account the non-Euclidean geometry of the shape spaces mentioned above. When variation in shape is sufficiently small it is possible to make a good linear approximation to the space and then use standard multivariate methods (Kent 1994) to test hypotheses. The resulting space is of the same dimensionality as Kendall shape space and may be viewed as tangent to it. The reference shape corresponds to the point of tangency. A linear approximation will, of course, be best when the point of tangency is taken as close as possible to the positions of the points that will be used in an analysis (that is why the average shape is usually used as the reference shape). The projections of the points corresponding to the observed shapes are used for subsequent statistical analyses. Thus, one of the first things to investigate in a practical application is whether the observed variation in shape is sufficiently small that the distribution of points in the tangent space may be used as a satisfactory approximation to their distribution in shape space. A direct method for investigating this is simply to plot Euclidean distances between all pairs of points in the linear tangent space against their Procrustes distances in curved shape space. An approximately linear relationship with a slope close to unity implies that one may satisfactorily use the tangent space to approximate shape space for these data. The tpsSmall software (Rohlf 1998b) performs these computations. In practice the fit is usually very good (I am not aware of any cases of a poor fit except when some specimens were inadvertently reflected).

Multivariate statistical analyses are usually performed using measurements on suites of variables rather than directly on points in a multidimensional space. There are several approaches that can be used to generate variables from shape spaces, two of which are described below.

Kendall tangent space coordinates (Kent 1994), $\mathbf{V}'$, are computed as

$$\mathbf{V}' = \mathbf{X}' - \mathbf{I}_n \mathbf{X}_c, \tag{9.1}$$

where $\mathbf{X}'$ is the projection of points in a space orthogonal to the reference using

$$\mathbf{X}' = \mathbf{X}(\mathbf{I}_{kp} - \mathbf{X}_c^t \mathbf{X}_c), \tag{9.2}$$

where $\mathbf{X}$ is the $n \times kp$ matrix of aligned specimens (each centered on the origin and scaled to unit centroid size), $\mathbf{I}_{kp}$ is a $kp \times kp$ identity matrix, $\mathbf{X}_c$ is the reference (also centered on the origin and scaled to unit centroid size) as a row vector of kp elements, and the superscript t indicates matrix transpose. Matrix $\mathbf{V}'$ will be at most of rank $kp - k - 1 - k(k-1)/2$.

Kent (1994) suggests that one may use these shape variables in standard multivariate analyses if the data are concentrated in a relatively small region of shape space. Shapes close to the reference shape map to points near the origin and maximally dissimilar shapes map to points at a distance of 1 from the origin. Multivariate analyses using these variables may run into difficulties because their covariance matrix will be singular. This singularity results from the rank of the matrix being less than the number of shape variables. If this is taken into account – for example by using generalized inverses – then the results will be identical to those obtained using the next approach (see Rohlf 1999a).

Partial warp scores including the uniform component (Bookstein 1991, 1996b) are the basis for another approach. These shape variables partition shape variation into uniform (infinite scale) and non-uniform (local deformation) components. The former has two dimensions for two-dimensional data and five dimensions for three-dimensional data. The latter have $2p - 6$ dimensions for two-dimensional data and $3p - 12$ for three-dimensional data. The uniform component is best estimated using the linearized Procrustes method of Bookstein (1996b). For two-dimensional data, the uniform component scores may be given by $U = V'T$, where

$$T^t = \begin{pmatrix} \sqrt{\frac{\alpha}{\gamma}}y_1 & \sqrt{\frac{\gamma}{\alpha}}x_1 & \sqrt{\frac{\alpha}{\gamma}}y_2 & \sqrt{\frac{\gamma}{\alpha}}x_2 & \cdots & \sqrt{\frac{\alpha}{\gamma}}y_p & \sqrt{\frac{\gamma}{\alpha}}x_p \\ -\sqrt{\frac{\gamma}{\alpha}}x_1 & \sqrt{\frac{\alpha}{\gamma}}y_1 & -\sqrt{\frac{\gamma}{\alpha}}x_2 & \sqrt{\frac{\alpha}{\gamma}}y_2 & \cdots & -\sqrt{\frac{\gamma}{\alpha}}x_p & \sqrt{\frac{\alpha}{\gamma}}y_p \end{pmatrix}, \qquad (9.3)$$

and x and y are the coordinates of the landmarks in the reference (which has been aligned to its principal axes so that $\sum x_i y_i = 0$, $\alpha = \sum x_i^2$, and $\gamma = \sum y_i^2$). The matrix U has n rows and two columns. The three-dimensional case is somewhat more complicated and explicit equations have not yet been fully worked out (see Bookstein 1996b).

The non-uniform shape component may be decomposed to partial warps (Bookstein 1991) and used as shape variables. These are based on the thin-plate spline and are described in Bookstein (1991), Rohlf (1993), and Rohlf (1998a). This spline can be used to represent shape differences as a smooth deformation of a reference shape into another shape. Partial warp scores (projections) are computed as

$$W = V(E \otimes I_k), \qquad (9.4)$$

an $n \times 2(p - 3)$ matrix where E contains the first $p - k - 1$ columns of the matrix of normalized eigenvectors of the bending energy matrix (see below), I_k is a $k \times k$ identity matrix, and $\otimes$ is the Kronecker tensor product operator. The order of operations differ from that given in Rohlf (1993) because it was assumed there that all the x coordinates were given first followed by the y coordinates. The bending energy matrix is the upper left $p \times p$ block of L^{-1}, where L is a $(p + 3) \times (p + 3)$ matrix which is a function of the reference and is defined in Bookstein (1991). The U and W matrices are orthogonal to each other. Together,

$$W' = (W \mid U), \qquad (9.5)$$

they have $2p - 4$ columns which span the tangent space.

While the decomposition of shape variation into components at different spatial scales is mathematically elegant, one should be careful in how one interprets it biologically. The decomposition is relative to the selection and configuration of landmarks in the reference shape. Unlike many types of multivariate ordination analyses, it is not based on any information about covariation among shape changes in the data. The addition of a landmark can result in what was a uniform shape change becoming a local shape change and a landmark deletion can transform a local shape change into a uniform shape change. Even differences in the relative positions of the landmarks in the reference result in changes in the spatial scales to which variation is assigned. One must also be careful not to interpret the partial warp variables individually (e.g., $1x$, $1y$, etc.) since a change in the orientation of the reference will cause a change in all of the partial warps (geometrically, they also rotate at each spatial scale). Despite these limitations, partial warps are very useful as a set of non-redundant geometrically orthogonal axes that can be used as shape variables that capture all possible shape variation for a given set of landmarks.

If one wishes to analyze landmarks located on more than one structure (e.g., on two structures that articulate), then one can perform the above computations on each structure separately and then append the resulting $\mathbf{W}'$ matrices (Adams 1999a,b).

Fitting shape data to a phylogeny

Given an estimated phylogeny, several methods could be used to estimate the shapes corresponding to the internal nodes of the tree (the hypothetical taxonomic units, HTUs). However, it is important that the methods produce estimates of shape that are invariant to the effects of variation in the orientation of the specimens or to rotations of the tangent space. Procrustes superimposition removes the effects of variation in orientation by superimposition of all specimens onto a reference shape that is set at some particular orientation.

Methods of statistical analysis should not give different results dependent upon different choices for the orientation of the reference shape. This means that the usual linear parsimony method (Farris 1970) should not be used to estimate ancestral states since computations minimizing Manhattan distances are not invariant to the effects of rotation (Rohlf 1998a). The squared-change parsimony method described by Huey and Bennett (1987) and Maddison (1991) is a simple method that satisfies this important constraint. The maximum-likelihood method for continuous characters (Felsenstein 1988) also has this property of invariance. For an evolutionary model based on normally distributed Brownian motion it yields the identical estimates for the ancestral states (Maddison 1991; Martins 1999).

In the squared-change parsimony method ancestral states are estimated such that the sum of the squared branch lengths are minimized over a phylogenetic tree. Huey and Bennett (1987) and Maddison (1991) noted that the estimates of the character values for an internal node is simply the average of the character values of the adjacent nodes. This is because a mean minimizes a sum of squared deviations (Sokal and Rohlf 1995). Of course, these computations are complicated by the fact that the character states of one or more of the connecting internal nodes will also have to be estimated so that an iterative algorithm has been used. However, McArdle and Rodrigo (1994) presented a convenient matrix-based algorithm to simultaneously estimate the character states for all the internal nodes on a tree. This algorithm for an unrooted tree is described below.

Following McArdle and Rodrigo (1994), the matrix of estimated ancestral states is computed as follows. First, define matrix $\mathbf{M}$ as a $(n + n_I) \times (n + n_I)$ connectance matrix, where n is the number of terminal nodes (operational taxonomic units, OTUs) and n is the number of internal nodes (maximally $n - 2$). If nodes i and j are directly connected in the tree then m_{ij} is equal to the reciprocal of the length of the branch connecting them (if estimates of branch lengths are not available then they are treated as all of unit length). All other elements of $\mathbf{M}$ are set to zero. The obvious problem with zero-length branches can be handled by replacing any zero-length branches with multifurcations. Then define $\mathbf{M}_A$ as the $n_I \times (n + n_I)$ matrix consisting of the last n_I

rows of matrix **M**. Matrix $\mathbf{M}_{A(T)}$ contains the first n columns of $\mathbf{M}_A$ (for the terminal nodes) and $\mathbf{M}_{A(I)}$ contains the last n_I columns of the $\mathbf{M}_A$ (for the internal nodes).

The diagonal elements of the $n_I \times n_I$ coefficient matrix **C** are defined as the row sums of matrix $\mathbf{M}_A$ and the off-diagonal entries are equal to the corresponding entries of $\mathbf{M}_{A(I)}$, but multiplied by -1 (i.e., they are the negative reciprocals of branch lengths between all pairs of internal nodes). Finally, a matrix of the estimates of the ancestral states is given by

$$\hat{\mathbf{Z}} = \mathbf{C}^{-1}\mathbf{M}_{A(T)}\mathbf{Z}, \tag{9.6}$$

where **Z** is a matrix with n rows and each column corresponding to a shape variable (the **V′** or **W′** matrices as defined above to yield the $\hat{\mathbf{V}}'$ or $\hat{\mathbf{W}}'$ matrices). Thus, the estimated states are computed as a weighted average of the states in the OTUs by pre-multiplying a matrix of shape variables by the $n_i \times n$ matrix $\mathbf{C}^{-1}\mathbf{M}_{A(T)}$.

The procedure used here differs from that used by McArdle and Rodrigo (1994) in how a rooted tree is treated. When estimates of branch lengths are not available (i.e., unit length branch lengths are used), their procedure estimated separate evolutionary steps along both branches connected to the root. In effect, this doubles the length of the branch in which the root is placed. As noted by Maddison (1991), this results in different estimates of the ancestral states that can change the length of the tree itself. This seems undesirable and is not consistent with the other methods of phylogenetic inference such as linear parsimony or maximum-likelihood. A simple solution is to follow Huey and Bennett (1987), use an unrooted tree in the computations, and then root the tree afterwards for display. This, then, yields the identical estimate for the root as obtained using the method of independent contrasts and GLS (see Garland and Ives 2000; Rohlf 2001). Estimates of the ancestral states for the root can then be computed using interpolation between the nodes at the two ends of the branch where the root is placed. This strategy also has the advantage that matrix **C** will be square and not require any adjustments to ensure that it will be non-singular. The matrix of coefficients, $\mathbf{C}^{-1}\mathbf{M}_{A(T)}$, is augmented to include an initial row corresponding to the root HTU.

Visualizations

Conventional multivariate statistical analyses usually provide scatter plots that allow one to visualize the patterns of variation and covariation in a dataset to the extent that they are adequately summarized in a few dimensions. Geometric morphometric methods enable additional visualizations of the results of multivariate analyses. Points can be visualized as shapes and vectors can be visualized as a sequence of shape changes. This is possible because points in the multivariate space can be mapped to a corresponding position in tangent space and from there back to a set of landmark coordinates in the physical space of an organism. The visualization computation is easy to do because the transformations are linear even though they correspond to non-linear shape deformations (Rohlf 1999a). In those analyses that include a projection into a lower-dimensional space (e.g., when one retains only the first few principal components analysis (PCA) or CVA axes) some information is lost, but the shape can still be estimated by assuming that the projection of a shape onto the discarded dimensions are equal to the mean (zero) for that dimension (Rohlf 1993). Similarly, one can visualize the shapes corresponding to the interior nodes of a tree from the estimated values for the shape variables for the interior nodes. Shapes corresponding to positions along a branch can then be visualized by interpolating the values of the shape variables for the nodes at each end of the branch (interpolating using the reciprocals of the distances to the endpoints of the branch).

Given an estimated shape expressed in terms of Kendall tangent space coordinates, $\mathbf{V}'$, the matrix, $\mathbf{X}$, of coordinates of the landmarks can be computed using the relationship

$$\mathbf{X} = \mathbf{V}' + \cos(\rho)\mathbf{X}_c, \qquad (9.7)$$

where ρ is the Procrustes distance from the shape to the reference. Because $\cos(\rho)$ is usually just slightly less than 1.0 and $\mathbf{V}'$ is approximately a deviation of an aligned shape from the reference, we are approximately just adding the reference back in.

If the shape was expressed in terms of partial warps, then the matrix of landmark coordinates is given by

$$\mathbf{X} = \mathbf{W}'(\mathbf{E} \otimes \mathbf{I}_k \,|\, \mathbf{T})^{-1} + \cos(\rho)\mathbf{X}_c. \qquad (9.8)$$

The columns of the $(\mathbf{E} \otimes \mathbf{I}_k \,|\, \mathbf{T})$ matrix are orthogonal and of unit length so the matrix inverse can be implemented as a simple matrix transposition. Because the $\mathbf{V}'$ and $\mathbf{W}'$ matrices differ by only a rotation and a projection to eliminate dimensions in which there is no variation, both approaches yield the same visualizations.

Because shape differences can often be subtle, it is sometimes helpful to include other information in a plot of the coordinates of an estimated shape. For example, a standard technique is to show the estimated shape as a thin-plate spline that warps the reference shape into the estimated shape. This can make it easier to detect regions of expansion, contraction, or other deformations of the landmarks needed to warp the reference into the estimated shape. Use of the thin-plate spline also allows one to exaggerate the differences if they are very small and hard to see. However, in phylogenetic studies the reference shape may not have any special significance and it may be much more interesting to show an estimated shape as a deformation from some ancestral starting form (i.e., show a thin-plate spline warping the estimated shape of an ancestor into that of a descendant).

A limitation of the plots described above is that a set of points representing the locations of the landmarks does not provide a very realistic depiction of the part of the organism being studied. This can sometimes make it difficult to remember the relationships between the configuration of points and the actual structures that they represent. A solution is to include a digitized image of the organism in the background of the plot.

Of course, the image must be registered with respect to the locations of the landmarks. Bookstein (1991) gives an algorithm to construct an average image corresponding to the landmark locations in the reference shape. This is done by transforming the image of each specimen to create images with landmarks that align with those in the reference. For the ith specimen, the pixels in an image of the reference are replaced by the pixels they correspond to in the image of the ith specimen. The correspondence is determined by the thin-plate spline that maps the location of each pixel in the reference to a unique location in the image of the ith specimen.

This is called 'image unwarping' because each specimen is treated as a transformation of the reference. The resultant registered images are then averaged pixel by pixel (averaging the RGB intensities separately) to create an average image. The technique can easily be generalized to produce images for any estimated shape such as that corresponding to an internal node of a tree. One simply unwarps the images to the estimated landmark configuration and then averages them. Since the landmark configuration for an internal node is computed as a weighted average of the OTUs (see Equation (9.6)), one could also use a weighted average of the pixels. This would mean that pixels for terminal nodes closer (shorter path length) to an internal node would receive greater weight.

Once one has estimates of the ancestral states of the interior nodes, one can perform ordination analyses (e.g., PCA, or non-metric multidimensional scaling analysis (NMMDSA) Kruskal 1964a,b), that includes for the OTUs and the HTUs. The phylogenetic tree can be represented in the plot by connecting points corresponding to ancestors and their descendants (analogous to the common practice of showing minimum spanning trees in ordinations, Rohlf 1977). This is a more direct approach than that of Rohlf (1981). Assuming that most of the variations can be expressed in a few dimensions, such plots should provide useful visualizations of the estimated evolutionary trajectory through shape space. If the tree is accurate then such plots should give a good overall impression of how the shapes evolved.

Examples of some of these visualizations are given in the next section for a small dataset.

An example

To illustrate the methods presented above, a small example dataset was created consisting of the x, y coordinates of 18 landmarks located on the wings of mosquitoes. One species was used from each of the genera in the study by Harbach and Kitching (1998) that included North American species. The list of 11 species used in the present study is provided in Table 9.1. Images of the wings were scanned from Carpenter and LaCasse (1955) and the locations of the landmarks were digitized using the tpsDig software

Table 9.1 List of species and their codes used in this study

Species	Code
Anopheles	Anop
Aedes	Aedes
Psorophora	Psor
Culex	Cule
Culiseta	Culi
Mansonia	Mans
Orthopodomyia	Orth
Wyeomyia	Wyeo
Uranotaenia	Uran
Toxorhynchites	Toxo
Deinocerites	Dein

184   F. James Rohlf

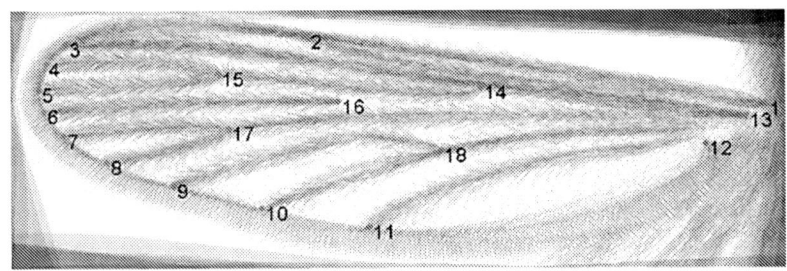

Figure 9.1 Average positions of the 18 landmarks (GLS consensus configuration) superimposed on the average unwarped image of a mosquito wing.

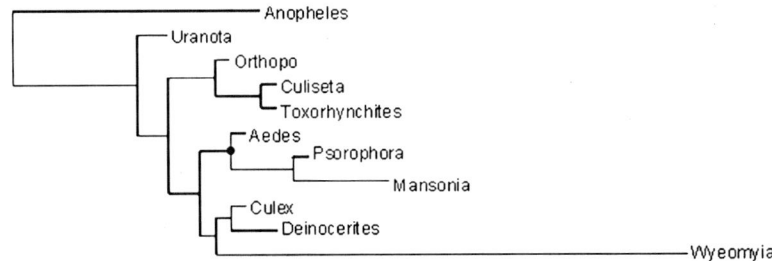

Figure 9.2 Phylogenetic tree extracted from figure 15A of Harbach and Kitching (1998). Branch lengths estimated as the number of character changes reported by them for each node. The location corresponding to the ancestor of the Aedini is indicated by the solid dot (see Figure 9.3).

(Rohlf 1999b). The GLS average locations of the 18 landmarks are shown in Figure 9.1 superimposed on the image of the average unwarped mosquito wing (computed using the tpsSuper software, Rohlf 2000d). This average configuration was used as the reference configuration for the subsequent statistical computations. Figure 9.2 shows a phylogenetic tree extracted from figure 15A of Harbach and Kitching (1998) for the 11 genera included in the present study. The branch lengths are shown proportional to the numbers of character changes they list for each node (taking into account genera included in their study but not included in the present one). Only two of their characters involved the landmarks in the present study. Their character no. 65 is related to the relative location of landmark 15 and character no. 67 is related to the relative location of landmark 11. The other 71 characters were from other parts of the adult and from the pupal and fourth-instar larval stages.

A matrix of partial warp scores (including the uniform component) was computed and the squared-change parsimony criterion was used to estimate the values of these shape variables for all of the interior nodes (except the root) in the tree given in Figure 9.2. The values for the root were then estimated as the weighted average of the values for the nodes at each end of the basal branch. The partial warp scores were then transformed to landmark coordinates using Equation (9.8) in order to display the landmark configurations for the interior nodes (as was done for the root of the tree, one can use interpolation to visualize shapes corresponding to intermediate positions along the branches). Figure 9.3 shows the estimated configuration of landmarks (○)

Geometric morphometrics and phylogeny 185

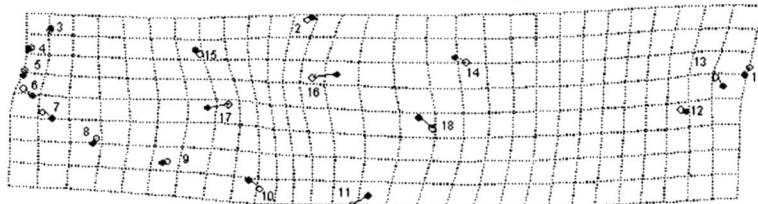

Figure 9.3 Visualization of the estimated shape for ancestor of the Aedini (*Aedes*, *Psorphora*, and *Mansonia*, see Figure 9.2). The grid shows the thin-plate spline transformation from the starting form (●) to the estimated configuration (○). The estimated shape at the root of the tree was used as the starting form.

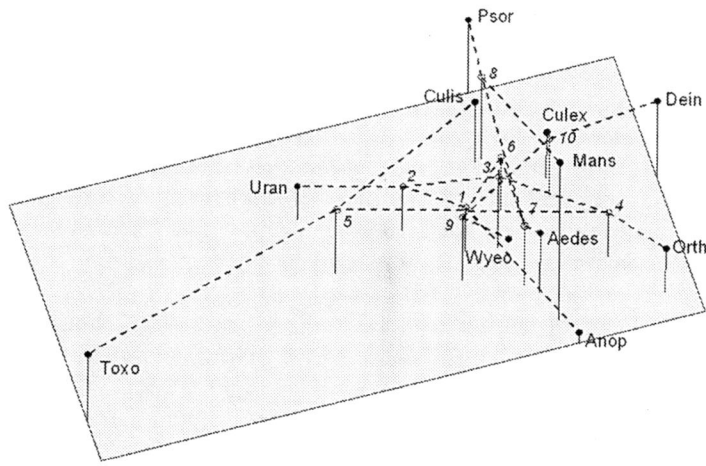

Figure 9.4 Ordination from a non-metric multidimensional scaling analysis of a matrix of distances between all species (●) and estimated internal nodes (○). Phylogenetic tree from Figure 9.2 superimposed using broken lines to link points. OTU codes are given in Table 9.1. Internal nodes numbered in a preorder traversal of the tree beginning with the root. Stress = 0.145. matrix correlation = 0.978.

for the ancestor of the Aedini (HTU 7, the node basal to *Aedes*, *Psorophora*, and *Mansonia*, see Figure 9.2). Since many of the differences among shapes are somewhat subtle, the estimated landmark configuration for the root of the tree is also shown and the differences in position indicated by vectors. The thin-plate spline grid indicates a region of compression with landmarks 2, 15, 16, and 17 moving closer together. These computations were performed using the tpsTree software (Rohlf 2000e).

Figure 9.4 shows a perspective view of a three-dimensional ordination of the relationships among both the OTUs and the estimated internal nodes. Even though the results were similar, a NMMDSA solution is presented rather than the results of a PCA since it achieved a better fit (a matrix correlation of 0.98 rather than 0.92). The PCA solution was used as the initial configuration for the NMMDSA computations. A disadvantage of using a non-metric ordination is that it is not possible to directly compute the shape corresponding to any position in the ordination. Fortunately, in our case PCA solution is similar enough so that it can be used as a guide to estimate the shape changes associated with different directions in the ordination. The OTUs and HTUs are linked together as in Figure 9.2 to show the estimated evolutionary trajectory in shape space. The HTUs are numbered in the order in which they would first be encountered in a pre-order traversal of the phylogenetic tree. These computations were performed using the NTSYSpc software (Rohlf 2000b) on the combined W' and $\hat{W}'$ matrix (partial warp scores for both the OTUs and the HTUs) produced by tpsTree.

Assuming the phylogenetic tree in Figure 9.2 is correct, the ordination indicates a rather complicated and un-parsimonious evolutionary trajectory through shape space. The root and the next two internal nodes (HTUs 1–3) are located near the center of the space, but then there is a large change to the right for HTU 4 (node 51 in Harbach and Kitching 1998) followed by a dramatic shift to the left in HTU 5. One of the descendants of HTU 5, *Toxorhynchites*, is at the extreme left but the other descendant, *Culiseta*, is located to the back right of the diagram. Harbach and Kitching (1998: 354) state that the monophyly of their node 51 is poorly supported though 'not inconceivable'. They also state that if the relationship is real then one must postulate a remarkable divergence of *Toxorhynchites* from its sister group. That conclusion is certainly supported by Figure 9.4. The length of the tree in our shape space would be much shorter if, for example, *Toxorhynchites* and *Uranotaenia* were sister groups. Interestingly, in their discussion of the difficulties of placing the *Toxorhynchites*, Harbach and Kitching (1998) list several characters that have independent occurrences in these two genera.

Wing shape in *Toxorhynchites* is shown in Figure 9.5 as a deviation from the estimated shape for the root of the tree. Much of the change can be described by a region of expansion moving landmarks 15 and 17 away from 16 and 18. The wing is also more elongate which is a uniform shape change. The estimated image for the ancestral mosquito is similar to that of the reference (shown in Figure 9.1). This is expected since the node corresponding to the root of the tree is somewhat centrally located in Figure 9.4 (if the reference were included in a PCA ordination it would be at the centroid of the distribution).

The other unexpected result is the placement of *Weomyia* near the center of the distribution rather than at the periphery as one might expect because of its very long branch in Figure 9.2. *Weomyia* is a derived member of the Sabethini and many Old

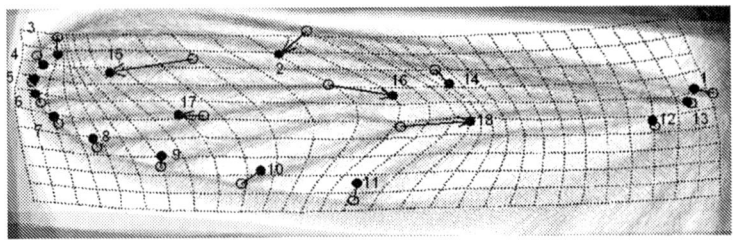

Figure 9.5 Wing shape in *Toxorhynchites* (○) expressed as a deformation of the estimated root (●). The grid represents a thin-plate spline from the root to *Toxorhynchites*. An image of the estimated root (computed as an unweighted average of images unwarped to the shape of the root) is shown in the background.

and New World genera that belong to this lineage were not included in the present study.

Estimating a phylogeny from shape data

There are two approaches that could be used. First, one could just use geometric morphometric methods as exploratory tools to look for conventional characters that could be used along with other information to infer a phylogeny. There is little reason to expect the individual tangent space coordinates or particular partial warps to correspond to the taxonomically most useful or biologically most meaningful variables since they are defined *a priori* and do not take into account any patterns of covariation in the data. It seems likely that visualizations of the results of multivariate techniques such as PCA, canonical variates analysis, or multivariate multiple regression could be quite helpful in discovering useful characters. To maximize the chance of finding useful features one would want to use multivariate techniques that fully search the multivariate shape space (including all possible rotations) and not limit one to studying variation along a single set of axes (Rohlf 1998a).

Alternatively, one could use morphometric shape variables directly as data for a method of phylogenetic inference. The continuous, maximum-likelihood (Felsenstein 1988), squared-change parsimony methods, and neighbor-joining methods are possible approaches because they are able to treat the shape variables as continuous and will produce the same results for different arbitrary orientations of the reference shape. However, it may not be worthwhile to try to estimate a tree using shape variables from only a single structure unless it is very complex and many landmarks can be identified. If the results shown in Figure 9.4 are representative and the phylogeny is correct then it will be very difficult to infer the correct phylogenetic tree from morphometric data.

One could also pool information from more than one structure. Partial warps can be combined from different structures by simply appending them as additional variables. However, it is not clear what their relative weights should be. Should structures with more partial warps (because they have more landmarks) be given greater weight? That would be the effect of simply combining the partial warps for different structures in a single data matrix. Note that one does not need to standardize the partial warp scores before one can combine them as they are already in the same units. This approach corresponds to the separate subsets method developed by Adams (1999a,b) for the analysis of articulated structures.

Discussion

There have been many papers concerned with how one can make use of the new techniques of geometric morphometrics in systematics – especially for phylogeny estimation. Zelditch *et al.* (1992, 1993, 1995), Zelditch and Fink (1995), Swiderski (1993), and Fink and Zelditch (1995) used partial warp scores to search for variables that could be used in cladistic studies. They compared regressions of individual partial warps to obtain discrete characters and used linear (Wagner) parsimony to estimate a phylogeny. This approach has the attraction that it treats shape variables as just additional characters that can be combined with other conventional characters.

Adams and Rosenberg (1998) and Rohlf (1998a) discuss some problems with their approach (see, however, Zelditch *et al.* 1998 for a rebuttal). One of the points which is relevant here is that by reducing continuous shape variables to discrete states one loses the ability to visualize shape change along an estimated clade and to estimate evolutionary trajectories in shape space. A technical point is that one should use methods that ensure that the results are free from artifacts due to arbitrary decisions about the orientation of the specimens. While the partial warps are invariant to variation in the orientation of the specimens, the partial warp scores are influenced by the orientation of the reference. One should not use statistical analyses whose results are sensitive to the arbitrary orientation of the reference.

The studies cited above did not use the mean configuration of landmarks as the reference configuration (which they called the starting form). They usually used the average landmark configuration for juveniles of an outgroup species. The effect of this choice is to degrade their approximation of shape space by the tangent space. This was done because the authors wished to express shape changes as deformations of a logical starting form rather than the overall average shape. As shown above, one can keep these roles separate. The mean shape can be used as the reference to define the tangent space that is used for statistical computations (e.g., regression analysis, estimation of ancestral states, phylogeny estimation, etc.). One is then free to use another shape as the starting form for visualizing shape differences as a deformation using the thin-plate spline. One need not have just a single starting form. Each node above the root could be expressed as a deformation of any of its ancestors. This flexibility may facilitate the biological interpretation of the results.

The studies cited above also emphasize analyses of the partial warps at each spatial scale (i.e., shape differences are described separately corresponding to each principal warp and uniform shape differences are explicitly excluded from consideration). Describing the overall shape differences only in terms of this particular decomposition implies that it should be more interpretable than any other decomposition. As Zelditch *et al.* (1992) point out, because the decomposition of shape variation by partial warps is complete any other complete decomposition would just be a rearrangement of the same information. However, rearrangements of information can be very useful. Just as analyses of individual Fourier harmonics can miss important patterns of covariance in outline shape because they do not happen to correspond to a single harmonic. That is, the pattern does not correspond exactly to a frequency that is $\frac{1}{2}$, $\frac{1}{3}$, $\frac{1}{4}$ etc. of the outline length. Examining harmonics or partial warps individually makes it difficult to detect patterns that do not happen to correspond to a single spatial scale in the decomposition.

The thin-plate spline decomposition is determined by the particular choice of the reference configuration of landmarks and a mathematical model related to the physics of bending infinite sheets of metal – not by any analysis of the empirical patterns of covariance in the data or any biological model of ontogeny or phylogeny. The warps do decompose the overall shape variation into components at different spatial scales, but there is nothing biologically special about any particular scale. Even the distinction between the uniform component and the non-uniform components is somewhat arbitrary. The addition of a landmark can change an infinite scale uniform shape into a local deformation. The solution is to use multivariate methods that take into consideration all possible rotations of the tangent space (e.g., multivariate analysis of variance, canonical variates analysis, etc.). Changes in a single feature or of a single developmental process could result in variation at several spatial scales depending on the selection of landmarks (i.e., it could involve more than one principal warp). Such changes may be difficult to detect if the partial warps are only examined individually.

Richtsmeier *et al.* (1992) distinguish between developmental and evolutionary *trajectories* (a path in a multidimensional morphometric shape space) and developmental and evolutionary *patterns* (the corresponding sequence of shape changes in the physical space of the organism). They stress that the importance of the ability to go from a trajectory to the corresponding pattern and back again because it ensures that "statistical rigor and biological meaning are components of the same approach" Richtsmeier *et al.* (1992: 298). The methods presented in this chapter are able to do this using simple equations to map points in shape space to coordinate configurations (or even estimated two-dimensional or three-dimensional images) in the physical space of an organism and then back again. However, they conclude that comparisons of two trajectories is problematic using coordinate-based approaches and they suggest the use of Euclidean distance matrix analysis (EDMA), methods (see Richtsmeier and Lele 1993, Richtsmeier *et al.* 1998, Lele and Richtsmeier 1991, Lele 1993, Lele and Cole 1996 for a description of these methods and examples of applications to developmental and evolutionary trajectories).

Rohlf (1999a, 2000a) point out a major problem with this approach – that trajectories in the EDMA shape space are greatly constrained by the curved geometry of the space itself. Richtsmeier *et al.* (1992) describe another major problem – points along an estimated trajectory may not correspond to configurations of landmarks that are physically possible when EDMA methods are used. Physical impossibility can be detected by the presence of more than two (for two-dimensional data) or three (for three-dimensional data) non-zero eigenvalues in the PCA used to transform an estimated inter-landmark distance matrix into a set of landmark coordinates. These authors argue (Richtsmeier *et al.* 1992: 299) that this property could be used to "determine the physical boundaries of ontogenetic and evolutionary changes and to answer questions concerning constraints on physical systems". However, the types of physical boundaries that EDMA methods could violate are properties (e.g., the length of a side of a triangle cannot exceed the sum of the lengths of the other two sides). The EDMA algorithm is not capable of detecting whether an estimated configuration of landmarks corresponds to a biomechanically unreasonable structure. It does not seem likely that the detection of physically impossible shapes along an estimated trajectory would lead to new biological insights. The generation of physically impossible shapes is simply an artifact of the EDMA approach. Physically impossible shapes cannot arise with the morphometric methods presented above.

Geometric morphometric shape variables can also be used with the comparative method to estimate correlations adjusted to take into account the effects of non-independence of OTUs in a phylogeny. This can be done by projecting each of the columns of the $\mathbf{V}'$ or $\mathbf{W}'$ matrices onto the independent contrasts (Felsenstein 1985) defined by the phylogeny. More generally, one could use GLS methods as described by Martins and Hansen (1997). Rohlf (2001) compares these approaches and discusses their interrelationships.

An important constraint on the application of these techniques to morphometric data is that the adjusted shape variables must be treated in a way that does not depend upon an arbitrary orientation of the reference and the analyses should be relatively insensitive to small changes in the configuration of landmarks in the reference. This means that one should not estimate correlations between some variable and individual partial warps or Kendall tangent space coordinates. However, one could estimate correlations between a variable and the entire suite of partial warps or perform a two-block partial least-squares analysis (Rohlf and Corti 2000) of the covariation between a set of variables and the entire set of partial warps after correcting for the effects of phylogeny.

Acknowledgements

The helpful critical comments by Leslie Marcus, Dennis Slice, and Michel Baylac are gratefully acknowledged. This work was supported in part by a grant from the Ecological and Evolutionary Physiology (IBN-9728160) program of the National Science Foundation. This paper is contribution no. 1059 from the Graduate Studies in Ecology and Evolution, State University of New York at Stony Brook.

References

Adams, D. C. (1999a) 'Ecological character displacement in *Plethodon* and methods for shape analysis of articulated structures', unpublished Ph. D. dissertation, *Ecology and evolution*, Stony Brook, New York, State University of New York, p. 159.

Adams, D. C. (1999b) 'Methods for shape analysis of landmark data from articulated structures', *Evolutionary Ecology Research*, **1**, 959–970.

Adams, D. C. and Rosenberg, M. S. (1998) 'Partial warps, phylogeny, and ontogeny: a comment on Fink and Zelditch (1995)', *Systematic Biology*, **47**, 168–173.

Bookstein, F. L. (1991) *Morphometric tools for landmark data: geometry and biology*, New York: Cambridge University Press.

Bookstein, F. L. (1994) 'Can biometrical shape be a homologous character?', in Hall, B. K. (ed.) *Homology: the hierarchical basis of comparative biology*, New York: Academic Press, pp. 197–227.

Bookstein, F. L. (1996a) 'Biometrics, biomathematics and the morphometric synthesis', *Bulletin of Mathematical Biology*, **58**, 313–365.

Bookstein, F. L. (1996b) 'A standard formula for the uniform shape component in landmark data', in Marcus, L. F., Corti, M., Loy, A., Naylor, G. J. P. and Slice, D. E. (eds) *Advances in morphometrics*, New York: Plenum, pp. 153–168.

Bookstein, F. L. (1996c) 'Visualizing group differences in outline shape: methods from biometrics of landmark points', in Höhne, K. and Kikinis, R. (eds) *Visualization in biomedical computing. Lecture notes in computer science*, Amsterdam: Springer-Verlag, pp. 405–410.

Carpenter, S. J. and LaCasse, W. J. (1955) *Mosquitoes of North America (north of Mexico)*, Los Angeles and California: University of California Press.

Dryden, I. L. and Mardia, K. V. (1993) 'Multivariate shape analysis', *Sankhya*, **55**, 460–480.

Dryden, I. L. and Mardia, K. V. (1998) *Statistical shape analysis*. New York: Wiley.

Farris, J. S. (1970) 'Methods for computing Wagner trees', *Systematic Zoology*, **19**, 83–92.

Felsenstein, J. (1985) 'Phylogenies and the comparative method', *American Naturalist*, **125**, 1–15.

Felsenstein, J. (1988) 'Phylogenies and quantitative characters', *Annual Review of Ecology and Systematics*, **192**, 445–471.

Fink, W. L. and Zelditch, M. L. (1995) 'Phylogenetic analysis of ontogenetic shape transformations: a reassessment of the piranha genus *Pygocentrus* (Teleostei)', *Systematic Biology*, 44, 344–361.

Garland, T. and Ives, A. R. (2000) 'Using the past to predict the present: confidence intervals for regression equations in phylogenetic comparative methods', *American Naturalist*, 155, 346–364.

Goodall, C. R. (1991) 'Procrustes methods in the statistical analysis of shape (with discussion and rejoinder)', *Journal of the Royal Statistical Society, Series B*, 53, 285–339.

Gower, J. C. (1971) 'Statistical methods of comparing different multivariate analyses of the same data', in Hodson, F. R., Kendall, D. G. and Tautu, P. (eds) *Mathematics in the archaeological and historical sciences*, Edinburgh: Edinburgh University Press, pp. 138–149.

Gower, J. C. (1975) 'Generalized Procrustes analysis', *Psychometrika*, 40, 33–51.

Harbach, R. E. and Kitching, I. J. (1998) 'Phylogeny and classification of the Culicidae (Diptera)', *Systematic Entomology*, 23, 327–370.

Huey, R. B. and Bennett, A. F. (1987) 'Phylogenetic studies of coadaptation: preferred temperature *versus* optimal performance temperatures of lizards', *Evolution*, 41, 1098–1115.

Kendall, D. G. (1984) 'Shape-manifolds, Procrustean metrics and complex projective spaces', *Bulletin of the London Mathematical Society*, 16, 81–121.

Kent, J. T. (1994) 'The complex Bingham distribution and shape analysis', *Journal of the Royal Statistical Society, Series B*, 56, 285–299.

Kruskal, J. B. (1964a) 'Multidimensional scaling by optimizing goodness of fit to a nonmetric hypothesis', *Psychometrika*, 29, 1–27.

Kruskal, J. B. (1964b) 'Nonmetric multidimensional scaling: a numerical method', *Psychometrika*, 29, 28–42.

Lele, S. (1993) 'Euclidean distance matrix analysis: estimation of mean form and form difference', *Mathematical Geology*, 25, 573–602.

Lele, S. and Cole, T. M. III. (1996) 'A new test for shape differences when variance–covariance matrices are unequal', *Journal of Human Evolution*, 31, 193–212.

Lele, S. and Richtsmeier, J. T. (1991) 'Euclidean distance matrix analysis: a coordinate free approach for comparing biological shapes using landmark data', *American Journal of Physical Anthropology*, 86, 415–427.

Maddison, W. P. (1991) 'Squared-change parsimony reconstructions of ancestral states for continuous-valued characters on a phylogenetic tree', *Systematic Zoology*, 40, 304–314.

Marcus, L. F., Corti, M., Loy, A., Naylor, G. J. P. and Slice, D. E. (1996) *Advances in morphometrics*, New York: Plenum.

Martins, E. P. (1999) 'Estimation of ancestral states of continuous characters: a computer simulation study', *Systematic Biology*, 48, 642–650.

Martins, E. P. and Hansen, T. F. (1997) 'Phylogenies and the comparative method: a general approach to incorporating phylogenetic information into the analysis of interspecific data', *American Naturalist*, 149, 646–667.

McArdle, B. and Rodrigo, A. G. (1994) 'Estimating the ancestral states of a continuous-valued character using squared-change parsimony: an analytical solution', *Systematic Biology*, 43, 573–578.

McCullagh, P. and Nelder, J. A. (1989) *Generalized linear models*, 2nd edn, London: CRC Press.

Richtsmeier, J. T., Cheverud, J. M. and Lele, S. (1992) 'Advances in anthropological morphometrics', *Annual Review of Anthropololgy*, 21, 283–305.

Richtsmeier, J. T., Cole, T. M. III, Valeri, C. J. and Lele, S. (1998) 'Preoperative morphology and development in sagittal synostosis', *Journal of Craniofacial Genetics and Developmental Biology*, 18, 64–78.

Richtsmeier, J. T. and Lele, S. (1993) 'A coordinate free approach to the analysis of growth patterns: models and theoretical considerations', *Biological Reviews*, 68, 381–411.

Rohlf, F. J. (1977) 'Classification of *Aedes* mosquitoes using statistical methods', *Mosquito Systematics*, 9, 372–388.

Rohlf, F. J. (1981) 'Spatial representation of phylogenetic trees computed from dissimilarity matrices', in Martinell, J. (ed.) *International Symposium on Concepts and Method in Paleontology*, Barcelona, Spain, Departament de Paleontologia, Univ. de Barcelona, pp. 303–311.

Rohlf, F. J. (1993) 'Relative warp analysis and an example of its application to mosquito wings', in Marcus, L. F., Bello, E. and Garcia-Valdecasas, A. (eds) *Contributions to morphometrics*, Madrid: Museo Nacional de Ciencias Naturales, pp. 131–159.

Rohlf, F. J. (1996) 'Morphometric spaces, shape components and the effects of linear transformations', in Marcus, L. F., Corti, M., Loy, A., Naylor, G. J. P. and Slice, D. E. (eds) *Advances in morphometrics*, New York: Plenum, pp. 117–129.

Rohlf, F. J. (1998a) 'On applications of geometric morphometrics to studies of ontogeny and phylogeny', *Systematic Biology*, 47, 147–158.

Rohlf, F. J. (1998b) 'tpsSmall: is shape variation small?, version 1.11', Department of Ecology and Evolution, State University of New York at Stony Brook. Online. Available <HTTP://life.bio.sunysb.edu/morph>.

Rohlf, F. J. (1999a) 'Shape statistics: procrustes superimpositions and tangent spaces', *Journal of Classification*, 16, 197–223.

Rohlf, F. J. (1999b) 'tpsDig, version 1.18', Department of Ecology and Evolution, State University of New York at Stony Brook. Online. Available <HTTP://life.bio.sunysb.edu/morph>

Rohlf, F. J. (2000a) 'On the use of shape spaces to compare morphometric methods', *Hystrix*, 11, 9–25.

Rohlf, F. J. (2000b) 'NTSYS-PC: numerical taxonomy and multivariate analysis system, version 2.02k', Setauket: Exeter Software.

Rohlf, F. J. (2000c) 'Statistical power comparisons among alternative morphometric methods', *American Journal of Physical Anthropology*, 111, 463–478.

Rohlf, F. J. (2000d) 'tpsSuper: superimposition, version 1.06', Department of Ecology and Evolution, State University of New York at Stony Brook. Online. Available <HTTP://life.bio.sunysb.edu/morph>

Rohlf, F. J. (2000e) 'tpsTree: fitting shapes to trees, version 1.12', Department of Ecology and Evolution, State University of New York at Stony Brook.

Rohlf, F. J. (2001) 'Comparative methods for the analysis of continuous variables: Geometric interpretations', *Evolution*, 55, 2143–2160.

Rohlf, F. J. and Corti, M. (2000) 'The use of partial least-squares to study covariation in shape', *Systematic Biology*, 49, 740–753.

Rohlf, F. J. and Marcus, L. F. (1993) 'A revolution in morphometrics', *Trends in Ecology and Evolution*, 8, 129–132.

Rohlf, F. J. and Slice, D. E. (1990) 'Extensions of the Procrustes method for the optimal superimposition of landmarks', *Systematic Zoology*, 39, 40–59.

Small, C. G. (1996) *The statistical theory of shape*, New York: Springer.

Sneath, P. H. A. (1967) 'Trend-surface analysis of transformation grids', *Journal of Zoology, London*, 151, 65–122.

Sokal, R. R. and Rohlf, F. J. (1995) *Biometry: the principles and practice of statistics in biological research*, 3rd edn, San Francisco: W. H. Freeman.

Swiderski, D. L. (1993) 'Morphological evolution of the scapula in the tree squirrels, chipmunks, and ground squirrels (Sciuridae): an analysis using thin-plate splines', *Evolution*, 47, 1854–1873.

Zelditch, M. L., Bookstein, F. L. and Lundrigan, B. L. (1992) 'Ontogeny of integrated skull growth in the cotton rat *Sigmodon fulviventer*', *Evolution*, 46, 1164–1180.

Zelditch, M. L., Bookstein, F. L. and Lundrigan, B. L. (1993) 'The ontogenetic complexity of developmental constraints', *Journal of Evolutionary Biology*, 6, 121–141.

Zelditch, M. L. and Fink, W. L. (1995) 'Allometry and developmental integration of body growth in a Piranha, *Pygocentrus nattereri* (Teleostei: Ostariophysi)', *Journal of Morphology*, 223, 341–355.

Zelditch, M. L., Fink, W. L. and Swiderski, D. L. (1995) 'Morphometrics, homology, and phylogenetics: quantified characters as synapomorphies', *Systematic Biology*, 44, 179–189.

Zelditch, M. L., Fink, W. L., Swiderski, D. L. and Lundrigan, B. L. (1998) 'On applications of geometric morphometrics to studies of ontogeny and phylogeny: a reply to Rohlf', *Systematic Biology*, 47, 159–167.

Chapter 10

A parametric bootstrap approach to the detection of phylogenetic signals in landmark data

Theodore M. Cole III, Subhash Lele, and Joan T. Richtsmeier

ABSTRACT

A phylogenetic signal is present in a morphometric data set if similarities in form reflect genealogical relationships. The degree to which such a reflection exists can be measured by comparing the topology of a morphometric-based hierarchical clustering with the topology of a cladogram that is specified a priori using other sources of data. A strong phylogenetic signal is indicated by a high degree of agreement between topologies. A lack of agreement is indicative either of data with a strong "alternative" signal (attributable to homoplasy) or of data with a lack of a signal of any kind. In considering the uncertainties inherent in morphometric data, we present a new method for detecting phylogenetic signals when form is described using landmark-coordinate data. We also provide a parametric bootstrapping algorithm that, while applied to landmarks, is general enough to be applied to any sort of morphometric data where a reasonable model of within-sample variation can be specified. We go on to demonstrate how the bootstrap data can be used to make topological comparisons between morphometric clusterings and the cladogram, using: (1) bootstrap proportions attached to cladogram nodes; (2) tree-comparison statistics; and (3) analysis of the frequencies of morphometric-based clusterings that occur when bootstrapping under the model. Our method is exemplified by examining phylogenetic patterning in midfacial shape for ateline primates. We conclude by discussing topics where more research is needed, concentrating on efforts to partition morphometric data into homologous and homoplasious components.

Introduction

Within the past quarter century, the science of comparative biology has undergone a substantial transformation, centered on the advocacy of an explicitly phylogenetic (historical) perspective in the study of evolution. Such a perspective is now considered essential for testing hypotheses about adaptation, the evolution of biological roles, evolutionary covariances among characters, and general principles of organismal design (Gould and Lewontin 1979; Lauder 1981, 1990; Coddington 1988; Wake and Larson 1987; Larson and Losos 1996; Huelsenbeck *et al.* 2000). In addition, a phylogenetic framework is required for the study of the mechanisms underlying evolutionary transformations in form, such as heterochrony (e.g., Fink 1982; Wake and Larson 1988; McKinney and McNamara 1991). Within the same quarter century, there has been a proliferation of new methods in morphometrics, particularly where analyses of landmark data are concerned (Rohlf and Marcus 1993; Bookstein 1996). While applications of morphometrics have varied widely, many of them address evolutionary questions, so that a combination of the "new comparative biology" and the "new morphometrics" seems natural. However, despite their simultaneous development, there has been little synthesis of the two fields, and we perceive substantial theoretical and methodological gaps between them.

The mandates of the new comparative biology should strongly influence the course of morphometric research, as the latter should be regarded as a tool for pursuing the goals of the former. Therefore, we are faced with the task of developing new methods (or retooling existing ones) so that morphometrics become more relevant to researchers who study the history of biological patterns and processes. The purpose of this chapter is to take an initial step in meeting this challenge. We begin with a basic question that has broad relevance: what does it mean for a data set to have (or to lack) a phylogenetic signal? We then propose a method for recognizing a phylogenetic signal in a set of morphometric data. We will be concerned specifically with the analysis of landmark data, although much of what we will present may be applied to other quantitative descriptors of biological form.

What are phylogenetic signals?

In the most basic sense, we say that there is a phylogenetic signal in morphometric data when closely-related taxa are more similar to one another than they are to more distantly-related taxa. Because phylogenies are hierarchically organized, it is useful to refine this basic definition in terms of a comparison between two nested, hierarchical trees. The first tree is a phylogenetic one, represented by a *cladogram* that provides a history of the speciations that gave rise to the clade's member taxa. The second tree is a *phenogram*, which is constructed using the landmark data, where the hierarchical structure is described using a clustering algorithm. The strength of the landmark data's phylogenetic signal is reflected in the degree to which these two trees match. If the topologies of the cladogram and the phenogram are very similar, we can say that there is evidence of a strong phylogenetic signal contained within the landmark data. The morphometric affinities based on the landmark data would therefore reflect the genealogical relationships among taxa. If the two topologies are very different, we can conclude that form has evolved in a way that does not reflect phylogeny.

Now that we have established (in basic terms) what phylogenetic signals are, we can begin to think about how they come about and what they mean. We can also think about what it means if there is a strong, nonrandom signal that does not reflect phylogeny. Finally, we can consider why there may be no apparent signal at all, so that morphometric variation is randomly distributed with respect to phylogeny.

How do phylogenetic signals originate and why are they interesting?

Before discussing phylogenetic signals further, we must state a fundamental assumption. Throughout this chapter, we will assume that there is a preexisting estimate of

196 Theodore M. Cole III et al.

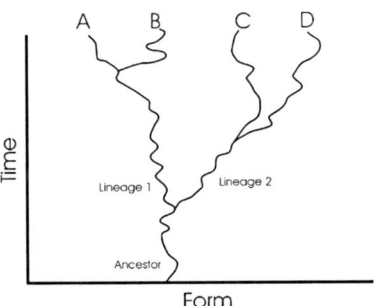

Figure 10.1 Evolution of phylogenetic signal through stochastic processes. The observed differences in form between taxa are functions of the time since they shared a common ancestor (see text).

the phylogenetic relationships among the study taxa (a cladogram) and that this estimate has been made without error (Pagel 1999; Huelsenbeck *et al.* 2000). To reduce the chance of making circular arguments, we also will assume that the cladogram has been estimated using data other than morphometric measurements (e.g., using molecular sequences, developmental patterns, aspects of behavior and life history, etc.). Naturally, as other researchers refine their estimates of the phylogenetic relationships among taxa, we will be faced with the task of revising our own morphometric analyses accordingly.

There are many possible scenarios where organisms evolve so that phylogenetically-patterned form variation is the result. As a first example, let us consider a scenario where natural selection plays no role, so that form evolves solely *via* stochastic (= random) processes (e.g., Felsenstein 1988). If we make some simplifying assumptions, we may find that the history of speciations largely determines the observed differences in form among terminal taxa. More specifically, the expected difference between any two taxa will be a function of the time that has passed since they last shared a common ancestor.[1] Consider the history of a four-taxon clade that is mapped onto a space where the horizontal axis is some measure of form (measured with morphometrics) and the vertical axis is time (Figure 10.1). The common ancestor for the radiation gives rise to two daughter lineages (labelled Lineage1 and Lineage 2), which, in turn, give rise to four terminal taxa. Taxa *A* and *B* arise from Lineage 1, while taxa *C* and *D* arise from Lineage 2. Whenever speciation occurs, we assume that the daughter taxa (whether lineages or terminal taxa) will evolve independently and at random from that point onward (Cavalli-Sforza and Piazza 1975; Cheverud *et al.* 1985; Felsenstein 1985, 1988). We therefore expect taxa *A* and *B* to be the most similar because their evolutionary histories are identical, up to the point where they diverge from their common ancestor. From that point onward, they evolve independently.

---

1 The assumptions we must make in this simple case are that all of the lineages evolve at the same rate and that the forms of sister taxa will tend to diverge following a speciation. From another point of view, these assumptions are the conditions that must hold if phenetic similarities are to be an accurate reflection of genealogical relationships (Colless 1970; Sneath and Sokal 1973; Cavlli-Sforza and Piazza 1975; Felsenstein 1982).

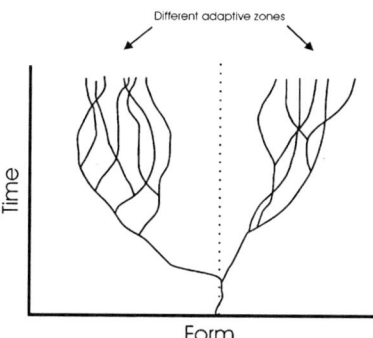

Figure 10.2 A phylogenetic signal that is produced and maintained by natural selection. The different clades radiate within their respective adaptive zones, but no taxa move from one zone into the other. Note that while a phylogenetic signal is present at a high taxonomic level, it becomes obscured at lower levels.

However, relative to the age of the entire clade, they have not had very much time for differences between them to evolve. In contrast, if we consider the difference between taxa *A* and *D*, we see that their last common ancestor is much closer to the base of the clade. Therefore, their common history is proportionately much shorter, and the time over which they have evolved independently is much longer. We thus expect them to have greater morphometric differences, simply because they have had more time to accumulate those differences through random processes.

We can also imagine scenarios where natural selection plays an active role in the production and maintenance of phylogenetic signals. As a simple example, suppose that the taxa in a clade have evolved different forms as specializations to different biological roles (Bock and van Wahlert 1965). Figure 10.2 shows a hypothetical situation where there is an association between organismal form and the occupation of two different adaptive zones. Now suppose that populations of the ancestral taxon for the clade encounter two novel sets of environmental conditions, so that a speciation occurs. Associated with the initial speciation, there may be a morphological divergence between the lineages, which may be especially pronounced if the lineages are entering novel adaptive zones where their respective forms can function as key adaptations to new biological roles (Harvey and Pagel 1991). Following the initial speciation, there may be a number of subsequent speciations in each clade (particularly if a key adaptation is involved), but the members of the two large clades may experience no further selective forces that would tend to force them outside of their respective adaptive zones. Their occupation of those zones is, therefore, very stable over time. As a result, the phylogenetic signal is maintained at a relatively high taxonomic level, even though it may become obscured at lower levels. Harvey and Pagel (1991) refer to this pattern as phylogenetic niche conservatism. They also point out that reversals in the evolution of some complex forms and adaptations may be very unlikely, providing further reinforcement of phylogenetic signals once they have originated. Finally, Simpson (1961) points out that the separation between clades in such a case might be further reinforced by extinctions in the boundary between adaptive zones (i.e., in the valleys between adaptive peaks). Therefore, we might consider extinctions and

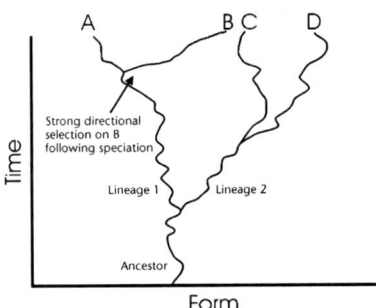

Figure 10.3 Homoplasy that obscures the phylogenetic signal, resulting in an "alternative" signal. While taxa B and C are distant relatives, they have converged on similar morphologies.

the incompleteness of the fossil record as other potential factors that influence the expression of a phylogenetic signal in a data set.

In addition, there may be interesting situations where there is a strong, nonrandom signal in the data that is not a phylogenetic signal. In this discussion, we will call such instances "alternative signals". Researchers who are interested in adaptation and the role of selection may find instances of alternative signals particularly valuable. For example, there may be instances where striking morphometric similarities between distantly-related animals are the result of convergence (Figure 10.3). If such similarities have arisen independently they can allow the construction of testable hypotheses about adaptation (Coddington 1988; Harvey and Pagel 1991; Wake 1991; Brooks 1996; Losos and Larson 1996). Homoplasies might also be frequent enough at lower taxonomic levels to be considered "rampant" in the measurements examined. While striking structural and functional similarities can evolve independently in distantly-related animals (e.g., the well-known similarities between some marsupial and placental carnivores), this pattern is also likely to occur in closely-related taxa with similar developmental programs (Sluys 1989; Brooks 1996; see also Alberch 1980, 1985). When taxon-specific developmental programs are essentially minor variations on the same theme, different taxa are likely to find similar morphological solutions to similar biological problems, obscuring evidence of shared history.

There may also be cases where there is no apparent signal of any kind. One familiar example is "star radiation" (Figure 10.4), where speciations occurred very rapidly, so that the cladogram's nodes are all concentrated near the root. If there have been no further selective forces to cause taxa to evolve in parallel, the taxa will have evolved independently for nearly all of the clade's history. As a result of this speciation pattern, nearly all of the morphometric differences that accumulate between taxa will be autapomorphic, so that "closely-related species are no more likely to be similar than any two species picked at random" (Mooers *et al.* 1999: 250). If autapomorphies are ubiquitous, phylogenetic signals will be very hard to recognize. This will be especially true if each of the lineages experiences very different selective pressures, causing them to follow highly divergent evolutionary pathways.

Finally, there may be difficulties in detecting a phylogenetic signal if morphometric traits exhibit high evolvabilities (*sensu* Houle 1992), so that they are evolutionarily labile. In our initial depiction of the origins of a phylogenetic signal (Figure 10.1),

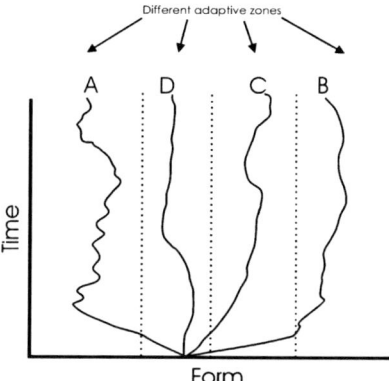

Figure 10.4 A "star radiation". Because the taxa separated very early and have been subjected to different patterns of selection, the accumulation of autapomorphies has completely obscured any type of phylogenetic signal.

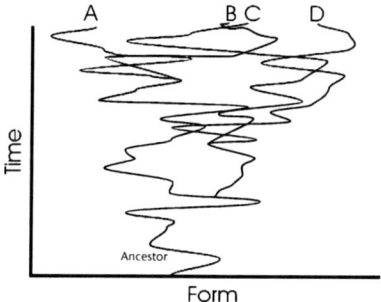

Figure 10.5 Evolutionary lability that has obscured a phylogenetic signal (see text for discussion). The phylogenetic relationships are the same as in Figures 10.1 and 10.3.

the evolution of the clade was very neat and orderly, with the evolutionary paths of the taxa staying well separated in space. However, suppose that each taxon covers a great deal of morphological ground as it evolves. If this is the case, the evolutionary pathways of the individual taxa may cross one another many times, so that form variation becomes randomly distributed with respect to the phylogenetic relationships between taxa (Figure 10.5). The reasons that morphometric data may be labile (so phylogenetic signals are therefore absent) are varied. On one hand, the lability may be an intrinsic characteristic of the organisms themselves. For example, high degrees of within-population genetic variation and low levels of stabilizing selection may both be contributing factors (Houle 1992). On the other, the apparent lability of the traits may not be a quality of the organisms themselves, but of a rapidly fluctuating environment, where the organisms are chasing fast-moving adaptive peaks through the morphological space. Whatever their cause, a particularly interesting aspect of labile data is that the patterns they produce may be very difficult to distinguish from the homoplasy that results from parallel responses to selection (compare the positions of the taxa in Figures 10.3 and 10.5).

To conclude this section, we would like to consider the implications of strong signals (whether phylogenetic or alternative) for the way that we think about morphometric variation in a phylogenetic framework. To begin, we must recognize that morphometric data are phenetic data, and that phenetic similarities are, by definition, mixtures of homologous and homoplasious similarities (Cain and Harrison 1960; Simpson 1961; Sneath and Sokal 1973; Felsenstein 1982; Cheverud et al. 1985). The difficulties of recognizing homologies in morphometric variables prior to construction of a phylogeny are widely recognized, and many investigators have been skeptical of the validity of continuous data as characters for estimating phylogenies (e.g., Pimentel and Riggins 1987; Cranston and Humphries 1988; Chappill 1989; Bookstein 1994). Much recent debate has focused on landmark data in particular (Zelditch et al. 1995, 2000; Fink and Zelditch 1995; Adams and Rosenberg 1998; Rohlf 1998). However, in looking at the distribution of morphometric variation relative to a phylogeny that has already been estimated using other data, we can approach questions about homology and homoplasy from a somewhat different perspective, because we already have a cladogram in place.

Discussions about homology frequently include detailed considerations of terminology, but for the purposes of our presentation we will opt for a fairly simple definition. We would consider a shared "state" of form or shape to be homologous if it is a shared-derived state that characterizes a monophyletic group; this is Patterson's (1982) criterion of congruence (Zelditch et al. 1995; Chang and Kim 1996). If a shared morphometric state is, in fact, homologous, we expect to see congruence between phenetic and cladistic topologies, that returns us to our original definitions of a phylogenetic signal. Therefore, the presence of a strong phylogenetic signal suggests that morphometric similarities tend (in an overall sense) to be homologous, rather than homoplasious. If we judge some shared aspect of form to be homologous, we gain the advantage of discussing that morphometric similarity in terms that are familiar to phylogenetic systematists, including "symplesiomorphy" and "synapomorphy". However, what is most important about the provisional identification of a morphometric homology is that it provides us with a starting point for better understanding the processes that generate evolutionary diversity. As Sanderson and Hufford (1996: 329) succinctly state: "At issue in the study of homology is how character states become different despite their common origin."

In contrast, a strong "alternative" signal tells us something very different about how morphometric similarities tend to evolve from different beginnings (Sanderson and Hufford 1996). To recognize homoplasy in discrete character states, Hennig (summarized by Brooks 1996) recommended that similar states should be provisionally considered as homologous. Following the construction of a cladogram, this assumption is reevaluated for each character, and similarities that are incongruent with the phylogeny are reclassified as homoplasies. For morphometric data, we could make the same assumption initially (i.e., that all morphometric similarities are homologous). However, if we then found a strong alternative signal, we would have evidence that our initial assumption was incorrect. We would then have to consider the possibility that some proportion of the morphometric similarities we observed were homoplasies. We might then be able to construct testable hypotheses about the biological roles of these similarities and the factors that would tend to produce homoplasies of form (e.g., adaptations to similar environments, developmental constraints).

Uncertainty and the bootstrap

Now that we have introduced some basic ideas about what phylogenetic signals are and how they might arise, we can turn our attention to how they can be studied using real data. As we stated earlier, we can evaluate the phylogenetic signal in morphometric data by comparing cladistic and phenetic hierarchical topologies. However, we first need to discuss the nature of the data that are used to construct them.

When we compare a phenogram to a cladogram, we assume that the cladogram is measured without error. However, this assumption cannot be made for morphometric phenograms, because we know that the measurements vary within populations. We also acknowledge that our sample sizes often may be small, particularly when we are studying rare organisms or fossils, so that sampling errors become an important consideration in estimating morphometric affinities. We therefore realize that *uncertainty* in statistical estimation threads its way throughout our study of phylogenetic signals from start to finish. The fact of this uncertainty has led us to use the bootstrap, which is a very versatile method for addressing issues of statistical uncertainty in interesting and informative ways.

The bootstrap technique was developed in the late 1970s and early 1980s by Efron and colleagues (see Efron and Tibshirani (1991, 1993) for reviews). It is a frequently used method for working with statistics that have either very complex or unknown distributions, where intensive computational effort can be used to address problems that might otherwise be intractable (Efron and Tibshirani 1991). Phylogenetic applications of the bootstrap and related methods were introduced soon after the development of the bootstrap itself. As with other bootstrap applications, the first phylogenetic uses were motivated by concerns about the uncertainties of statistical estimation.

Felsenstein (1985) was the first to use the nonparametric bootstrap to address concerns about the uncertainty of sampling discrete characters for use in phylogeny estimation. His method of attaching bootstrap proportions to cladogram nodes is now widely used in the systematics literature (see below). At roughly the same time, Lanyon (1985) used the jackknife (a related method) as a means of dealing with the uncertainties of estimating genetic distances and the phylogenies inferred from them. Mueller and Ayala (1982) had previously suggested the use of the jackknife for this same purpose. More recently, Huelsenbeck *et al.* (1996) used the parametric bootstrap to model variation in DNA sequence data, demonstrating the versatility of the method. Their aims were to examine bias in phylogenetic estimation, to compare the support for competing phylogenetic hypotheses, to conduct power analyses, and to measure the repeatability of a tree's subclades. This last aim is the same as Felsenstein's (1985) and is probably the most frequent application of the bootstrap in phylogenetics.

Before we describe how we have applied the bootstrap in this study, we will provide a brief illustration of how the method generally works in applications other than phylogenetics. Suppose we have measured a sample of organisms using a continuously-distributed variable (e.g., body mass or length), and we want to use the data to estimate a parameter θ, which is a smooth function of the data. In addition to obtaining a point estimate of θ (a statistic called T), we want to say something about our uncertainty in making that estimate. The uncertainty involved in making point estimates is usually quantified using standard errors and confidence intervals. For many familiar statistics (e.g., means, regression coefficients, and correlation coefficients), there are analytical formulae for calculating these uncertainty measures. However, if there are no available analytical methods for the statistics that interest us, we can apply the bootstrap. If the sample has n observations, we can construct a *pseudosample* by drawing n observations from the sample randomly and with replacement. By "randomly", we mean that all observations have the same probability of being selected. By "with replacement", we mean that any given observation can be sampled more than once and that some observations may not be sampled at all. From the pseudosample, we compute a bootstrap estimate of T and call it T^*. This process is then repeated for M independent pseudosamples, where M is a large number (usually between 200 and 1,000), so that we get a *bootstrap distribution* of T^*. Once the bootstrap distribution is obtained, we calculate its mean $\hat{T}^* = \Sigma(T^*)/M$, which is the bootstrap estimate of T. The standard *error* of T is estimated by the standard *deviation* of the T^* estimates, and there are several ways that bootstrap confidence intervals for T can be computed (Efron and Tibshirani 1991, 1993; Davison and Hinkley 1997).

The type of resampling just described is called nonparametric bootstrapping, because no assumptions are made about the distribution of the data. However, suppose we can make reasonable assumptions about how the data are distributed (although the distribution of the statistic of interest may remain very complex or unknown). In that case, we can use parametric bootstrapping, where a fitted parametric model serves as the basis for generating random data sets that can serve as pseudosamples (Efron and Tibshirani 1991, 1993; Huelsenbeck et al. 1996; Davison and Hinkley 1997). Returning to our simple example, suppose we can assume that the measurement data are distributed as $N(\mu, \sigma^2)$; that is, the data are normally distributed with mean μ and variance σ^2. To perform a parametric bootstrap, we first obtain sample estimates of μ and σ^2, called $\bar{x}$ and s^2, respectively. We then generate M independent pseudosamples, each with n random observations that are distributed as $N(\bar{x}, s^2)$. The parametric bootstrap estimates of the mean and standard error of T are then computed as with the nonparametric method. The primary advantage of parametric bootstrapping is that the standard-error and confidence-interval estimates are generally more accurate than nonparametric estimators. This is an especially important concern when studying multivariate data, where the sample sizes required for precise nonparametric estimates increase exponentially with the number of variables (Silverman 1986).

A model for describing morphometric variation using landmarks

An explicit model of within-sample variation is necessary for any application of the parametric bootstrap. Before we describe the statistical model and computations that we use in this study, we will present a more general picture of biological variation in landmark data, which is largely based on Lele (1999). Suppose we are interested in a sample of n organisms and we measure them using a series of K landmarks in D (= 2 or 3) dimensions. The mean for the population is described by a $K \times D$ matrix called **M**, where each row represents the D-dimensional coordinates of a landmark. While **M** is not directly observable, we can imagine the mean configuration of landmarks in a "Nature Space" (Figure 10.6), where within-sample variation arises. No single individual is likely to be identical in form to the mean, nor are any two individuals likely to be identical. This phenotypic variability is due to both genetic and environmental

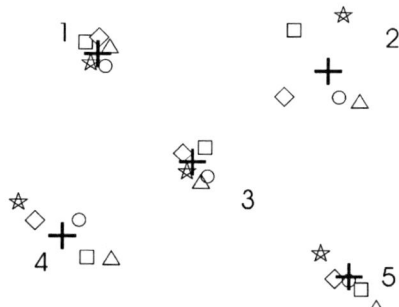

Figure 10.6 The "Nature Space", where individual differences in form originate. The parametric mean configuration for a hypothetical five-landmark organism is indicated by the crosses (+). The filled symbols represent the landmark locations of different specimens (where like symbols belong to the same specimen). These locations are phenotypic perturbations of the mean, that reflect underlying genetic and environmental variations. Note that the dispersion patterns differ from one landmark to the next. Some landmarks have roughly circular distributions (1, 2, and 3), while others are elliptical. Some landmarks (1 and 3) have relatively small dispersions, while others (2) are large. In addition, some of the perturbations may be correlated – note the similarity in the rank-order of perturbations (from upper left to lower right) for landmarks 4 and 5. As described in the text, the positions of the perturbations in Nature Space cannot be reconstructed; however, some descriptors of the dispersion patterns can be estimated.

variation (Falconer and Mackay 1996). In the Nature Space, phenotypic variation is manifested as perturbations around **M** (Figure 10.6). Note that the dispersion patterns of these perturbations can vary in size and shape from landmark to landmark. Some landmarks may be more variable than others and this will be indicated by the relative sizes of their dispersions. The perturbation scatters also can vary in shape from one landmark to another, with some being round and others elliptical. Finally, there may be covariances between the landmarks, so that the relative positions of the observations at one landmark may be correlated with their positions at other landmarks.

We describe the phenotypic variation statistically with a *general perturbation model*. This model was used by Goodall (1991) in the development of superimposition (Procrustes) methods and by Lele (1993; Lele and Richtsmeier 1990; Lele and McCulloch 2001) in the development of Euclidean distance matrix analysis (EDMA). We can describe the landmark data for each observation in a sample with a $K \times D$ matrix called $\mathbf{X}_i$. Each $\mathbf{X}_i$ is related to **M** as follows:

$$\mathbf{X}_i = (\mathbf{M} + \mathbf{E}_i)\Gamma_i + \mathbf{t}_i$$

$\mathbf{E}_i$ is a $K \times D$ matrix of perturbations that describe how $\mathbf{X}_i$ differs from **M** in the Nature Space. For the population, these perturbations are assumed to have a multivariate normal distribution with a $K \times D$ mean matrix **0** and a covariance structure $\Sigma_K \otimes \Sigma_D$, where the $\otimes$ operator denotes a Kronecker product. Σ_K is a $K \times K$ matrix that describes the variances and covariances of the landmarks, while Σ_D is a $D \times D$ matrix that describes the variances and covariances of the perturbations with respect to the

Nature Space's coordinate-system axes (i.e., they describe the eccentricity and orientation of the perturbation scatters). The mean and the variance–covariance matrices are obviously of great biological interest. Unfortunately, they are not estimable because of the presence of the other terms in the equation, which are called *nuisance parameters* (Neyman and Scott 1948; Lele and Richtsmeier 1990; Lele 1993). The orthogonal $K \times K$ matrix Γ_i describes the rotation of $\mathbf{X}_i$ (as we measure it) relative to $\mathbf{M}$ (as it lies in the Nature Space), while the $K \times D$ matrix $\mathbf{t}_i$ describes the translation of $\mathbf{X}_i$ relative to $\mathbf{M}$. Unfortunately, these entirely arbitrary parameters are unobservable, which means that reconstruction of the Nature Space from empirical data is impossible (Lele 1993, 1999; Lele and McCulloch 2001).

Fortunately, there are some biologically interesting components of the model that are identifiable and can be estimated using method-of-moments techniques developed by Lele (1993) and Lele and McCulloch (2001). While we cannot observe directly the coordinates of the population mean form ($\mathbf{M}$), we can compute the coordinates of a consistent sample estimate of the mean, called $\hat{\mathbf{M}}$, up to translation, rotation, and reflection. Similarly, while we cannot estimate the sample among-landmarks variance–covariance matrix (Σ_K), we can compute a consistent estimate of a singular version of it, called Σ_K^*. Finally, while neither the among-axes variance–covariance matrix (Σ_D) nor its eigenvectors is estimable, its eigenvalues can be estimated (Lele and McCulloch 2001), describing the overall eccentricity of the perturbation scatters. Alternatively, we can make the simplifying assumption that $\Sigma_D = \mathbf{I}$ (Lele and Cole 1996). We should emphasize that all of these estimators are *coordinate-system invariant*, so that they are not affected by the positions and orientations of the observations in any arbitrary coordinate system (Lele 1993). If we assume that the landmark perturbations about $\mathbf{M}$ approximate a multivariate normal distribution, we can use $\hat{\mathbf{M}}$ and Σ_K^* to randomly generate pseudosamples under the model (Lele and Cole 1996). This data-generating procedure is at the heart of our parametric bootstrapping method, as described in the following section.

Parametric bootstrapping under the model

We now provide a description of the parametric bootstrapping algorithm that we use to assess a phylogenetic signal. The method is illustrated schematically in Figure 10.7. The particular details of scale adjustments, dissimilarity metrics, and clustering algorithms may vary from one application to another, so we will only speak of them in general terms for the time being.

1 Using the sample-specific estimates of the mean form ($\hat{\mathbf{M}}$ for each sample), compute a matrix of the pairwise dissimilarities in form between taxa. As explained below, this matrix is called $\mathbf{F}_\Omega$ (or $\mathbf{S}_\Omega$ if the data are scale-adjusted). $\mathbf{F}_\Omega$ is used as the basis of a hierarchical cluster analysis, yielding a morphometric-based clustering that is referred to as the "empirical phenogram". If the goal is to study shape, rather than form (where information about scale is retained), the mean forms should be scaled first (see Lele and Cole 1996 for a discussion).

2 Again using the sample-specific estimates of $\hat{\mathbf{M}}$ and Σ_K^*, generate a set of pseudosamples of the appropriate sample sizes. An algorithm for generating random

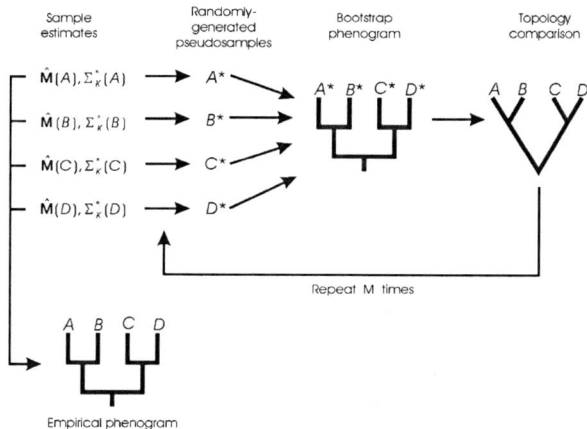

Figure 10.7 Schematic of the parametric bootstrapping algorithm used in this study. Data from four taxa (A, B, C, and D) are used to estimate mean forms ($\hat{M}$) and among-landmark variance–covariance matrices (Σ_K^*). The means are first used to estimate the empirical tree, shown at the bottom of the figure. The means and covariance matrices are then used to generate a large number (e.g., 500) of pseudosamples, assuming multivariate Gaussian perturbations. Each set of pseudosamples is, in turn, used to estimate a bootstrap tree. The intermediate step of computing a dissimilarity matrix is not shown. Finally, each of the bootstrap trees is compared with the cladogram, using bootstrap proportions, tree-comparison statistics, or some other measure of topological similarity.

data under the model is provided by Lele and Cole (1996). After calculating the pseudosample means, construct a matrix of pairwise dissimilarities called F_Ω^* (or S_Ω^*), where the asterisk indicates that it is computed from a set of pseudosamples.

3   Generate *M* sets of pseudosamples by repeating Step 2 a large number of times (e.g., 200–1,000). Use the resulting bootstrap distribution of F_Ω^* (or S_Ω^*) matrices and a hierarchical clustering algorithm to obtain a distribution of bootstrap phenograms.
4   Compare the bootstrap phenograms to the cladogram, examining the bootstrap proportions attached to cladogram nodes or using tree-comparison statistics. We might also examine the frequencies with which different bootstrap phenograms occur when resampling under the model. Each of these methods for comparison is described below.

Constructing phenograms based on morphometric dissimilarity

Given the collection of sample mean forms, we must determine which of the means are most similar and which are most different (Steps 1 and 2 of the bootstrapping algorithm). These similarities and differences serve as the basis for constructing our morphometric clusterings (the empirical and bootstrap phenograms). For landmark data, we can use EDMA (Lele and Richtsmeier 1991, 2001; Lele 1993; Lele and McCulloch 2001), which is a coordinate-system-invariant method of describing and comparing forms. The basis of all EDMA applications is the *form matrix* (**FM**). Suppose we have the mean coordinates of a taxon A, measured with K landmarks. The form matrix **FM**(A) is defined as:

$$\mathbf{FM}(A) = \begin{bmatrix} 0 & d(1,2) & \cdots & d(1,K) \\ d(2,1) & \ddots & & \\ \vdots & & \ddots & d(K-1,K) \\ d(K,1) & & d(K,1-K) & 0 \end{bmatrix}$$

where $d(i,j)$ is the Euclidean distance between landmarks i and j. This representation of form is coordinate-system invariant because the elements of **FM**(A) are always the same, no matter how A is positioned (= translation) or oriented (= rotation and reflection). Now, suppose we have a second taxon B, with its own form matrix **FM**(B), and we want to compare the forms. We can define a *form-difference matrix*, called **FDM**(A, B), as follows:

$$\mathbf{FDM}(A, B)_{ij} = \frac{\mathbf{FM}(A)_{ij}}{\mathbf{FM}(B)_{ij}}$$

where $i, j = 1, \ldots, K$ and with the convention that $0/0 = 0$. Each element of the **FDM** is the ratio of like distances in A and B. If A and B are identical in form, all of the off-diagonal elements of the **FDM** will be one. If a given distance is greater in A than in B, the corresponding element of the **FDM** will be greater than one. Similarly, an instance where the distance in B is greater will be indicated by an element that is less than one. If all of the off-diagonal elements are equal and are different from one, A and B will have the same *shape*, but will differ in *scale*. Finally, if the off-diagonal elements are heterogeneous, A and B will differ in *shape*. Importantly, the **FDM** shares the property of coordinate-system invariance with the form matrices, meaning that its elements are always the same, no matter how either A or B are translated, rotated, or reflected.

In the algorithm described above, we discussed dissimilarity measures in very general terms, but we can now define them more precisely. The form-difference matrix can be used as the basis for a dissimilarity measure, called F_Ω, between two taxa (Richtsmeier et al. 1998). Given **FDM**(A, B), we can calculate:

$$F_\Omega(A, B) = \sqrt{\sum [\ln(\mathbf{FDM}(A, B))]^2}$$

where the summation is over all of the below-diagonal elements. $F_\Omega(A, B)$ is a metric and is equivalent to the Q-mode Euclidean distance between A and B (Sneath and Sokal 1973: 124), when that distance is calculated using all ln-transformed interlandmark distances. If A and B are identical, then F_Ω will equal zero; otherwise, F_Ω will become increasingly positive as A and B become more different in form. If the taxa we are comparing vary substantially in scale, we may want to concentrate on variation in shape, rather than in form. Scaling of each sample mean is accomplished simply by computing the FM and then dividing each element by an appropriate scaling factor (derived from the elements themselves), yielding a shape matrix called **SM** (Lele and

Cole 1996). Some examples of possible scaling factors include any single interlandmark distance (e.g., one that measures maximum length or breadth), the maximum distance, the median distance, or the geometric mean of all of the distances. Whichever scaling factor is chosen, we strongly recommend that the choice should be based solely on biological grounds (Lele and Cole 1996).

Given the means of multiple taxa, we collect all of the pairwise F_Ω statistics in a symmetric dissimilarity matrix called $\mathbf{F}_\Omega$. For example, if there are three taxa called A, B, and C:

$$\mathbf{F}_\Omega = \begin{bmatrix} 0 & F_\Omega(B,A) & F_\Omega(C,A) \\ F_\Omega(A,B) & 0 & F_\Omega(C,B) \\ F_\Omega(A,C) & F_\Omega(B,C) & 0 \end{bmatrix}$$

Alternatively, a dissimilarity matrix based on shape matrices would be called $\mathbf{S}_\Omega$. With such a dissimilarity matrix in hand, we can use it as the basis for hierarchical cluster analysis to obtain both the empirical phenogram and the distribution of bootstrap phenograms.

In outlining our algorithm, we have spoken of hierarchical clustering in very general terms. However, the choice of a clustering algorithm should be carefully considered. There are many different algorithms available for constructing hierarchical clusters and these may yield different results, even when based on the same dissimilarity matrix (Sokal and Rohlf 1962; Sneath and Sokal 1973; Johnson and Wichern 1982). In deciding which method to use, we note that hierarchical clustering methods cannot summarize morphological relations between taxa in a multivariate space without introducing some kind of distortion, and some of the information about pairwise dissimilarities is invariably lost (Sokal and Rohlf 1962; Sneath and Sokal 1973). Therefore, the best choice of a clustering algorithm might be the method that introduces the least distortion and provides the most faithful summary of the information in the dissimilarity matrix. One way of measuring this accuracy is through use of the cophenetic correlation (Sokal and Rohlf 1962), which is the correlation between the elements of the original dissimilarity matrix and those implied by the hierarchical clustering. The "best" algorithmic result, using this criterion, will be the one with the cophenetic correlation that is closest to 1.0. Finally, we would strongly discourage selecting the method that produces the clustering that is most similar in topology to the cladogram; this practice would defeat the purpose of our method by biasing the results toward the detection of a strong signal, where a weaker one may exist in reality.

Measuring the signal

After generating the empirical phenogram and a distribution of bootstrap phenograms, we can compare their topologies to that of the cladogram. Because we are interested in comparing the topologies of two different hierarchical trees, we have a choice of several different approaches. First, we can calculate bootstrap proportions (colloquially referred to as "bootstrap support") for each of the nodes of the cladogram. Second, we can employ any number of tree-comparison statistics for evaluating the agreement between cladistic and phenetic topologies. Finally, we can examine the frequencies of the different topologies that occur under when resampling under the model. Although this last approach does not involve actual tree comparisons, it can be very interesting and informative.

Bootstrap proportions: The distribution of bootstrap phenograms can be used to assign a bootstrap proportion (Felsenstein 1985; Efron *et al.* 1996) to each internal node (= subclade) of the cladogram. Suppose we are interested in the node corresponding to a subclade that includes three taxa *A*, *B*, and *C*. The associated bootstrap proportion is the percentage of bootstrap phenograms where *A*, *B*, and *C* cluster together to the exclusion of all other taxa. Note that there is no consideration of the internal structure of the clade, only of the identity of its member taxa. Note also that bootstrap proportions are *marginal* proportions (Felsenstein 1985), meaning that the proportions for different nodes are calculated independently. To relate bootstrap proportions to a phylogenetic signal's strength, we would say that a bootstrap proportion of 100 percent would indicate a perfect phylogenetic signal for the subclade, while 0 percent would indicate that the subclade's members never cluster together. We can attach bootstrap proportions not only to the cladogram, but to the empirical phenogram as well. In doing so, we get an idea of the repeatability in the data (*sensu* Hillis and Bull 1993), so that we have an explicit picture of the uncertainty in estimating the phenetic clustering. This is similar to the concept of the "robust validity" of distance and correlation matrices that was used by Cheverud *et al.* (1989).

Finally, and as an aside, we would like to make a comment regarding terminology. While bootstrap proportions are popularly referred to as measures of "bootstrap support", we have avoided using that term. "Support" has a very specific statistical definition that is related to principles of likelihood (Edwards 1992), and this definition is different from what we measure using bootstrap proportions. Hillis and Bull (1993) similarly favor the use of "bootstrap proportions", while discouraging the use of "bootstrap *P* values".

Tree-comparison statistics: Tree-comparison statistics are commonly used to measure the degree of difference between two hierarchical structures, expressing this degree as a single number. There are many different statistics available (see reviews by Rohlf (1974, 1982), Hubert (1978), Penny and Hendy (1985), and Lapointe and Legendre (1990)), and bootstrapping can be used to generate estimates of their standard errors and confidence intervals. For this study, we have developed a simple tree-comparison statistic that we find useful because its interpretation is very straightforward. To compare a cladogram with a phenogram, we first represent their topologies in matrix form using *cardinality matrices* (Lapointe and Legendre 1992), where all of the information about branching sequences is retained, but where branch lengths are ignored. The dissimilarity between two taxa *A* and *B* is defined as the total number of taxa in the smallest clade/cluster containing both *A* and *B*. Therefore, if *A* is closest to *B*, the dissimilarity between them will be the minimum possible value of 2 (the size of the smallest possible clade/cluster). If *A* is most distant from *B*, the dissimilarity will be the maximal value, which is equal to the total number of taxa in the study (which is the largest possible clade/cluster). Figure 10.8 provides three examples of how cardinality matrices are constructed. In the third example, note how cardinality matrices can be used with cladograms that have unresolved multifurcations.

To compare the cladogram with the empirical phenogram, suppose we represent the topology of the former with a cardinality matrix C_C, and we represent the topology of the latter with a second cardinality matrix C_P. We can compare the

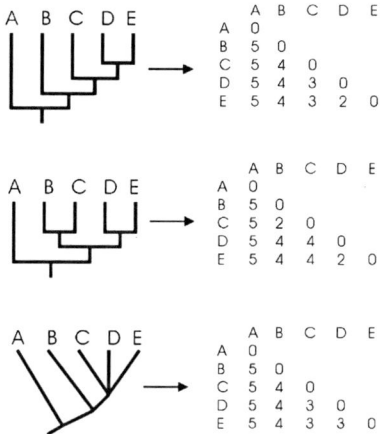

Figure 10.8 Examples of how the topologies of phenograms and cladograms are expressed using cardinality matrices. See text for an explanation of how the matrix elements are defined. For the cladogram at the bottom of the figure, note how unresolved multifurcations can be accommodated.

two cardinality matrices by subtracting one from the other, forming a *cardinality difference matrix* (**CDM**). Suppose we subtract the cladogram topology from the phenogram topology:

$$\mathbf{CDM} = \mathbf{C}_P - \mathbf{C}_C$$

If the two topologies are identical, then all of the elements of **CDM** will equal zero. If the topologies differ, then some or all of the off-diagonals of **CDM** will be non-zero. The elements of **CDM** are useful for defining an overall measure of topological dissimilarity. As one of a number of possibilities, we can use the absolute value of the off-diagonal element of **CDM** that is furthest from zero, calling that number C: $C = \max(\text{abs}(\mathbf{CDM}))$. The minimal value that C can take is 0, indicating that the topologies are identical (a perfect phylogenetic signal). The maximal value that C can take is the total number of taxa minus 2. An advantage of this particular statistic is that it gives us an idea of the *depth* of the topological dissimilarity (moving from the branch tips toward the root). If $C = 1$, then we know that the greatest difference between the topologies is concentrated near the branch tips. More specifically, if $C = 1$, we know that the greatest differences occur within one or more three-taxon subclades. (Note that C does not tell us how many three-taxon subclades differ, only that there is at least one difference of that magnitude.) If C is maximal, we know that the disagreement between topologies extends all the way to their roots.

Suppose we have used the parametric bootstrap algorithm to estimate M bootstrap estimates of C (each called C^*). When we look at the distribution of C^*, we see that it contains two types of information (Figure 10.9). First, the *mode* of the distribution is a measure of the *agreement* between the cladogram and the bootstrap phenograms. The closer the mode is to zero, the greater the agreement tends to be, as described above. Second, the *dispersion* of the distribution is a measure of the *precision* (= repeatability)

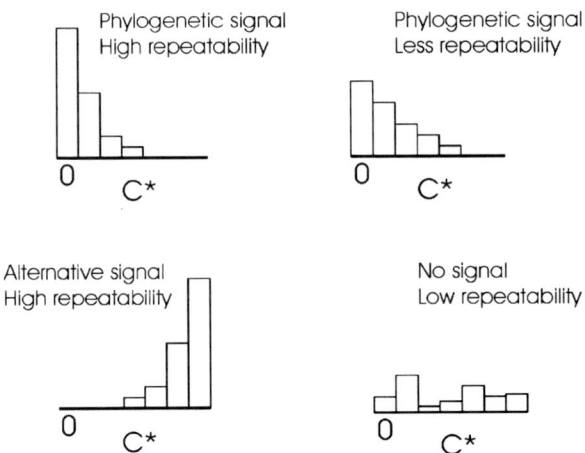

Figure 10.9 Hypothetical bootstrap distributions of the C^* statistic. A perfect agreement between topologies is indicated by a value of zero. At the upper left, there is a case where there is a strong phylogenetic signal (mode of zero) and high repeatability, indicated by the small dispersion. At the upper right, there is still evidence of a phylogenetic signal (mode of zero), but it is less repeatable, indicated by the greater dispersion. At the lower left is a strong alternative signal, indicated by the combination of a lack of agreement (mode very different from zero) and high repeatability. At the lower right, there is no apparent signal, with a large dispersion and no distinct mode.

of C^*. Here, we define precision in a way that is similar to Hillis and Bull's (1993: 183–184) definition: "... the correspondence between multiple sets of bootstrap pseudosamples taken from the same initial sample". In our case, we are looking at the correspondence between bootstrap phenograms generated from the same set of sample estimates. Note that while the mode of the distribution can be seen as an indication of a phylogenetic signal, the dispersion does not necessarily lead us to that conclusion by itself. C^* can be highly precise (= repeatable) when bootstrapping under the model without the presence of a phylogenetic signal; it may instead be indicative of a strong alternative signal. By itself, a small dispersion of C^* simply tells us that the hierarchical clusterings that are based on the bootstrap data are highly repeatable. In contrast, if the bootstrap clusterings are highly variable, the precision may be too low to distinguish any repeatable structure in the morphometric data, so that no signal of any kind is detected.

Topology frequencies: We have found that it is instructive to look at the frequencies with which different topologies occur under bootstrap resampling. Here, we are looking at the same data that are used to compute bootstrap proportions, but without condensing them in that way. We simply count the number of times that each topology is observed and express that number as a percentage of the total number of bootstrap resamples. We then know whether some topologies are more likely to be observed than others, and we can make observations about how these topologies tend to be similar or different.

Given a set of taxa, how many different topologies do we expect to observe when bootstrapping? Initially, we might guess that each possible topology will be observed with the same frequency, given enough resamples. However in practice, we tend to observe far fewer topologies than all those possible. There are two reasons for this. The first is that we would only expect to see each of the possible topologies if the sample means were all the same. If the means differ (as they usually do – otherwise, we would probably not be interested in carrying out the study in the first place), then the number of topologies observed will necessarily be limited. The second reason is that the among-taxon variation observed in interspecific studies tends to be substantially larger than within-taxon variation. Therefore, the relationships between well-separated means in multidimensional space will tend to remain stable under bootstrap sampling. As a result, most of the bootstrap phenograms will probably lie close together in the space of all possible phenogram topologies (see Efron *et al.* 1996), unless within-taxon variances are relatively very large.

In summary, we recommend the use of a combination of all three of the methods just described here. Taken together, they can give a measurement of the repeatability of the data when resampling under the model, they can provide a picture of the taxonomic levels where homoplasies occur, and (primarily in the case of bootstrap proportions) they allow us to "localize" parts of the cladogram where the phylogenetic signal may be particularly strong or may be nonexistent.

A simple example

To demonstrate our methods, we will examine morphometric variation in the facial skeletons of ateline primates. The subfamily Atelinae is a small clade of Neotropical monkeys that are characterized by large body size and possession of prehensile tails. Despite their close phylogenetic relationships, there is a large degree of anatomical diversity within the clade, especially in skull form. This diversity makes them a particularly interesting group for studies of comparative functional anatomy. There are four living ateline genera: *Ateles* (spider monkeys), *Alouatta* (howler monkeys), *Lagothrix* (woolly monkeys), and *Brachyteles* (muriquis or woolly spider monkeys). Figure 10.10 shows the hypothesized genealogical relationships among the genera, following Rosenberger and Strier (1989). *Ateles* and *Brachyteles* are the two most closely-related genera. The *Ateles–Brachyteles* clade is then joined by *Lagothrix*, followed by *Alouatta*, that joins as the sister-taxon to the other three genera. This phylogenetic estimate is based on a variety of data (e.g., anatomy, genetics, life history, and behavior), but is not based on skull form.

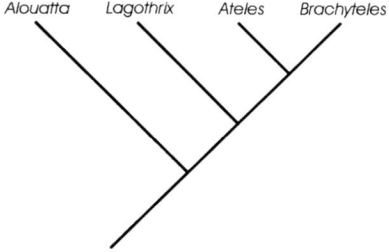

Figure 10.10 Phylogenetic relationships of living ateline genera, following Rosenberger and Strier (1989).

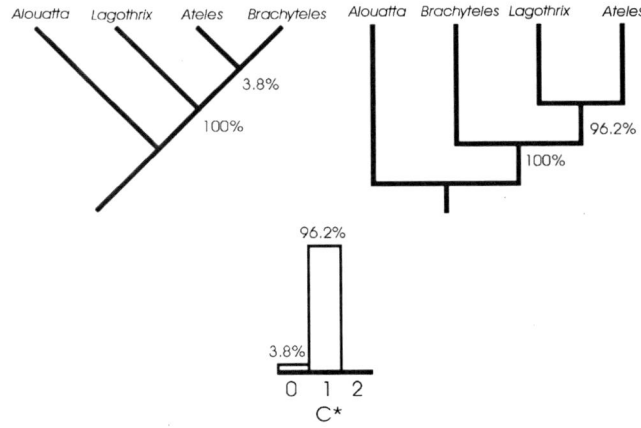

Figure 10.11 Results of the bootstrap analysis. The cladogram (left) is shown with bootstrap proportions that measure the agreement between it and the structures of the bootstrap phenograms. The UPGMA phenogram (right) is shown with bootstrap proportions that measure its repeatability (see text). The bootstrap distribution for the C^* statistic is shown in the center.

Our sample consists of adult specimens of *Ateles geoffroyi* ($N = 10$); *Alouatta seniculus* ($N = 6$); *Lagothrix lagothricha* ($N = 14$), and *Brachyteles arachnoides* ($N = 7$). To avoid the potentially confounding effects of sexual dimorphism, only females were examined. Variation in midfacial form was initially quantified by recording the three-dimensional positions of six homologous landmarks on the facial skeleton: (1) rhinion; (2) prosthion; (3) premaxilla-maxilla suture at alveolus; (4) inferior end of the zygomaxillary suture; (5) maxillary tuberosity; and (6) posterior nasal spine. Because ateline taxa vary in their adult sizes, we scaled the mean form matrix of each genus by the geometric mean of all interlandmark distances. This scaling measure seems to be a reasonable representation of the overall "volume" of the midfacial skeleton.

Figure 10.11 shows the empirical phenogram that results when S_Ω (the matrix of shape dissimilarities) is subjected to UPGMA clustering, with a cophenetic correlation > 0.99. Other clustering methods (including single linkage, complete linkage, and neighbor joining) yielded the same phenogram. While this particular data set seems robust to differences in the choice of clustering algorithm, it should not be assumed that this will always be the case. When the empirical phenogram is compared to the cladogram, we see a similarity in the position of *Alouatta*, relative to the *Lagothrix–Ateles–Brachyteles* cluster and clade. So, not only is *Alouatta* more distantly related to *Lagothrix*, *Ateles*, and *Brachyteles* than any of those taxa are to each other, it exhibits the most distinctly different shape. The difference between the topologies lies within the *Lagothrix–Ateles–Brachyteles* cluster and clade. In terms of shape, *Lagothrix* and *Ateles* are the most similar taxa, despite of the fact that *Ateles* and *Brachyteles* are the most closely related.

To quantify the effects of uncertainty in measuring shape differences, we generated 500 sets of parametric pseudosamples that were used to obtain a distribution of bootstrap phenograms. Within the bootstrap phenograms, only two topologies were observed (out of the 15 that are possible with four taxa). The most frequently observed topology (481/500 or 96.2%) matched the empirical phenogram, while the remaining topology (19/500 or 3.8%) matched the cladogram. Because we observed one topology an overwhelming majority of the time, and because that topology does not match the cladogram, we have convincing evidence for a strong alternative signal in the morphometric data. At the same time, we have evidence for a high degree of repeatability in the morphometric data, because the overwhelming majority of bootstrap phenograms are identical, matching the empirical phenogram.

Figure 10.11 shows the bootstrap proportions superimposed on the nodes of the cladogram. *Alouatta* was always the most distinct in shape, so that the bootstrap proportion associated with the *Lagothrix–Ateles–Brachyteles* clade is 100 percent. Within that clade, we see the very low proportion associated with the *Ateles-Brachyteles* clade (3.8%), indicating that this close phylogenetic relationship was not reflected in the strongly-patterned morphometric data. Finally, we come to the same conclusion when we look at the bootstrap distribution of the C^* statistic. C^* equals zero 3.8 percent of the time, indicating that perfect matches between the bootstrap phenograms and the cladogram were rare. C^* equals one 96.2 percent of the time and never equaled two (the maximum value). This indicates that the mismatches were always restricted to the structure of the *Lagothrix–Ateles–Brachyteles* cluster and never extended any deeper toward the cladogram root (so that *Alouatta* was always the most distinct).

Since our ultimate goal is to better understand how organisms evolve, we would like to suggest an evolutionary scenario that accounts for the strong alternative signal. As we mentioned previously, there is considerable intergeneric variation in ateline facial morphology, and much of this variation has been interpreted in terms of biomechanical adaptation to different diets (e.g., Rosenberger and Strier 1989; Anapol and Lee 1994). *Ateles* and *Lagothrix* are primarily frugivorous, with relatively small and low-crowned postcanine teeth and facial skeletons that are somewhat more gracile, especially for *Ateles*. In contrast, *Alouatta* and *Brachyteles* incorporate a far greater proportion of mature leaves in their diets. As a correlate, they possess very large molars with high shearing crests. Their faces are also substantially larger (relative to the rest of the skull) than in their frugivorous relatives. It is generally thought that the ancestral ateline was probably a generalized frugivore and that the most parsimonious hypothesis for the evolution of dietary specializations in the clade is a parallel acquisition of folivory in *Alouatta* and *Brachyteles* (Rosenberger and Strier 1989; Rosenberger 1992). Therefore, when we return to the empirical phenogram, we might hypothesize that the strong alternative signal is due in part to the symplesiomorphic (shared primitive) retention of a frugivorous diet by both *Ateles* and *Lagothrix*, with a correlated retention of a primitive facial structure.

An obvious question is why *Alouatta* and *Brachyteles* are not more similar, given our hypothesis that they have adopted similar dietary strategies in parallel. We can speculate, that the lack of similarity may be due to *Alouatta*'s unique and bizarre modifications in its hyolaryngeal apparatus (Rosenberger and Strier 1989), so that many of the diet-related morphological similarities that might have arisen between *Alouatta* and *Brachyteles* have simply been "swamped" by *Alouatta*'s many morphological autapomorphies that are not related to its diet. It is interesting to note that, as the analysis now stands, we cannot tell how much of the clustering of *Lagothrix–Ateles* and *Brachyteles* might be attributed to synapomorphies for those data and how much is due to the fact that *Alouatta* is simply so different. To make such a distinction, we would require an outgroup and a way to decompose the morphometric similarities into homologous and homoplasious components (see *Further research*).

Finally, we should emphasize again that our interpretations of evolutionary patterns in the atelines is strongly dependent on our assumptions of how the taxa are related. It is clear that our ideas could require substantial revisions if we were confronted with new evidence that would lead us to accept a different cladogram.

Discussion

Uncertainty in constructing the cladogram: As we have mentioned several times, the results that we gain in applying our methods are always contingent on our *a priori* specification of a cladogram. As a result, our interpretations could be impacted substantially if the cladogram were different. In general, we will choose the cladogram that we believe to be correct, given our current knowledge, and we assume that there has been no uncertainty in its construction (Pagel 1999; Huelsenbeck *et al.* 2000). We realize, of course, that uncertainties also play a role in estimating phylogenetic relationships and that our assumption is ultimately (though necessarily) a simplification. While considerations of cladistic uncertainties are beyond the scope of this study, we are intrigued by the recent work of Huelsenbeck *et al.* (2000), who have taken a Bayesian perspective toward that problem. Using Markov chain Monte Carlo methods with DNA sequence data, these authors have presented a method of assigning weights to different cladograms, in an effort to determine which are more likely to be correct, given the data. For researchers who use morphometrics, such methods may ultimately prove to be extremely useful for evaluating the distribution of form variation relative to a set of "credible" cladograms, so that the implications of cladogram differences can be evaluated.

Parametric versus *nonparametric bootstrapping:* We have stressed parametric methods for pseudosample generation in this chapter, but our methods are easily used with nonparametric bootstrapping. There are relative advantages and disadvantages to both resampling methods. As we mentioned previously, one of the primary advantages to using a fitted parametric model is that estimators of uncertainty will generally be more accurate, especially for modest sample sizes. For this reason, we recommend the parametric bootstrap for most applications. However, if we have doubts about the suitability of the parametric model, the nonparametric bootstrap may be preferable. In our ateline example, we believe that the assumption of normally-distributed perturbations is reasonable for modeling within-sample variation. If this assumption proved unreasonable, a nonparametric bootstrap would be a better choice, as the advantages of parametric bootstrapping only hold when the model is valid. Unfortunately, because the Nature Space is unobservable, we cannot test the adequacy of the parametric model directly. Therefore, it might be worthwhile to try nonparametric bootstrapping as a test of the parametric model's suitability. This is because we expect the results to converge (as the number of pseudosamples increases) if the parametric model is a good descriptor of the observed variation. This type of comparison measures the "robustness of specification" for the parametric model (Davison and Hinkley 1997). If the results differ considerably, it might be advisable either to use the nonparametric results or to use another parametric model.

Generality of the method: While we have emphasized the use of parametric bootstrapping with landmark data and EDMA, the basic bootstrapping strategy that we have outlined is very general. It can be applied with any type of data where the data vary within samples and, preferably, where a reasonable model of within-sample variation can be specified. For example, a researcher might be interested in determining whether a phylogenetic signal is present in the postcranial skeleton for a group of organisms, and the data may consist of the maximum lengths of the limb bones. In that case, within-sample variations can be described by assuming multivariate normal distributions and using the sample mean vectors and variance–covariances matrices to generate pseudosamples. This approach can be taken with either unscaled (form) or scaled (shape) data. The measurements do not even have to be continuously distributed. Meristic (= count) data can also be used, providing they exhibit within-sample variation. For example, Mosimann *et al.* (1978) show how a multivariate lognormal model can be used to describe the relative proportions of counts. This model could conceivably serve as the basis for parametric bootstrapping to examine phylogenetic signals in scale counts for reptiles or in fin-ray counts for fishes. In contrast to landmark-based applications, we can apply tests for multivariate normality in these cases to determine whether our model choices are reasonable; otherwise, nonparametric bootstrapping is always an option.

Further research: While signal detection is an important first step in the phylogenetic analysis of morphometric data, it is clearly not an end in itself. Rather, it is an indication that the data contain interesting information that should be investigated further. One particularly interesting area for future research is the development of methods for studying "mosaic" evolution in a phylogenetic context. Our method for detecting phylogenetic signals is based on measures of morphometric dissimilarity (e.g., F_Ω), where differences in many interlandmark distances are summarized in terms of a single number that describes the "overall" degree of difference in form. However, a signal in overall form does not necessarily imply that there is the same pattern or strength of signal for each of the contributing measurements. In reality, complex organisms tend to evolve as "mosaics", where an organism's components or parts can potentially evolve at different rates and in response to different selective pressures (e.g., Lande and Arnold 1983). So, the distribution of the "whole" with respect to the phylogeny can be thought of as a consensus of the distribution of the "parts". We would naturally like to decompose this consensus into its component parts. For example, we would like to recognize those parts that are homologous and those that are homoplasious. We would also like the ability to distinguish different patterns and events of homoplasy (e.g., different episodes of convergence, parallelism, and reversal). While there are existing methods of "optimizing" changes in discrete character states over a cladogram (e.g., Swofford and Maddison 1987), there are no analogous methods for studying the evolution of multivariate, continuously-distributed "characters" in a phylogenetic context. We believe that the development of such methods should be a high priority in further efforts to bridge the gaps between the "new comparative biology" and the "new morphometrics".

Acknowledgements

Thanks to Norman MacLeod and Peter Forey for inviting us to contribute to their Systematics Association symposium and to this volume. Much of this research was supported by NSF Grant DBS 9209083 to JTR and SL, as well as by Project II of NIH Grant 1 P50 DE11131-03 to JTR. Support for collection of the ateline data was provided to TMC by NSF Grant BNS 9020562, by Wenner-Gren Foundation Grant 5303, and by the Field Museum of Natural History. All of the analyses presented here were performed using the *WinEDMA* software package (Cole 2001), available via the Internet at no cost (faith.med.jhmi.edu or c.faculty.umkc.edu/colet).

References

Adams, D. C. and Rosenberg, M. S. (1998) 'Partial warps, phylogeny, and ontogeny: a comment on Fink and Zelditch (1995)', *Systematic Biology*, **47**, 167–172.

Alberch, P. (1980) 'Ontogenesis and morphological diversification', *American Zoologist*, **20**, 653–667.

Alberch, P. (1985) 'Developmental constraints: why St. Bernards often have an extra digit and poodles never do', *American Naturalist*, **126**, 430–433.

Anapol, F. and Lee, S. (1994) 'Morphological adaptation to diet in platyrrhine primates', *American Journal of Physical Anthropology*, **94**, 239–261.

Bock, W. J. and von Wahlert, G. (1965) 'Adaptation and the form-function complex', *Evolution*, **19**, 269–299.

Bookstein, F. L. (1994) 'Can biometrical shape be a homologous character?', in Hall, B. K. (ed.) *Homology: the hierarchical basis of comparative biology*, San Diego: Academic Press.

Bookstein, F. L. (1996) 'Biometrics, biomathematics and the morphometric synthesis', *Bulletin of Mathmatical Biology*, **58**, 313–365.

Brooks, D. R. (1996) 'Explanations of homoplasy at different levels of biological organization', in Sanderson, M. J. and Hufford, L. (eds) *Homoplasy: the recurrence of similarity in evolution*, San Diego: Academic Press, pp. 3–36.

Cain, A. J. and Harrison, G. A. (1960) 'Phyletic weighting', *Proceedings of the Zoological Society of London*, **135**, 1–31.

Cavalli-Sforza, L. L. and Piazza, A. (1975) 'Analysis of evolution: evolutionary rates, independence and treeness', *Theoretical Population Biology*, **8**, 127–165.

Chang, J. T. and Kim, J. (1996) 'The measurement of homoplasy: a stochastic view', in Sanderson, M. J. and Hufford, L. (eds) *Homoplasy: the recurrence of similarity in evolution*, San Diego: Academic Press, pp. 189–203.

Chappill, J. A. (1989) 'Quantitative characters in phylogenetic analysis', *Cladistics*, **5**, 217–234.

Cheverud, J. M., Dow, M. M. and Leutenegger, W. (1985) 'The quantitative assessment of phylogenetic constraints in comparative analyses: sexual dimorphism in body weight among primates', *Evolution*, **39**, 1335–1351.

Cheverud, J. M., Wagner, G. and Dow, M. M. (1989) 'Methods for the comparative analysis of variation patterns', *Systematic Zoology*, **38**, 201–213.

Coddington, J. A. (1988) 'Cladistic tests of adaptational hypotheses', *Cladistics*, **4**, 3–22.

Cole, T. M. III (2001) *WinEDMA: Windows-based software for Euclidean distance matrix analysis*, Kansas City: School of Medicine, University of Missouri–Kansas City.

Colless, D. H. (1970) 'The phenogram as an estimate of phylogeny', *Systematic Zoology*, **19**, 352–362.

Cranston, P. S. and Humphries, C. J. (1988) 'Cladistics and computers: a chironomid conundrum?', *Cladistics*, 4, 72–92.

Davison, A. C. and Hinkley, D. V. (1997) *Bootstrap methods and their application*, Cambridge: Cambridge University Press.

Edwards, A. W. F. (1992) *Likelihood*, Expanded edition, Baltimore: The Johns Hopkins University Press.

Efron, B., Halloran, E. and Holmes, S. (1996) 'Bootstrap confidence levels for phylogenetic trees', *Proceedings of the National Academy of Sciences USA*, 93, 13429–13434.

Efron, B. and Tibshirani, R. S. (1991) 'Bootstrap methods for standard errors, confidence intervals, and other measures of statistical accuracy', *Statistical Science*, 1, 54–77.

Efron, B. and Tibshirani, R. S. (1993) *An introduction to the bootstrap*, New York: Chapman & Hall.

Falconer, D. S. and Mackay, T. F. C. (1996) *Introduction to quantitative genetics*. 4th edn, Harlow, England: Longman.

Felsenstein, J. (1982) 'Numerical methods for inferring evolutionary trees', *Quarterly Review of Biology*, 57, 379–404.

Felsenstein, J. (1985) 'Confidence limits on phylogenies: an approach using the bootstrap', *Evolution*, 39, 783–791.

Felsenstein, J. (1988) 'Phylogenies and quantitative characters', *Annual Review of Ecology and Systematics*, 19, 445–471.

Fink, W. L. (1982) 'The conceptual relationship between ontogeny and phylogeny', *Paleobiology*, 8, 254–264.

Fink, W. L. and Zelditch, M. L. (1995) 'Phylogenetic analysis of ontogenetic shape transformations: a reassessment of the piranha genus *Pygocentrus* (Teleostei)', *Systematic Biology*, 44, 343–360.

Goodall, C. R. (1991) 'Procrustes methods in the statistical analysis of shape', *Journal of the Royal Statistical Society, Series B*, 53, 285–339.

Gould, S. J. and Lewontin, R. (1979) 'The spandrels of San Marcos and the Panglossian paradigm: a critique of the adaptationist programme', *Proceedings of the Royal Society of London Series B*, 205, 581–598.

Harvey, P. H. and Pagel, M. D. (1991) *The comparative method in evolutionary biology*, Oxford: Oxford University Press.

Hillis, D. M. and Bull J. J. (1993) 'An empirical test of bootstrapping as a method for assessing confidence in phylogenetic analysis', *Systematic Biology*, 42, 182–192.

Houle, D. (1992) 'Comparing evolvability and variability in quantitative traits', *Genetics*, 130, 195–204.

Hubert, L. J. (1978) 'Generalized proximity function comparison', *British Journal of Mathematical and Statistical Psychology*, 31, 179–192.

Huelsenbeck, J. P., Hillis, D. M. and Jones, R. (1996) 'Parametric bootstrapping in molecular phylogenetics: applications and performance', in Ferraris, J. D. and Palumbi, S. R. (eds) *Molecular zoology: advances, strategies, and protocols*, New York: Wiley–Liss, pp. 19–45.

Huelsenbeck, J. P., Rannala, B. and Masly, J. P. (2000) 'Accommodating phylogenetic uncertainty in evolutionary studies', *Science*, 288, 2349–2350.

Johnson, R. A. and Wichern, D. W. (1982) *Applied multivariate statistical analysis*, Englewood Cliffs, New Jersey: Prentice-Hall.

Lande, R. and Arnold, S. J. (1983) 'The measurement of selection on correlated characters', *Evolution*, 37, 1210–1226.

Lanyon, S. (1985) 'Detecting internal inconsistencies in distance data', *Systematic Zoology*, 34, 397–403.

Lapointe, F.-J. and Legendre, P. (1990) 'A statistical framework to test the consensus of two nested classifications', *Systematic Zoology*, 39, 1–13.

Lapointe, F.-J. and Legendre, P. (1992) 'Statistical significance of the matrix correlation coefficient for comparing independent phylogenetic trees', *Systematic Biology*, 41, 378–384.

Larson, A. and Losos, J. B. (1996) 'Phylogenetic systematics of adaptation', in Rose, M. R. and Lauder, G. V. (eds) *Adaptation*, San Diego: Academic Press, pp. 187–220.

Lauder, G. V. (1981) 'Form and function: structural analysis in evolutionary morphology', *Paleobiology*, 7, 430–442.

Lauder, G. V. (1990) 'Functional morphology and systematics: studying functional patterns in an historical context', *Annual Review of Ecology and Systematics*, 21, 317–340.

Lele, S. (1993) 'Euclidean distance matrix analysis (EDMA): estimation of mean form and mean form difference', *Mathematical Geology*, 25, 573–602.

Lele, S. (1999) 'Invariance and morphometrics: a critical appraisal of statistical techniques for landmark data', in Chaplain, M. A. J., Singh, G. D. and MacLachlan, J. C. (eds) *On growth and form: spatio-temporal pattern formation in biology*, Chichester: Wiley, pp. 325–336.

Lele, S. and Cole, T. M. III (1996) 'A new test for shape differences when variance–covariance matrices are unequal', *Journal of Human Evolution*, 31, 193–212.

Lele, S. and McCulloch, C. E. (2001) 'Invariance and morphometrics', *Journal of the American Statistical Association*, submitted manuscript.

Lele, S. and Richtsmeier, J. T. (1990) 'Statistical models in morphometrics: are they realistic?', *Systematic Zoology*, 39, 60–69.

Lele, S. and Richtsmeier, J. T. (1991) 'Euclidean distance matrix analysis: a coordinate free approach to comparing biological shapes using landmark data', *American Journal of Physical Anthropology*, 86, 415–428.

Lele, S. and Richtsmeier, J. T. (2001) *An invariant approach to the statistical analysis of form*, London: CRC Press.

Losos, A. and Larson, J. B. (1996) 'Phylogenetic systematics of adaptation', in Rose, M. R. and Lauder, G. V. (eds) *Adaptation*, San Diego: Academic Press, pp. 187–220.

McKinney, M. L. and McNamara, K. J. (1991) *Heterochrony: the evolution of ontogeny*, New York: Plenum.

Mooers, A. Ø., Vamosi, S. M. and Schluter, D. (1999) 'Using phylogenies to test macroevolutionary hypotheses of trait evolution in cranes (Gruinae)', *American Naturalist*, 154, 249–259.

Mosimann, J. E., Malley, J. D., Cheever, A. W. and Clark, C. B. (1978) 'Size and shape analysis of schistosome egg-counts in Egyptian autopsy data', *Biometrics*, 34, 341–356.

Mueller, L. D. and Ayala, F. J. (1982) 'Estimation and interpretation of genetic distance in empirical studies', *Genetical Research (Cambridge)*, 40, 127–137.

Neyman, J. and Scott, E. (1948) 'Consistent estimates based on partially consistent observations', *Econometrika*, 16, 1–32.

Pagel, M. (1999) 'Inferring the historical patterns of biological evolution', *Nature*, 401, 877–884.

Patterson, C. (1982) 'Morphological characters and homology', in Joysey, K. A. and Friday, A. E. (eds) *Problems of phylogenetic reconstruction*, London: Academic Press, pp. 21–74.

Penny, D. and Hendy, M. D. (1985) 'The use of tree comparison metrics', *Systematic Zoology*, 34, 75–82.

Pimentel, R.A. and Riggins, R. (1987) 'The nature of cladistic data', *Cladistics*, 3, 201–209.

Richtsmeier, J. T., Cole, T. M. III, Krovitz, G. E., Valeri, C. J. and Lele, S. (1998) 'Preoperative morphology and development in sagittal synostosis', *Journal of Craniofacial Genetics and Developmental Biology*, 18, 64–78.

Rohlf, F. J. (1974) 'Methods of comparing classifications', *Annual Review of Ecology and Systematics*, 5, 101–113.

Rohlf, F. J. (1982) 'Consensus indices for comparing classifications', *Mathematical Biosciences*, 59, 131–144.

Rohlf, F. J. and Marcus, L. F. (1993) 'A revolution in morphometrics', *Trends in Ecology and Evolution*, 8, 129–132.

Rohlf, F. J. (1998) 'On applications of geometric morphometrics to studies of ontogeny and phylogeny', *Systematic Biology*, 47, 147–158.

Rosenberger, A. L. (1992) 'Evolution of feeding niches in New World monkeys', *American Journal of Physical Anthropology*, 88, 525–562.

Rosenberger, A. L. and Strier, K. B. (1989) 'Adaptive radiation of the ateline primates', *Journal of Human Evolution*, 18, 717–750.

Sanderson, M. L. and Hufford, L. (1996) 'Homoplasy and the evolutionary process: an afterword', in Sanderson, M. J. and Hufford, L. (eds) *Homoplasy: the recurrence of similarity in evolution*, San Diego: Academic Press, pp. 327–330.

Simpson, G. G. (1961) *Principles of animal taxonomy*, New York: Columbia University Press.

Silverman, B. W. (1986) *Density estimation for statistics and data analysis*, London: Chapman and Hall.

Sluys, R. (1989) 'Rampant parallelism: an appraisal of the use of nonuniversal derived states in phylogenetic reconstruction', *Systematic Zoology*, 38, 350–370.

Sneath, P. H. A. and Sokal, R. R. (1973) *Numerical taxonomy*, San Francisco: Freeman.

Sokal, R. R. and Rohlf, F. J. (1962) 'The comparison of dendrograms by objective methods', *Taxon*, 11, 33–40.

Swofford, D. L. and Maddison, W. P. (1987) 'Reconstructing ancestral character states under Wagner parsimony', *Mathematical Bioscience*, 87, 199–229.

Wake, D. B. (1991) 'Homoplasy: the result of natural selection, or evidence of design limitations?', *American Naturalist*, 138, 543–567.

Wake, D. B. and Larson, A. (1987) 'Multidimensional analysis of an evolving lineage', *Science*, 238, 42–48.

Zelditch, M. L., Fink, W. L. and Swiderski, D. L. (1995) 'Morphometrics, homology, and phylogenetics: quantified characters as synapomorphies', *Systematic Biology*, 44, 179–189.

Zelditch, M. L., Swiderski, D. L. and Fink, W. L. (2000) 'Discovery of phylogenetic characters in morphometric data', in Wiens, J. J. (ed.) *Phylogenetic analysis of morphological data*, Washington, DC: Smithsonian Institution Press, pp. 37–83.

Chapter 11

Phylogenetic tests for differences in shape and the importance of divergence times
Eldredge's enigma explored

P. David Polly

ABSTRACT

When the shapes of phylogenetically divergent samples are compared, standard tests of significance (such as Hotelling's T^2) are not applicable. Instead shape differences between taxa should be assessed relative to the amount of time since they last shared a common ancestor. This can be accomplished using the same methods applied to univariate data from an evolutionary time-series. Thirteen samples of viverravid carnivorans (Viverravidae, Carnivora) from the Paleogene of the Bighorn Basin, Wyoming were compared in this way. The amount of divergence time separating samples was determined from their phylogenetic tree and their stratigraphic setting. Branch lengths were first estimated in millions of years, then converted to generations using an allometric equation relating body-mass and generation-time in extant mammals. A Log-Rate–Log-Interval (LRI) distribution was used to estimate the per-generation rate of shape change, the latter of which was used to calculate the expected shape divergence for each pair of taxa. The amount of shape divergence tended to be less than that expected, given the number of generations separating the samples. Shape change was close to a statistical definition of stasis, indicating that long-term rates of molar shape evolution are slow relative to their potential. This may be due to functional constraints on the configuration of crown features and the complexity of the developmental processes controlling their topographical relationships. Phylogenetic comparisons of shape are dependent on accurate estimates of branch lengths, that may be complicated by data and methodological considerations. In particular, many phylogenetic methods are ambiguous when it comes to determining the time interval separating two stratigraphically distinct samples.

Introduction

Two goals of morphometrics are the description of differences in size and shape and the statistical testing of those differences. There are many kinds of differences, however, each with its own appropriate test. In this chapter, I consider shape differences between phylogenetically divergent populations, showing that an estimate of the time since their common ancestry is necessary to assess their significance.

Morphological shape, like any other heritable trait, covaries with phylogeny. The amount of divergence between two biological populations is likely to be proportional to the time elapsed since they last shared a common ancestor. Standard tests for differences in sample means (e.g., T or T^2 tests) are therefore not particularly meaningful. Rather, differences between populations should be compared to the amount of divergence expected given the time interval since their common ancestry. Populations with an ancient ancestry are expected to be more divergent than those with a recent one. The amount of divergence expected under a null hypothesis of random change can be estimated from data on among-population shape difference and a phylogenetic hypothesis that includes divergence times. Here I apply this procedure to molar shape differences, as measured by Procrustes distance, in early Eocene viverravid carnivorans.

Why are standard tests for differences in mean not appropriate? A t-test, or its equivalent, is normally used when two samples are expected to have the same mean and variance. An example of its application is a pharmaceutical test in which a drug is applied to an experimental group and its effects compared to an untreated control group. If the drug has no effect, we expect the two groups to be alike except for differences due to sampling error (Figure 11.1A). The probability that the two samples will have different means is a function of the number of individuals in each and the variance of the parent population (Sokal and Rohlf 1995). In contrast, when two populations have diverged over evolutionary time, we cannot assume they will have the same mean. An example is a morphometric study of the scapulae of mammals with various locomotory habits. We might, on the one hand, expect the scapulae of fossorial species to be more like one another than they are to those of cursorial species. On the other hand, we might expect the scapulae of canids to be more like one another than they are to felids. In either case, we cannot presume that any of them will be exactly alike because differences will have accumulated, generation by generation, over time (Figure11.1B). The assumption of a t-test – that means are equal except due to sampling error – does not pertain. Rather, the difference in means between two biological populations is a function of the number of individuals in each sample, the additive variance of their respective populations, and the number of sampling generations between them and the original parent population (Felsenstein 1973; Bookstein 1991; Falconer and Mackay 1996). Any statistical test of between-species differences, including shape differences measured by landmark techniques, must take these factors into account. Techniques for comparing phylogenetically divergent samples are not new. Independent contrasts, autocorrelation analysis, squared-change parsimony, and maximum-likelihood methods are now commonly used in inter-specific studies of trait correlation (Felsenstein 1985; Cheverud and Dow 1985; Maddison 1991) and ancestral trait reconstruction (Martins and Hansen 1997; Schluter et al. 1997; Garland et al. 1999).

The amount of change expected in a numeric variable under a random walk model is a function of the square root of time (Bookstein 1991; Berg 1993; Gingerich 1993a). The single most probable value at the end of the walk is the same as its starting value. However, the distribution of other possible end values gets broader the more steps there are in the walk. The breadth of that distribution increases, not directly with the number of steps, but with the square root of the number of steps. If two closely related populations are compared, their difference is therefore likely to be proportionally greater than that of two distantly related ones. This is because over a longer time interval the random fluctuations of the walk are more likely to return to their starting value than they are over a short interval. We are familiar with this process in molecular evolution with its 'multiple hits'.

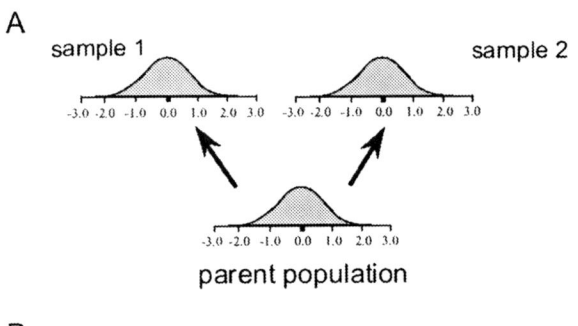
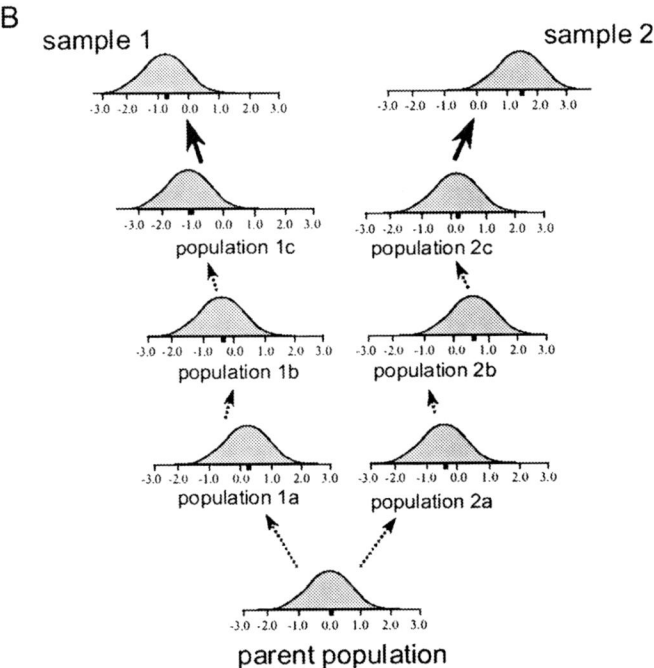

Figure 11.1 Populations, sampling, and phylogenetic divergence. (A) Standtard *t*-test scenario in which the two samples to be compared were randomly drawn from the same parent population. The mean and variance of each are expected to be identical to one another and to the parent population except for differences due to chance. (B) The scenario considered in this chapter in which the two samples being compared are descended from a common ancestor. Differences between the samples are the result of the cumulative effects of a number of cycles of generational resampling. Tests of difference or similarity must take into account the number of generations separating the two samples.

Given enough time, random evolution has a high probability of erasing itself. This means that rates calculated over short intervals are likely to be higher than ones calculated over long intervals if the evolutionary process is random. Time in a random walk is measured in the small steps over which change actually occurs. In sexually reproducing organisms, those steps are generations. For our comparison, then, the critical values change (1) with the square root of the number of generations separating

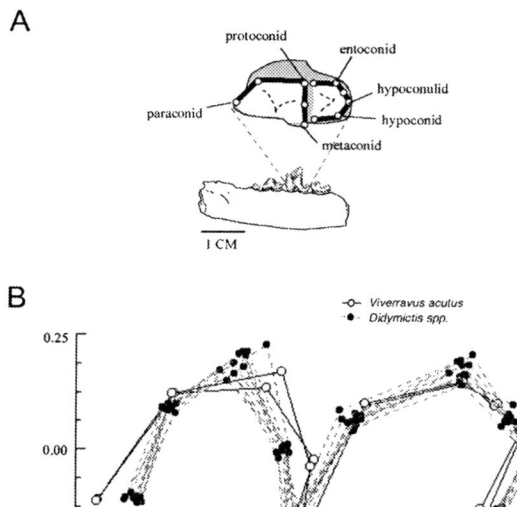

Figure 11.2 Procrustes superposition of 13 lower molar shapes. (A) Diagram showing the landmarks used in this study. (B) Generalized Least Squares superposition of the consensus landmark configurations from 13 population samples. Eleven of the shapes are from the genus *Didymictis* (filled circles) and two are from the genus *Viverravus* (open circles). Lines connecting the landmarks correspond to the heavy lines in part A.

the two populations and (2) in proportion with the amount of change expected per generation.

The problem is applicable to the data shown in Figure 11.2, which illustrate differences in the arrangement of molar cusp landmarks in some Paleogene carnivorans. The molars of *Didymictis* (filled circles, Figure 11.2B) are quite different in shape than those of *Viverravus* (open circles), although there is some shape variation among the samples of both genera. In many landmark-based studies Hotelling's T^2 statistic could be applied to test the significance of these differences. However, all of the samples in Figure 11.2 are phylogenetically related, some more closely to others, some more distantly. All of the *Didymictis* samples, for example, are more closely related to one another than they are to either of the *Viverravus* samples. In this chapter shape differences among these samples are assessed in terms of their phylogenetic relationships. Rather than arriving at a probability that the samples are different (as in t-tests), this investigation will assume that the samples are different and will instead assess whether the observed differences are consistent with a random walk in shape change, given an estimated time separating the samples.

Materials and methods

The data used in this study are from late Paleocene and early Eocene viverravid carnivorans from North America (Polly 1997). Viverravids were small carnivores – estimated body masses ranging from the size of the living Least Weasel to that of the Coyote (Appendix 1) – that were part of the early radiation of the Order Carnivora colloquially known as 'miacids'. Viverravids were primarily terrestrial in habit (Heinrich and Rose 1997) and some (e.g., *Viverravus acutus*) were as sexually dimorphic as are many extant weasels (Polly 1997). Viverravids are the earliest known members of Carnivora, first appearing in the Paleocene of North America about 60 million years ago (mya) (Fox and Youzwyshyn 1994). Their relationship to extant Carnivora is controversial. Flynn and Galiano (1982) argued that they are the sister group of extant felids, hyaenids, viverrids, and herpestids. Gingerich and Winkler (1985) argued that they are a completely extinct carnivoran lineage with no close relationship to any extant groups, a view echoed by Wyss and Flynn (1993) who considered them to be the sister group to all extant carnivorans. Hunt and Tedford (1993) suggested that Carnivora itself may be polyphyletic and viverravids related only to the aeluroid half.

The phylogenetic hypothesis used here (Figure 11.3) is from Polly (1997). It was generated using stratocladistics, a parsimony-based phylogeny reconstruction technique that simultaneously minimizes *ad hoc* hypotheses of homoplasy and stratigraphic non-preservation (Fisher 1992, 1994; Bodenbender 1995; Fox *et al.* 1999). Stratocladistic phylogenies are fully resolved in that taxa are placed at branch tips, at nodes, or along branches as dictated by character-state distributions and temporal ordering. The basic units of analysis are restricted to discrete stratigraphic intervals and represent 'segments' of evolutionary species (Simpson 1951). Thirty-nine discrete morphological characters and one stratigraphic ordering variable were used in the analysis. Figure 11.3A is a strict consensus of 1,470 morphologically most parsimonious cladograms generated in that analysis (each with a consistency index of 0.784 and a retention index of 0.914). Figure 11.3B shows the two most stratocladistically parsimonious phylogenetic trees (see Polly 1997 for details).

This study is restricted to viverravid species from the late Paleocene and early Eocene Bighorn Basin deposits (Wyoming, USA). Restricting consideration to Bighorn Basin material allowed tight control of sampling and age estimates. The Bighorn Basin in northwestern Wyoming represents perhaps the longest continuous record – more than eight million years – of mammalian life in the world (Gingerich 1980). More than 200,000 vertebrate specimens have been collected from the more than 2,000 meters of sediment that accumulated during the uplift of the northern Rocky Mountains. The combination of careful collecting, measured sections (Rose 1981; Gingerich and Klitz 1985; Bown *et al.* 1994; Clyde 1997), biostratigraphy (Gingerich 1991), and magnetostratigraphic dating (Butler *et al.* 1981; Clyde *et al.* 1994) make it possible to assign remarkably precise absolute ages to all 37 samples considered here, probably accurate to within 100,000 years. Samples relevant to this study are from the Torrejonian, Clarkforkian, and Wasatchian North American Land Mammal Ages (NALMAs), that span the late Paleocene and earliest Eocene. Absolute age estimates range from 58.8 to 52.7 mya. Figure 11.4 shows the stratigraphic and phylogenetic relation context of the samples used in this study.

Landmark coordinates were collected from the lower first molars of the 13 viverravid populations indicated with an asterisk in Appendix 1. These include two samples of *Viverravus acutus* (Wa-1 and Wa-2), two samples of *Didymictis leptomylus* (Wa-1 and Wa-2), three samples of *D. proteus* (Cf-2, Cf-3, and Wa-0), and six samples

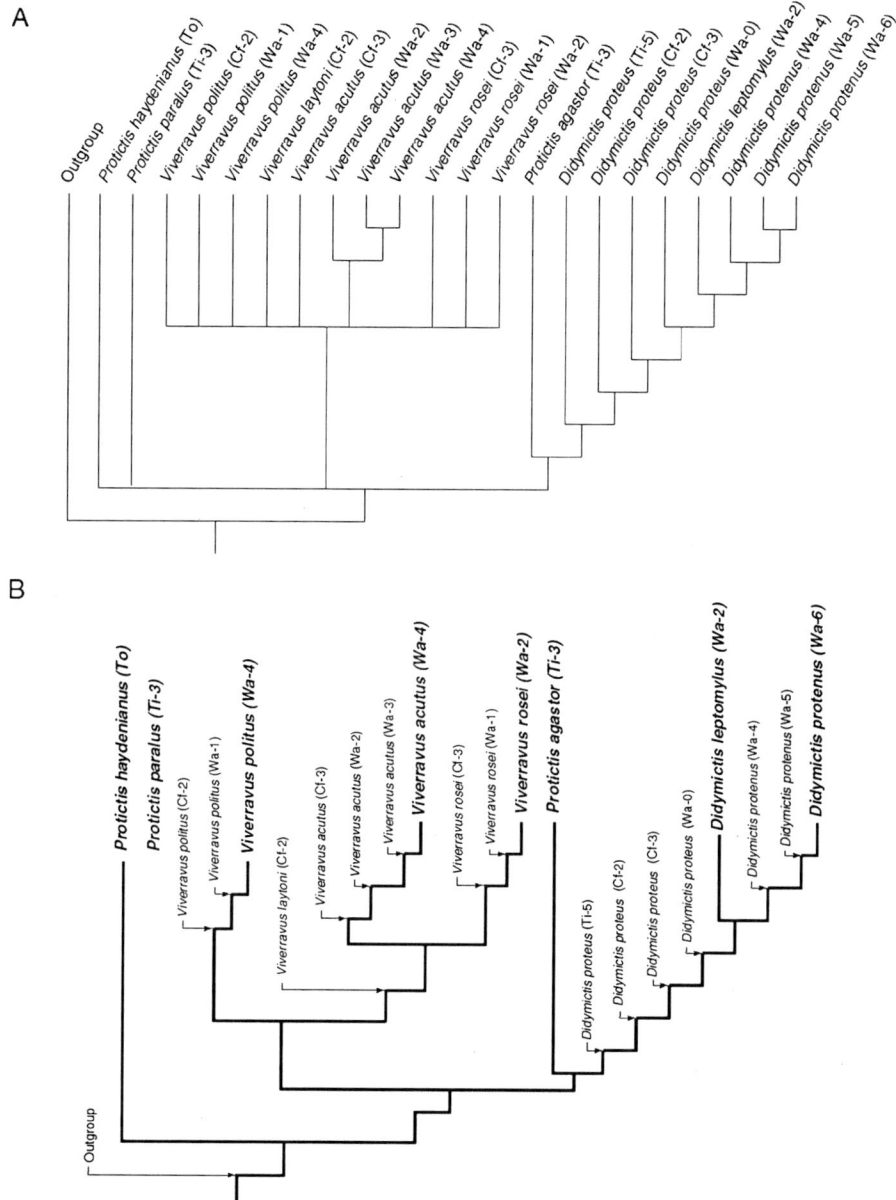

Figure 11.3 Cladogram and phylogenetic tree. (A) Consensus cladogram of viverravid samples from the early Paleogene of the Bighorn Basin, Wyoming. (B) Phylogenetic tree (stratocladogram) of the same taxa. Thin arrows indicate that a sample falls unequivocally along a branch or at a node; thicker lines indicate that the position of a sample is unequivocal, falling either at the node or as a sister-taxon. Both trees taken from Polly (1997).

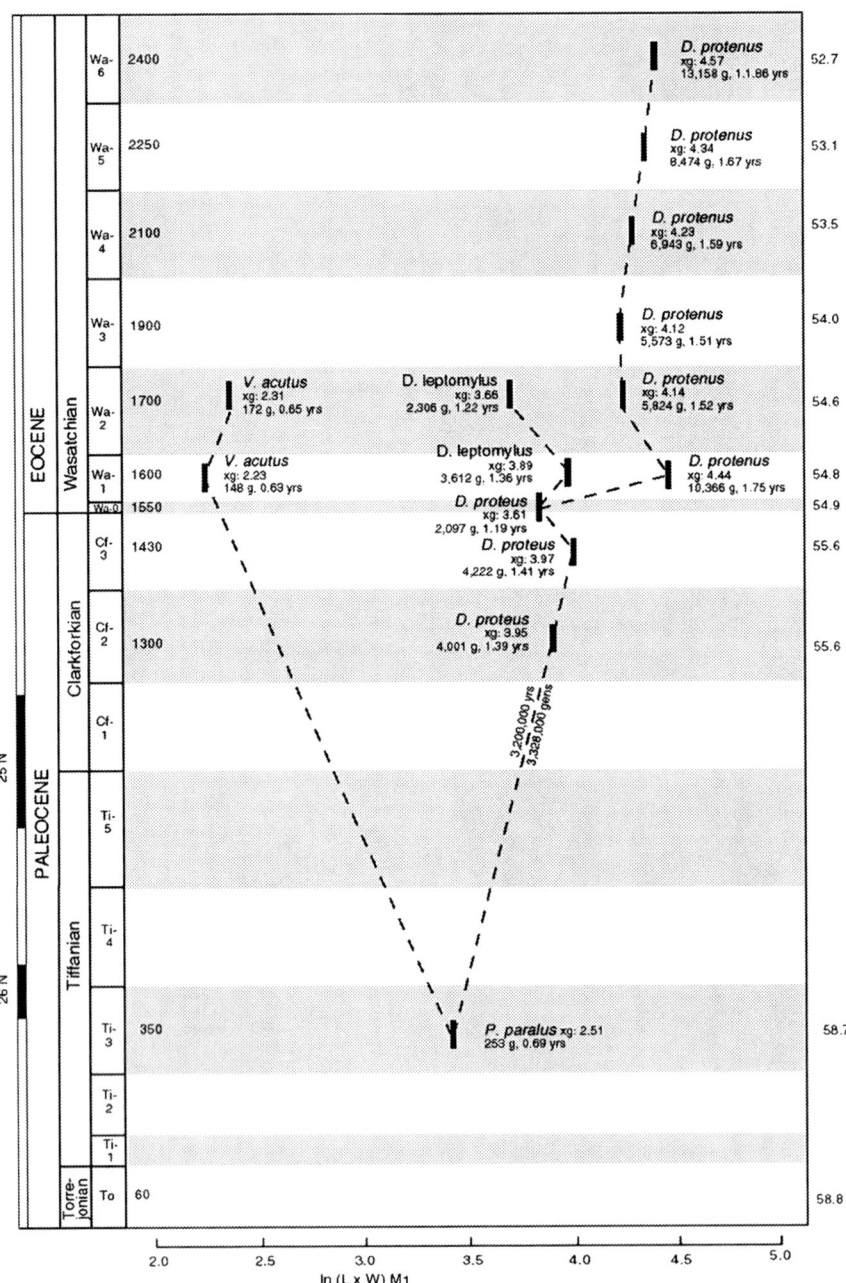

Figure 11.4 Phylogeny of samples used. The phylogenetic and stratigraphic relationships of the 13 samples considered in this study. The vertical axis represents time, the horizontal axis is the natural log of first lower molar occlusal area. The taxonomic name, mean molar area, estimated body mass, and generation length are indicated for each sample. Lines connecting samples are drawn according to the tree in Figure 11.3B. Along the left of the graph are magnetostratigraphic intervals, ages, and meter levels; to the right are absolute ages.

of *D. protenus* (Wa-1, Wa-2, Wa-3, Wa-4, Wa-5, and Wa-6). Eleven landmarks were chosen (Figure 11.2A) and their coordinates were digitized using a Reflex® microscope. A single specimen was redigitized 10 times to test measurement accuracy and reproducibility. Less than 10 per cent of within-sample variation could be attributed to irreproducibility of measurements, which was approximately equal at all landmarks. See Polly (1998) for further details about landmark selection and their biological significance.

Data were only collected from those teeth in which wear did not obscure cusp tips. Consensus configurations were calculated for all 13 samples using generalized least squares (GLS) fitting and these were used for the between-population comparisons in this study. The standard deviation of the Procrustes tangent distances among the specimens in each sample was calculated and these were pooled for use in calculating rates of morphological change as explained below. The pooled standard deviation was 0.025.

The shape metric used in this study is a variant of the Procrustes distance. In this case, the Euclidean distance, d, between points projected into a two-dimensional tangent space from a multidimensional Procrustes shape space was used. Partial Procrustes fitting using the GLS method was used to align the landmark sets. In partial Procrustes fitting, specimens are first centered, then scaled to unit centroid size, and then superimposed by minimizing d, the square-root of the sum of the squared distances between all corresponding landmarks (Rohlf 1999). In GLS the average configuration of the specimen shapes is used as the point of contact between the tangent and shape spaces. The tpsSmall program (Rohlf 1998) was used to calculate the Euclidean tangent distances, that are reported in Appendix 2. If shape distances are small enough that projection does not distort relationships among objects, tangent distances can be substituted for Procrustes distances for multivariate analysis (Rohlf 1998).

For these data there is a good correlation between the distances in shape space and the tangent distances (Figure 11.5). Because different superposition methods produce different distances, two alternatives were tried. These were full Procrustes fitting (d_F), in which one of the specimens is scaled to $\cos(\rho)$ and pairwise fits between specimens using the least squares (LS) method. These metrics did not produce different results and are not reported further.

The lengths of tree branches in years were calculated by subtracting the estimated end age from the estimated beginning age of each branch. Absolute ages in mya were assigned to each sample by extrapolating from known age tie-in points in the Big Horn Basin sequence. Linear regression of meters of sediment onto absolute age estimates for seven magnetostratigraphic and geochemical tie points yielded a prediction equation for converting meter level into absolute age. The tie points used included: 390 m = 58.2 mya, 500 m = 57.8 mya, 830 m = 56.5 mya, 1,080 m = 56.0 mya, 2,400 m = 52.8 mya, 2,200 m = 53.3 mya, 1,520 m = 55.5 mya (Koch *et al.* 1995; Clyde *et al.* 1994; Butler *et al.* 1981). The resulting prediction equation is:

$$y = -0.0026 x + 58.98 \tag{11.1}$$

where y is millions of years before present and x is meter level in the Big Horn Basin sequence. Both the meter levels and estimated ages for each sample are reported in Appendix 1. It should be noted that regression has the curious effect of 'reassigning' absolute ages to known tie points (e.g., the 1,520 m horizon which represents the Paleocene–Eocene boundary is assigned an age of 54.9 mya rather than 55.5 mya).

228   P. David Polly

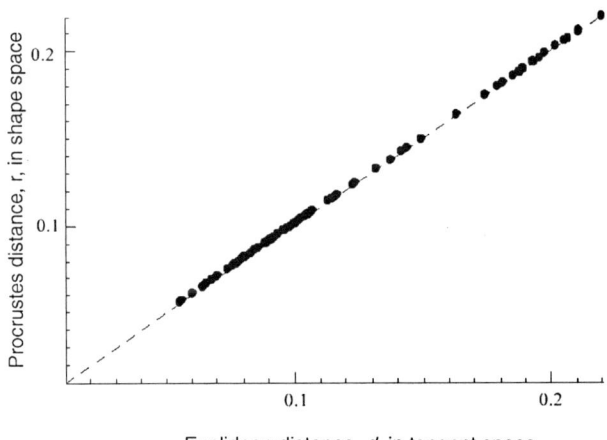

Figure 11.5 Plot of Euclidean distances, *d*, in tangent space against Procrustes distance ρ^2, in shape space for 13 of viverrid first lower molar consensuses. The slope of the regression is 0.998430 and the correlation is 0.999999. Distances between shapes are small enough to justify using tangent space distances in this analysis. Plot generated by tpsSmall (Rohlf 1998).

There are other interpolation methods that do not distort the original data points (e.g., Fricke *et al.* 1998). Since the end goal is not to assign absolute ages but rather to determine branch lengths these methodological discrepancies are unimportant. It should also be noted that the meter levels and the absolute ages estimated from them are subject to several other sources of uncertainty. Specimens were grouped into samples by subage (that introduces some time averaging) and there is uncertainty around the magnetostratigraphic tie points and the absolute ages assigned to them. Furthermore, the sedimentation rate in the Big Horn Basin was not constant. Many different facies are present in the sequence, each with its own sedimentary regime and rate, and the global sedimentary rate apparently decreased through the section as the Laramide orogeny ended (Clyde 1997). This error does not substantially affect the results of this chapter, because the uncertainty in branch lengths due to the error is only a small percentage of the total branch lengths.

Branch lengths were then converted from years to generations. The number of generations per branch was estimated by first estimating the body mass for the samples being considered here and then by estimating the relationship between generation time and body mass in mammals. To estimate body mass from the area of the first lower molar the prediction equation for Carnivora from Legendré (1986) was used:

$$y = 1.922 x + 0.709 \tag{11.2}$$

where *x* is the natural log of the area of the first lower molar and *y* is the natural log of body mass in grams. Body mass estimates for each sample considered here based on this equation are presented in Appendix 1. Generation times were estimated from body mass by using a prediction equation derived from regression generation time in days on body mass in grams. Generation times of 89 species were calculated from data in

Appendix 4 of Eisenberg (1981) by adding 'Gestation' and 'Age at First Mating'. Body masses for those species were taken from Eisenberg (1981) and Silva and Downing (1995). The natural log of generation time was regressed on body mass, yielding the following prediction equation:

$$y = 0.24x + 4.23 \tag{11.3}$$

where y is the natural log of generation time in days and x is the natural log of body mass in grams ($r^2 = 0.62$). This equation was used to estimate the generation time of each sample considered here (Table 11.1). Branch lengths between all possible sample pairs, both in years and generations, are reported in Appendix 3. Note that the conversion of branch length units to generations makes the tree additive rather than

Table 11.1 Rates of evolutionary change in mammalian dental traits

Trait	Species	Per-generation rate (haldanes)	LRI slope	Median log interval	References
Lower first molar length	*Kanisamys* spp.	0.000	−0.3	7.07	Gingerich and Gunnell 1995
Lower first molar length	*Cantius* spp.	0.000	−0.3	5.716	Clyde and Gingerich 1994
Upper third molar length	*Microtus pennsylvanicus*	0.003	−0.5	4.519	Gingerich 1993b
Lower first molar length	*Hyopsodus* spp.	0.011	−0.6	5.681	Gingerich and Gunnell 1995
Wing length	American house sparrow	0.024	—	—	Hendry and Kinnison 1999
Eye diameter	Norwegian stickleback	0.043	—	—	Hendry and Kinnison 1999
Lower first molar length	*Cosomys primus*	0.067	−0.8	5.299	Gingerich 1993b
Lower first molar length	*Giraffokery punjabiensis*	0.071	−0.8	5.80	Gingerich and Gunnell 1995
37 pooled dental traits	*Equus germanicus*	0.097	−0.8	3.392	Gingerich 1993b
Lower first molar	Viverravid spp.	0.132	−0.8	6.02	This study
Lower first molar length	*Phenacolemur praecox*	0.173	−0.9	5.58	Gingerich and Gunnell 1995
Body mass	*Mus musculus*	0.180	−0.2	0.836	Gingerich 1993a
Lower first molar length	*Progonomys* sp.	0.574	−1.0	6.58	Gingerich and Gunnell 1995
Lower first molar	*Cantius* spp.	0.653	−1.0	5.693	Clyde and Gingerich 1994
Spot number	Trinidadian guppies	0.742	—	—	Hendry and Kinnison 1999
Lower first molar length	*Hyracotherium grangeri*	0.225	−0.9	5.033	Gingerich 1993a
Shell spire height	*Littorina obtusata*	1.905	−0.6	1.711	Gingerich 1993a

ultrametric. The length of branches connecting smaller animals with shorter estimated generation times are reduced relative to those connecting larger animals with longer estimated generation times.

The per-generation rate of shape change was estimated as the y-intercept of a Log-Rate–Log-Interval (LRI) distribution of rates (Gingerich 1983, 1993a; Gingerich and Gunnell 1995; Hendry and Kinnison 1999). In many comparative methods, the rate of character change is estimated from the variance of tree-tip values scaled by their branch lengths (Felsenstein 1985). In addition to tree tips, within-branch and node samples are available in the present study. The LRI method allows all of these data to be combined in estimating the per-generation rate. This method employs regression through a log-rate *versus* log-interval distribution.

The slope of the regression indicates the extent to which the data fit a distribution expected under an evolutionary random walk (Brownian motion). The regression intercept provides an estimate of the per-generation rate of change. All possible rates of shape change were calculated in haldanes (standard deviations per generation), a dimension-independent unit that is convenient for comparing rates of evolutionary change (Gingerich 1993a; Hendry and Kinnison 1999). The Procrustes tangent distances, d, from Appendix 2 were first standardized by dividing each by the pooled standard deviation of the 13 samples (0.025). The standardized distances were then divided by the corresponding branch length (in generations) from Appendix 3. An LRI distribution of these rates was created by plotting the $\log_{10}$ of each rate by the $\log_{10}$ of the interval over which it was calculated (Figure 11.6). A LS regression line was calculated to determine the slope and the intercept. The slope indicates the extent to which the data are consistent with a Brownian-motion process, 0.5 representing a truly random process, 1.0 representing stasis, and 0.0 representing complete directionality (Gingerich 1993a). These slopes, projected through log-transformed data,

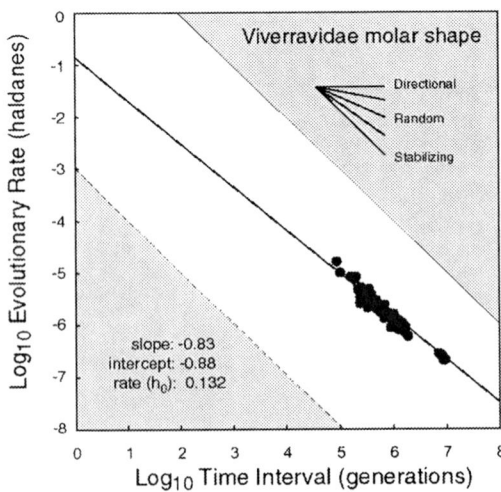

Figure 11.6 Log-Rate–Log-Interval plot. Data are molar crown shape of viverravid carnivores. The vertical axis is the $\log_{10}$ of rate of shape change (in haldanes), the horizontal axis is the interval over which the rate was calculated. The residual variation around the regression line is unusually low for an LRI distribution.

correspond respectively to three walk-modes in the original data space: Brownian-motion divergence (that accumulates as the square root of time, or time 0.5), no divergence (or which does not accumulate, or time 0.0), and constant, unidirectional divergence (that accumulates linearly with time, or time 1.0). The inverse log of the y-intercept is an estimate of the per-generation rate of the process, h_0 (the rate of change in haldanes over a single generation). The per-generation rate h_0 is equivalent to the square root of the rate β, a variance term, that is often used in phylogenetic comparative methods (e.g., Felsenstein 1985; Harvey and Pagel 1991).

To test whether two populations are more divergent than expected, a probability distribution must be estimated against which observed differences can be compared. Evolution is a time-series process in which change accumulates generation by generation; the end point of an evolutionary sequence depends on its initial starting point, the number of generations in the sequence, and the amount and 'direction' of change during each generation. A Brownian-motion random walk is a time-series process whose end point depends on its initial starting point, the number of steps in the walk, and the change at each step; furthermore, in Brownian motion the direction of change at any step is random and is not biased by change at other steps. There is no intrinsic directionality or trend in a Brownian-motion random walk, making it the appropriate model for random evolutionary change.

A probability distribution of series end points can be calculated using the Brownian-motion model, the per-generation rate of change, and the number of generations separating two populations on a phylogenetic tree. That distribution reflects the frequency of end points if the same random walk is 'walked' many times from the same initial starting point, with the same per-step rate of change, and for the same number of steps. Because there is no bias in the change at each step, the outcomes of a Brownian-motion walk are normally distributed. In this chapter shape differences are measured in units on a tangent plane to Procrustes space. Because that plane is two dimensional the variance of a random walk on it is calculated as

$$r^2 = 4Dt \tag{11.4}$$

where r^2 is the variance of the distribution, D is the squared per-step rate of change (or the per-step variance), and t is time (Berg 1993). In this study, D is the squared per-generation rate, h_0 as estimated from the LRI distribution, and t is the number of generations separating two populations on a phylogenetic tree. The probability distribution for test used here is, thus, a normal distribution with a variance of $4h_0 t$ centered on the value of one of the two populations being compared. Note that the standard deviation of that distribution is the square root of the variance, or $2\sqrt{h_0 t}$. The square root of time in this equation corresponds to the scaling discussed in the previous paragraph in relation to the slope of the LRI distribution. Regardless of the dimensionality of the data, the variance of a Brownian-motion process scales linearly with time.

Critical values for the test are now easily calculated. In evolutionary biology two phenomena are often of interest: more divergence than expected (e.g., because of directional selection) and less than expected (e.g., stasis). These suggest two sets of critical values, one for values at the tails of the normal distribution and one for values very near its mean. To accommodate this, the standard 5 per cent value can be divided in two so that 2.5 per cent of the curve is allocated to the tails of the distribution and 2.5 per cent around the mean (Bookstein 1991; Gingerich 1993a). Accordingly, 97.5 per cent of the space under a normal curve lies between points 2.241 standard deviations above and below its center and 2.5 per cent between 0.031 standard deviations (Sokal and Rohlf 1995). The standard deviation of the distribution is the square root of its variance (given by Equation 11.4). Thus, given a total α of 0.05, any value greater than $2.241 \times 2h_0\sqrt{t}$ can be considered non-random directional change and any value less than $0.031 \times 2h_0\sqrt{t}$ is non-random stasis.

When a large number of tests are made, it is likely that a fraction of them will result in rejection of the null hypothesis by chance (Type I error). In this study 72 pairwise comparisons are made (Appendix 2) to assess departures from a random-walk model. If the critical value (α) of each test is set at 0.05 then 5 per cent of the tests will falsely reject the null model. Out of 72 tests, more than three could fall into this category. Bonferroni-style corrections adjust the critical values to make the test more robust in the face of this possibility. Here an amended critical value was calculated as:

$$\alpha' = \frac{\alpha}{k} \qquad (11.5)$$

where α' is the Bonferroni-corrected critical value, α is the original critical value (0.05), and k is the number of comparisons made (72). The adjusted critical value is thus 0.0007, and the corresponding critical values would be $3.58 \times 2h_0\sqrt{t}$ for non-random directional change and $0.0004 \times 2h_0\sqrt{t}$ for non-random stasis. For comparison, consistency with the null Brownian-motion model are flagged at critical levels of both 0.05 and 0.0007.

Results

Euclidean shape distances, d, in tangent space ranged from 0.05 (found in several stratigraphically adjacent comparisons) to 0.21 (between *V. acutus* Wa-2 and *D. proteus* Wa-0). Measured in standard deviations, the range is 1.83–8.40. Both Euclidean shape distances and their standard deviation equivalents are reported in Appendix 2. All of the pairwise comparisons between a *Viverravus* and *Didymictis* population have much larger shape distances than do comparisons within either group. This is not surprising since those comparisons are separated by more than eight million years (Appendix 3). Nevertheless, the longest branch length, though (between *V. acutus* Wa-2 and *D. protenus* Wa-6), does not have the greatest shape divergence. Neither does the shortest branch length (*D. proteus* Wa-0 to either *D. leptomylus* Wa-1 or *D. protenus* Wa-1) have the smallest shape divergence.

Shape divergence and branch length are strongly correlated, however. The LRI distribution and the parameters estimated from it are summarized in Figure 11.6. The pairwise rates range from −6.68 to −4.79 log units and the intervals from 4.95 to 6.97 log units. The distribution is unusually linear compared to similar plots of univariate size traits (cf., Gingerich 1993a; Gingerich and Gunnell 1995). The slope of the regression through the points is −0.83, suggesting a process tending towards stasis, but with some random component (a perfectly random walk has a slope of −0.50, perfect stasis produces a slope of −1.00). The y-intercept gives the per-generation rate of change for the distribution (remembering that $\log_{10}$ of one generation is zero). It is −0.88 log units. Log detransformation yields an h_0 of 0.132 haldanes (standard deviations per generation). This rate is not exceptional, falling well within the range of other studied traits (Table 11.1).

The amount of shape change that can be produced by a random walk when the per-generation rate is scaled to the time intervals found between samples in this study is large. The upper 5 per cent critical values (above which a random walk is rejected in favor of directional selection) ranged from 175.9 to 1,811.8 standard deviations (Appendix 4). None of the observed values in Appendix 2 approach these magnitudes. The lower critical values (below which a random walk is rejected in favor of stabilizing selection) range from 2.4 to 25.1 standard deviations. All but six of the observed distances fall below their respective critical values at the 5 per cent level, meaning that they are not consistent with a Brownian-motion model. A stabilizing process must be invoked to explain the small changes in molar shape observed over evolutionary intervals of this length. The six comparisons that fall between the two critical values and are thus consistent with Brownian-motion random change are all comparisons made over very short intervals: *Didymictis proteus* Wa-0; *D. protenus* Wa-1, *D. proteus* Wa-0; *D. leptomylus* Wa-2, *D. protenus* Wa-1; *D. protenus* Wa-2, *D. protenus* Wa-2; *D. leptomylus* Wa-1, *D. protenus* Wa-1; *D. leptomylus* Wa-1, and *D. leptomylus* Wa-2; *D. leptomylus* Wa-1 (Appendix 2). When the Bonferroni adjustments are made to the critical values – which make the stasis interval narrower and the directional interval wider – all of the distances fall within the area consistent with Brownian motion. This emphasizes that the shape divergences observed in this study are small relative to the amount of divergence time, falling close to, but just outside of, the statistical limits of stasis. This finding fits with the LRI slope of −0.83 reported above.

Discussion

Accurate estimates of phylogenetic divergence times are an important part of this study. Calculation of evolutionary rates, estimation of per-generation rates, and estimation of phylogenetic variance all incorporate branch lengths calibrated in absolute time units. Accurately calibrated times of divergence are crucial. These are not unique to this study. Many standard methods in evolutionary biology require time-of-divergence estimates, including investigations of trait correlation (e.g., independent contrasts; Felsenstein 1985), molecular clock studies (Zuckerkandl and Pauling 1962; Wilson *et al.* 1987; Kumar and Hedges 1998), and ancestral node values reconstructions (Martins and Hansen 1997; Garland *et al.* 1999). These investigations can make due with coarse estimates of divergence time. Indeed, many have simply set each branch to unit length. However, the more precise (and accurate) the time-of-divergence estimate, the more precise the results. But precise branch lengths in units of absolute time are seldom available, sometimes because authors make little attempt to procure relevant data, but more often because paleontological data are either not available or have not been incorporated into a phylogenetic hypothesis in a way that makes branch lengths easy to determine. Traditional phylogenetic reconstruction algorithms – including parsimony cladistics, maximum-likelihood, neighbor-joining, and UPGMA clustering – all treat their operation taxonomic units (OTUs) as terminal taxa, meaning that only minimum divergence times can be estimated. However, that situation is quickly changing. In addition to discrete-character-parsimony-based stratocladistics there are now distance methods that place OTUs along branches and at nodes. Serial sample UPGMA (sUPGMA) reconstructs a phylogeny from data sampled serially in time using pairwise distances and a list of the time-order of the samples (Rodrigo 2000). Maximum-likelihood methods can also be adapted for dealing with non-contemporaneous OTUs (Huelsenbeck and Rannala 2000).

The difficulty of calibrating divergence times from paleontological data is compounded by vagueness introduced by standard phylogenetic reconstruction methods. Cladograms and synapomorphies diagnose sister-group relationships, but are unable to distinguish whether known taxa fall at tree nodes or lie at the tips. When closely related species are being considered, this difference can be crucial for estimating divergence times. Niles Eldredge pointed out that a cladogram showing the sister-group relationships of two taxa is logically consistent with at least three true phylogenies (Eldredge 1979; Eldredge and Cracraft 1980). Figure 11.7 illustrates this. Taxa A and B share a derived feature (Figure 11.7A), and A occurs stratigraphically lower than B. But the cladogram does not tell us whether both are descended from some unknown common ancestor (Figure 11.7B), whether A is the ancestor of B (Figure 11.7C), or whether B is the ancestor of A (Figure 11.7D). These three phylogenies imply three very different sets of phylogenetic time intervals. Let us presume that A is older than B by one million years and has a shape difference of one unit. If A and B are true sister groups – each descended from some unknown common ancestor that existed earlier in time – then total time separating them may be something like three million years

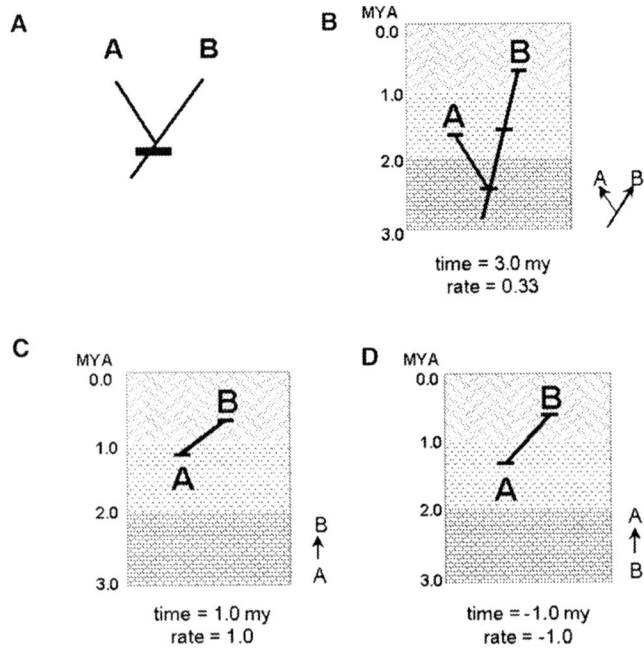

Figure 11.7 Eldredge's enigma. (A) A cladogram of two sister-taxa, united by a synapomorphy. (B, C, D) The three phylogenetic relationships consistent with the cladogram in A. The branch lengths connected the two taxa and the rate of change given that interval are shown below each scenario. See text for details.

(it could be more or less depending on how ancient the common ancestor is). The rate of shape change we estimate from them would thus be 1/3 (million) or 0.33 units per million years. If A is ancestral to B, however, then only one million years separates them and we estimate a shape-change rate of 1.0 units per million years. For this scenario the third alternative is actually impossible: B could not really be ancestral to A. But if it was, there would be an interval of negative one million years separating them and a shape-change rate of −1.0. Because of the different branch lengths and shape-change rates in each scenario, different per-generation rates would be estimated and different comparison intervals would be used in assessing the significance of shape change. The rate in Figure 11.7C is three times that in Figure 11.7B, and might be judged directional while the latter might be static. Eldredge's enigma should be given careful attention when applying significance tests to phylogenetic data.

If branch lengths are incorrectly estimated they may or may not affect tests of significance as used in this chapter. If all branches are uniformly misestimated, then the significance of all pairwise tests are likely to change. In this case, all of the pairwise rate comparisons using in the LRI distribution would be changed, which would shift the entire distribution and result in a different estimate of the per-generation rate. However, the interval over which each test is made and the variance associated with that interval would both change, too. Uniformly underestimated branch lengths would result in uniformly overestimated rates and an underestimated interval of comparison. A high rate over a short interval is likely to be judged directional. Conversely, uniformly overestimated lengths result in underestimated rates and a long interval of comparison. A false 'stasis' is likely to occur in this case. If only one or two branch lengths are misjudged then only comparisons between taxa connected by those branches are likely to be affected.

The incorrect branch lengths will affect a few of the points in the LRI distribution, probably making it less linear, but not significantly changing its slope and intercept. This means that the same per-generation rate would be estimated. Only those taxa connected by the incorrect branches will have incorrect comparison intervals, the rest will be unaffected. This study could be affected either by uniform misestimation or by individually incorrect branches. The former could result from the regression used to assign absolute ages to meter levels (Equation 11.1), from the regression used to estimate body mass (Equation 11.2), or from the regression used to estimate generation times (Equation 11.3).

Individual branch lengths could also be misestimated only if the phylogenetic hypothesis is incorrect. One such possibility is the basal node of Figure 11.4. *Protictis paralus* was considered to be the last common ancestor of *Viverravus* and *Didymictis* for the purposes of this study, but the original phylogenetic analysis (Figure 11.2B) found its position equivocal – it could be ancestral or it could be the sister group to *Didymictis* and *Viverravus*. If the latter, then the branches leading from *P. paralus* could either be overestimated or underestimated. This would alter all of the rates calculated between *Viverravus* and *Didymictis* sample and their intervals of comparison.

This study indicates that long-term molar shape evolution is slow compared to its estimated per-generation rate. Over periods of hundreds of thousands to millions of years molar shape change is slow, falling close to the statistical definition of stasis adopted here. The per-generation rate of change (0.132 h), however, is similar to that of other mammalian dental traits, many of which exhibit quite rapid long-term change (Table 11.1). Interestingly, most previously studied traits were univariate measures of size rather than shape. The only previous analysis of rates of dental shape change is Clyde and Gingerich (1994), who looked at rates of change in molar crown shape using the x and y components of Bookstein shape variables as their metric. These authors found a similar pattern: long-term shape change was near stasis, but the per-generation rate was like that of other dental size variables (Table 11.1).

Why might the evolution of molar shape have a normal per-generation rate of change (indicating that, on a short time scale, shape can change as rapidly as other variables), yet evolve slowly over long periods? One factor may be that occlusal relationships introduce functional constraints which allow small scale variation, but prevent major topographical rearrangements of cusps. The functional relationships between upper and lower tooth cusps, cristae, and basins have been well studied, indicating that there is often a precise match in the morphology of occluding structures (Herring 1993; Fortelius 1985; Kay and Hiiemae 1974; Crompton and Hiiemae 1970). In this study, the landmarks all represent functional components of the molar crown, including the carnassial blade, that is part of a tightly fitting shear surface, and the margins of the talonid basin, that helps guide the teeth into centric occlusion (Polly 1998; Savage 1976; Crompton and Hiiemae 1970). Changes in the relative positions of these cusps would require coordinated changes on the occluding upper teeth (P4 and M1). Another factor that may constrain long-term shape change is that the developmental and genetic control of molar cusp position is quite complex. The cusps on a single molar are initiated by a cascade of gene expression centers – the enamel knots – beginning with a single primary enamel knot, which coordinate a series of secondary enamel knots, each of which is associated with one of the tooth's cusps (Jernvall et al. 1998; Maas and Bei 1997; Thesleff and Sahlberg 1996; Jernvall et al. 1994). A number of genes and gene products are involved in the cascade, including SHH, BMP-2, BMP-4, BMP-7, FGF-4, p21, Msx-1, Msx-2, and Lef-1. The cascade of gene interactions suggests that changes in the primary enamel knot or early secondary knots have cumulative effects on the cusps associated with later knots (Jernvall 1995, 2000), although some studies have shown that cusps associated with early cascade events are more variable (Polly 1998). Keränen et al. (1998) found that species-specific morphological differences in tooth shape were associated with changes in the expression of regulatory genes in secondary knots.

Correspondingly, the early stages of molar crown ontogeny appear to be more conserved than late stages (Popowics 1998; Wolsan 1989) and genetic changes that reduce the signaling of the primary enamel knot result in drastically malformed teeth (Pispa et al. 1999). Futhermore, complexly occluding teeth have enamel microstructural patterns that are integrated with the gross crown morphology of the tooth (Koenigswald 1982, 1997a,b; Koenigswald and Clemens 1992). It has been found that the number of amelogenin products that go into producing microstructural patterns increases along with the complexity of the tooth (Mathur and Polly 2000; Girondot and Sire 1998). Evolution of molar crown shape is, thus, a complex process involving many functional and developmental variables. Long-term shape evolution may therefore require a large number of developmentally compatible changes.

One reviewer of this chapter noted that the pairwise rate comparisons used in the LRI analysis are not corrected for phylogenetic covariance and are, therefore, not statistically independent (Felsenstein 1985; Harvey and Pagel 1991). Phylogenetic effects could affect the LRI distribution – and the per-generation rate estimated from it – if the rate of shape change is significantly different in one part of the tree relative to another or if particular branches have a much higher or lower rate of change than the rest of the tree. For example, if the divergence of *Viverravus* and *Didymictis* were associated with a greater degree of change than is found within either genus, then between-genus rate comparisons would be higher relative to the interval over which they were measured than would within-genus rate comparisons. The longest-interval rates (the points at the lower right of the LRI distribution, Figure 11.6) would, in this case, be shifted upwards on the LRI plot, changing both the slope of the regression line and the per-generation rate estimate. However, the LRI distribution itself would become notably non-linear because the shorter-interval rates would not be changed. Any distortion caused by phylogenetic non-independence ought to manifest itself in an LRI plot as a curvilinear distribution of points. Because the data in this study form a tight linear distribution, it is presumed that phylogenetic non-independence is not a confounding factor.

Conclusions

Morphometric data, or any quantitative data, demonstrably have phylogenetic signal in them (Felsenstein 1985; Cheverud and Dow 1985; Schluter *et al.* 1997). Organisms that share a recent common ancestor are expected to be more similar than those whose ancestry is more ancient. For this reason, phylogenetic distance is an important consideration in interpreting morphometric data, and geometric landmark morphometrics is no exception. The standard T^2 test is probably not appropriate if the shapes being compared are from phylogenetically divergent populations or species. Instead, shape differences should be compared in light of the divergence expected given the length of time since the two shared a common ancestor. An appropriate null model is a Brownian-motion random walk, which makes no assumptions about directionality or magnitude of change. To apply such a test, several parameters must be estimated, including the phylogeny connecting the taxa to be compared, the branch lengths of that phylogeny (preferably in either absolute time or generations), the within-population variance of the shape, between-sample distances in the shape, and the per-unit-time rate of change in shape.

The fact that quantitative data contain phylogenetic signal implies that phylogeny can be recovered from morphometric data (Felsenstein 1973, 1988). Despite the simplicity of that statement, it has been a key controversy in the systematics literature for more than a quarter of a century, first in the debates over phenetics (overall similarity) and cladistics as phylogeny reconstruction methods (Williams 1971; Kluge and Farris 1969), and more recently in debates over maximum-likelihood and parsimony criteria (Felsenstein 1978; Hillis *et al.* 1994; Kluge 1997; Siddall and Kluge 1997). One interesting aspect of the question is that morphometrics usually involves a quantitative assessment of differences (or similarities) in the same homologous feature among taxa. The properties of random walks tell us that divergence in quantitative traits should accumulate, on average, with the square root of time elapsed since they last shared a common ancestor (Berg 1993). This suggests that, like molecular sequence data, morphometric data ought to contain information that can, with quantifiable error, be used to reconstruct phylogeny, and that morphometric data will have a 'saturation point', or a maximum divergence time beyond which they are incapable of accurately resolving a phylogeny. Different traits or variables will have different parameters in this regard. Character-based phylogeny reconstruction, unlike morphometrics, is usually concerned with the origin of new homologous traits (or the identification of homologies and homoplasies). Unlike random walks in quantitative variables, there is no obvious null expectation associated with the origin of new characters (but see Schluter *et al.* 1997). And unlike morphometrics, a multitude of characters can easily be combined into a single analysis using parsimony-based methods. These differences are often at the heart of debates on the use of morphometrics in systematics. It is entirely conceivable, though, that the choice of data and method may best be made by consideration of the scope of the phylogenetic problem, in terms of the period over which the taxa have diverged. Relationships among closely related taxa with little discrete character divergence may best be studied using quantitative data, while relationships among more distantly related taxa may be best resolved using parsimony analysis of discrete data.

Acknowledgements

Special thanks to Norm MacLeod and Peter L. Forey for inviting me to participate in the Systematic Association symposium on morphometrics and systematics. Philip Gingerich and Gregg Gunnell provided access to material. Joe Felsenstein provided critical comments that improved the manuscript. This research was supported by a grant from the Natural Environment Research Council of the United Kingdom (NERC GR8/03692).

References

Berg, H. C. (1993) *Random walks in biology*, Expanded edition, Princeton: Princeton University Press.

Bodenbender, B. E. (1995) 'Morphological, crystallographic, and stratigraphic data in cladistic analyses of blastoid phylogeny', *Contributions from the Museum of Paleontology, University of Michigan*, 29, 201–257.

Bookstein, F. L. (1991) *Morphometric tools for landmark data: geometry and biology*, Cambridge: Cambridge University Press.

Bown, T. M., Rose, K. D., Simons, E. L. and Wing, S. L. (1994) 'Distribution and stratigraphic correlation of Upper Paleocene and Lower Eocene fossil mammal and plant localities of the Fort Union, Willwood, and Tatman Formations, southern Bighorn Basin, Wyoming', *US Geological Survey Professional Paper*, 1540, 1–103.

Butler, R. F., Gingerich, P. D. and Lindsay, E. H. (1981) 'Magnetic polarity stratigraphy and biostratigraphy of Paleocene and lower Eocene continental deposits Clark's Fork Basin, Wyoming', *Journal of Geology*, 102, 367–377.

Cheverud, J. M. and Dow, M. M. (1985) 'An autocorrelation analysis of genetic variation due to lineal fission in social groups of rhesus macaques', *American Journal of Physical Anthropology*, 67, 113–121.

Clyde, W. C. (1997) 'Stratigraphy and mammalian paleontology of the McCullough Peaks, northern Bighorn Basin, Wyoming: implications for biochronology, basin development, and community reorganization across the Paleocene–Eocene boundary', Ph.D. thesis, The University of Michigan.

Clyde, W. C. and Gingerich, P. D. (1994) 'Rates of evolution in the dentition of early Eocene *Cantius*: comparison of size and shape', *Paleobiology*, 20, 506–522.

Clyde, W. C., Stamatakos, J. and Gingerich, P. D. (1994) 'Chronology of the Wasatchian Land-Mammal Age (Early Eocene): magnetostratigraphic results form the McCullough Peaks Section, Northern Bighorn Basin, Wyoming', *Journal of Geology*, 102, 367–377.

Crompton, A. W. and Hiiemae, K. (1970) 'Molar occlusion and mandibular movements during occlusion in American opossum, *Didelphis marsupialis*', *Zoological Journal of the Linnean Society*, 49, 21–47.

Eisenberg, J. F. (1981) The *mammalian radiations: an analysis of trends in evolution, adaptation, and behavior*, Chicago: Chicago University Press.

Eldredge, N. (1979) 'Cladism and common sense', in Cracraft, J. and Eldredge, N. (eds) *Phylogenetic analysis and paleontology*, New York: Columbia University Press, pp. 165–198.

Eldredge, N. and Cracraft, J. (1980) *Phylogenetic patterns and the evolutionary process: method and theory in comparative biology*, New York: Columbia University Press.

Falconer, D. S. and Mackay, T. F. C. (1996) *Introduction to quantitative genetics*, 4th edn, Harlow: Addison Wesley Longman, Ltd.

Felsenstein, J. (1973) 'Maximum-likelihood estimation of evolutionary trees from continuous characters', *American Journal of Human Genetics*, 25, 471–492.

Felsenstein, J. (1978) 'Cases in which parsimony or compatibility methods will be positively misleading', *Systematic Zoology*, 27, 401–410.

Felsenstein, J. (1985) 'Phylogenies and the comparative method', *The American Naturalist*, 125, 1–15.

Felsenstein, J. (1988) 'Phylogenies and quantitative methods', *Annual Review of Ecology and Systematics*, 19, 445–471.

Fisher, D. C. (1992) 'Stratigraphic parsimony', in Maddison, W. P. and Maddison, D. R. (eds) *MacClade: analysis of phylogeny and character evolution, version 3*, Sunderland, Mass.: Sinauer and Associates, Inc., pp. 124–129.

Fisher, D. C. (1994) 'Stratocladistics: morphological and temporal patterns and their relation to phylogenetic process', in Grande, L. and Rieppel, O. (eds) *Interpreting the hierarchy of nature*, New York: Academic Press, pp. 133–171.

Flynn, J. J. and Galiano, H. (1982) 'Phylogeny of early Tertiary Carnivora, with a description of a new species of *Protictis* from the middle Eocene of northwestern Wyoming', *American Museum Novitates*, 2725, 1–64.

Fortelius, M. (1985) 'Ungulate cheek teeth: development, function, and evolutionary interactions', *Acta Zoologica Fennica*, 180, 1–76.

Fox, D. L., Fisher, D. C. and Leighton, L. R. (1999) 'Reconstructing phylogeny with and without temporal data', *Science*, 284, 1816–1819.

Fox, R. C. and Youzwyshyn, G. P. (1994) 'New primitive carnivorans (Mammalia) from the Paleocene of western Canada, and their bearing on the relationships of the order', *Journal of Vertebrate Paleontology*, 14, 382–404.

Fricke, H. C., Clyde, W. C., O'Neil, J. R. and Gingerich, P. D. (1998) 'Evidence for rapid climate change in North America during the latest Paleocene thermal maximum: oxygen isotope compositions of biogenic phosphate from the Bighorn Basin (Wyoming)', *Earth and Planetary Science Letters*, 160, 193–208.

Garland, T. Jr., Midford, P. E. and Ives, A. R. (1999) 'An introduction to phylogenetically based statistical methods, with a new method for confidence intervals on ancestral values', *American Zoologist*, 39, 374–388.

Gingerich, P. D. (1980) 'The Bighorn Basin – Why is it so important?', *Papers on Paleontology, University of Michigan*, 24, 1–5.

Gingerich, P. D. (1983) 'Rates of evolution: effects of time and temporal scaling', *Science*, 222, 159–161.

Gingerich, P. D. (1991) 'Systematics and evolution of early Eocene Perissodactyla (Mammalia) in the Clarks Fork Basin, Wyoming', *Contributions from the Museum of Paleontology, The University of Michigan*, 28, 181–213.

Gingerich, P. D. (1993a) 'Quantification and comparison of evolutionary rates', *American Journal of Science*, 293-A, 453–478.

Gingerich, P. D. (1993b) 'Rates of evolution in Plio-Pleistocene mammals: six case studies', in Martin, R. A. and Barnosky, A. D. (eds) *Morphological change in quaternary mammals of North America*, Cambridge: Cambridge University Press, pp. 84–106.

Gingerich, P. D. and Gunnell, G. F. (1995) 'Rates of evolution in Paleocene–Eocene mammals of the Clarks Fork Basin, Wyoming, and a comparison with Neogene Siwalik lineages of Pakistan', *Palaeogeography, Palaeoclimatology, Palaeoecology*, 115, 227–247.

Gingerich, P. D. and Klitz, K. (1985) 'University of Michigan Paleocene and Eocene fossil localities in the Fort Union and Willwood Formations, Clark's Fork Basin, Wyoming', *University of Michigan Museum of Paleontology*, map, 1 sheet.

Gingerich, P. D. and Winkler, D. A. (1985) 'Systematics of Paleocene Viverravidae (Mammalia, Carnivora) in the Bighorn and Clark's Fork Basin, Wyoming', *Contributions from the Museum of Paleontology, University of Michigan*, 27, 87–128.

Girondot, M. and Sire, J.-Y. (1998) 'Evolution of the amelogenin gene in toothed and toothless vertebrates', *European Journal of Oral Science*, 106, 501–508.

Harvey, P. H. and Pagel, M. D. (1991) *The comparative method in evolutionary biology*, Oxford: Oxford University Press.

Heinrich, R. E. and Rose, K. D. (1997) 'Postcranial morphology and locomotor behavior of two early Eocene miacoid carnivorans, *Vulpavus* and *Didymictis*', *Palaeontology*, 40, 279–305.

Hendry, A. P. and Kinnison, M. T. (1999) 'Perspective: the pace of modern life: measuring rates of contemporary microevolution', *Evolution*, 53, 1637–1653.

Herring, S. W. (1993) 'Functional-morphology of mammalian mastication', *American Zoologist*, 33, 289–299.

Hillis, D. M., Huelsenbeck, J. P. and Swofford, D. L. (1994) 'Consistency: hobgoblin of phylogenetics?', *Nature*, 369, 363–364.

Huelsenbeck, J. P. and Rannala, B. (2000) 'Using stratigraphic information in phylogenetics', in Wiens, J. (ed.) *Phylogenetic analysis of morphological data*, Washington, D.C.: Smithsonian Institution Press, pp. 165–191.

Hunt, R. M. and Tedford, R. H. (1993) 'Phylogenetic relationships within the aeluroid carnivora and implications of their temporal and geographic distribution', in Szalay, F. S., Novacek, M. J. and McKenna, M. C. (eds) *Mammal phylogeny: placentals*, New York: Springer-Verlag, pp. 53–73.

Jernvall, J. (2000) 'Linking development with generation of novelty in mammalian teeth', *Proceedings of the National Academy of Sciences of the United States of America*, 97, 2641–2645.

Jernvall, J. (1995) 'Mammalian molar cusp patterns: developmental mechanisms of diversity', *Acta Zoologica Fennica*, 198, 1–61.

Jernvall, J., Åberg, T., Kettunen, P., Keränen, S. and Thesleff, I. (1998) 'The life history of an embryonic signaling center: BMP-4 induces p21 and is associated with apoptosis in the mouse tooth enamel knot', *Development*, 125, 161–169.

Jernvall, J., Kettunen, P., Karavanova, I., Martin, L. B. and Thesleff, I. (1994) 'Evidence for the role of the enamel knot as a control center in mammalian tooth cusp formation: non-dividing cells express growth stimulating Fgf-4 gene', *International Journal of Developmental Biology*, 38, 463–469.

Kay, R. F. and Hiiemae, K. M. (1974) 'Jaw movement and tooth use in recent and fossil primates', *American Journal of Physical Anthropology*, 40, 227–256.

Keränen, S. V. E., Berg, T., Kettunen, P., Thesleff, I. and Jernvall, J. (1998) 'Association of developmental regulatory genes with the development of different molar shapes in two species of rodent', *Development, Genes, and Evolution*, 208, 477–486.

Kluge, A. G. (1997) 'Testability and the refutation and corroboration of cladistic hypotheses', *Cladistics*, 13, 81–96.

Kluge, A. G. and Farris, J. S. (1969) 'Quantitative phyletics and the evolution of anurans', *Systematic Zoology*, 8, 1–32.

Koch, P. L., Zachos, J. C. and Dettman, D. L. (1995) 'Stable isotope stratigraphy and paleoclimatology of the Paleogene Bighorn Basin (Wyoming, USA)', *Palaeogeography, Palaeoclimatology, Palaeoecology*, 115, 61–89.

Koenigswald, W. von. (1982) 'Enamel structure in the molars of Arvicolidae (Rodentia, Mammalia), a key to functional morphology and phylogeny', in Kurtén, B. (ed.) *Teeth: form, function, and evolution*, New York: Columbia University Press, pp. 109–122.

Koenigswald, W. von. (1997a) 'The variability of enamel structure at the dentition level', in von Koenigswald, W. and Sander, P. M. (eds) *Tooth enamel microstructure*, Rotterdam: Balkema, pp. 193–202.

Koenigswald, W. von (1997b) 'Brief survey of enamel diversity at the schmelzmuster level in Cenozoic placental mammals', in von Koenigswald, W. and Sander, P. M. (eds) *Tooth enamel microstructure*, Rotterdam: Balkema, pp. 137–162.

Koenigswald, W. von and Clemens, W. A. (1992) 'Levels of complexity in the microstructure of mammalian enamel and their application in studies of systematics', *Scanning Microscopy*, 6, 195–218.

Kumar, S. and Hedges, S. B. (1998) 'A molecular timescale for vertebrate evolution', *Nature*, 392, 917–920.

Legendré, S. (1986) 'Analysis of mammalian communities from the late Eocene and Oligocene of southern France', *Palaeovertebrata*, 16, 191–212.

Maas, R. and Bei, M. (1997) 'The genetic control of early tooth development', *Critical Reviews in Oral Biology and Medicine*, 8, 4–39.

Maddison, W. P. (1991) 'Squared-change parsimony reconstructions of ancestral states for continuous-valued characters on a phylogenetic tree', *Systematic Zoology*, 40, 304–314.

Martins, E. P. and Hansen, T. F. (1997) 'Phylogenies and the comparative method: a general approach to incorporating phylogenetic information into the analysis of interspecific data', *American Naturalist*, 149, 646–667.

Mathur, A. K. and Polly, P. D. (2000) 'The evolution of enamel microstructure: how important is amelogenin?', *Journal of Mammalian Evolution*, 7, 23–42.

Pispa, J., Jung, H. S., Jernvall, J., Kettunen, P., Mustonen, T., Tabata, M. J., Kere, J. and Thesleff, I. (1999) 'Cusp patterning defect in Tabby mouse teeth and its partial rescue by FGF', *Developmental Biology*, 216, 521–534.

Polly, P. D. (1997) 'Ancestry and species definition in paleontology: a stratocladistic analysis of Paleocene–Eocene Viverravidae (Mammalia, Carnivora) from Wyoming', *Contributions from the Museum of Paleontology, The University of Michigan*, 30, 1–53.

Polly, P. D. (1998) 'Variability, selection, and constraints: development and evolution in viverravid (Carnivora, Mammalia) molar morphology', *Paleobiology*, 24, 409–429.

Popowics, T. E. (1998) 'Ontogeny of postcanine tooth form in the Ferret, *Mustela putorius* (Carnivora, Mammalia), and the evolution of dental diversity within the Mustelidae', *Journal of Morphology*, 237, 69–90.

Rodrigo, A. (2000) 'Serial sample UPGMA', hyperlink "http://www.cebl.auckland.ac.nz/" http://www.cebl.auckland.ac.nz/.

Rohlf, F. J. (1998) 'tpsSmall: is shape variation small?', Stony Brook, New York: Department of Ecology and Evolution, State University of New York.

Rohlf, F. J. (1999) 'Shape statistics: Procrustes superimpositions and tangent spaces', *Journal of Classification*, 16, 197–223.

Rose, K. D. (1981) 'The Clarkforkian Land-Mammal Age and mammalian faunal composition across the Paleocene–Eocene boundary', *University of Michigan Papers on Paleontology*, 26: 1–196.

Savage, R. J. G. (1976) 'Evolution in carnivorous mammals', *Palaeontology*, 20, 237–271.

Schluter, D., Price, T., Mooers, A. O. and Ludwig, D. (1997) 'Likelihood of ancestor states in adaptive radiation', *Evolution*, 51, 1699–1711.

Siddall, M. E. and Kluge, A. G. (1997) 'Probabilism and phylogenetic inference', *Cladistics*, 13, 313–336.

Silva, M. and Downing, J. A. (1995) *CRC handbook of mammalian body masses*, Boca Raton: CRC Press.

Simpson, G. G. (1951) 'The species concept', *Evolution*, 5, 285–298.

Sokal, R. R. and Rohlf, F. J. (1995) *Biometry*, 3rd edn, New York: W. H. Freeman and Co.

Thesleff, I. and Sahlberg, C. (1996) 'Growth factors as inductive signals regulating tooth morphogenesis', *Seminars in Cell and Developmental Biology*, 7, 185–193.

Williams, W. T. (1971) 'Principles of clustering', *Annual Review of Ecology and Systematics*, 2, 303–326.

Wilson, A. C., Ochman, H. and Prager, E. M. (1987) 'Molecular time scale for evolution', *Trends in Genetics*, 3, 241–247.

Wolsan, M. (1989) 'Dental polymorphism in the genus *Martes* (Carnivora, Mustelidae) and its evolutionary significance', *Acta Theriologica*, 34, 545–593.

Wyss, A. R. and Flynn, J. J. (1993) 'A phylogenetic analysis and definition of the Carnivora', in Szalay, F. S., Novacek, M. J. and McKenna, M. C. (eds) *Mammal phylogeny: placentals*, New York: Springer-Verlag, pp. 32–52.

Zuckerkandl, E. and Pauling, L. (1962) 'Evolutionary divergence and convergence in proteins', in Bryson, V. and Vogel, H. J. (eds) *Evolving genes and proteins*, New York: Academic Press.

Appendices

Appendix 1 Age, meter level, sample size, mean area and standard deviation of first lower molar, estimated body mass, and estimated generation times for samples relevant to this study. Age details include North America Land Mammal Subages (NALMS), meter level in the Bighorn Basin, and absolute age in millions of years before present. Mean molar area in the natural log of the length times the width of the first lower molar. Body mass estimates are given in grams. Estimated generation times are reported in years

Species	NALMS	Meters	Age[1]	N	Mean	SD	Mass[2]	Generation[3]
Didymictis proteus	Ti-4	550	57.5	2	3.85	0.071	3,323	1.3
Didymictis proteus	Ti-5	700	57.2	1	3.60	—	2,055	1.2
Didymictis proteus	Cf-1	1,050	56.2	3	3.69	0.057	2,455	1.2
*Didymictis proteus**	Cf-2	1,300	55.6	20	3.95	0.092	4,001	1.4
*Didymictis proteus**	Cf-3	1,430	55.3	9	3.97	0.064	4,222	1.4
Didymictis proteus,**	Wa-0	1,550	54.9	3	3.61	0.103	2,097	1.2
*Didymictis leptomylus**	Wa-1	1,600	54.8	7	3.89	0.128	3,612	1.4
*Didymictis leptomylus**	Wa-2	1,700	54.6	6	3.66	0.142	2,306	1.2
Didymictis leptomylus	Wa-3	1,900	54.0	1	3.38	—	1,347	1.1
*Didymictis protenus**	Wa-1	1,600	54.8	3	4.44	0.158	10,336	1.8
*Didymictis protenus**	Wa-2	1,700	54.6	2	4.14	0.007	5,824	1.5
*Didymictis protenus**	Wa-3	1,900	54.0	6	4.12	0.197	5,573	1.5
*Didymictis protenus**	Wa-4	2,100	53.5	14	4.23	0.105	6,943	1.6
*Didymictis protenus**	Wa-5	2,250	53.1	12	4.34	0.145	8,474	1.7
*Didymictis protenus**	Wa-6	2,400	52.7	8	4.57	0.261	13,158	1.9
Viverravus laytoni	Ti-5	700	57.2	2	2.05	0.057	104	0.6
*Viverravus laytoni***	Cf-2	1,300	55.6	2	2.11	0.129	117	0.6
Viverravus acutus	Cf-3	1,430	55.3	1	1.92	—	82	0.5
*Viverravus acutus**	Wa-1	1,600	54.8	20	2.23	0.176	148	0.6
*Viverravus acutus**	Wa-2	1,700	54.6	24	2.31	0.164	172	0.7
Viverravus acutus	Wa-3	1,900	54.0	27	2.55	0.151	272	0.7
Viverravus acutus	Wa-4	2,100	53.5	18	2.53	0.176	262	0.7
Viverravus acutus	Wa-5	2,250	53.1	12	2.61	0.136	309	0.7
Viverravus acutus	Wa-6	2,400	52.7	9	2.79	0.135	435	0.8
Viverravus politus	Ti-5	700	57.2	8	2.92	0.035	556	0.9
Viverravus politus	Cf-2	1,300	55.6	2	2.52	0.064	259	0.7
Viverravus politus	Cf-3	1,430	55.3	4	2.97	0.089	615	0.9
Viverravus politus	Wa-1	1,600	54.8	4	3.29	0.093	1,128	1.0
Viverravus politus	Wa-2	1,700	54.6	1	3.65	—	2,263	1.2
Viverravus politus	Wa-3	1,900	54.0	1	3.73	—	2,639	1.3
Viverravus rosei	Wa-2	1,700	54.6	1	1.81	—	66	0.5
Viverravus rosei	Wa-3	1,900	54.0	2	1.63	0.046	47	0.5
Viverravus rosei	Wa-4	2,100	53.5	2	1.63	0.031	47	0.5
*Viverravus paralus***	Ti-3	350	58.1	10	2.51	0.053	253	0.7
Protictis agastor	Ti-3	350	58.1	5	3.54	0.124	1,831	1.2
Protictis haydenianus	To	60	58.8	5	3.49	0.129	1,664	1.1

1 Age (millions of years before present) estimated from a linear LS regression of meters of accumulated sediment (Gingerich 1991) onto absolute age in millions of years before present (Clyde et al. 1994): $y = -0.0025x + 58.8$.
2 Body mass (grams) derived from linear LS regression line of body mass onto the area (length × width) of the first lower molars of extant Carnivores (Legendré 1986): $y = 1.922x + 0.709$.
3 Generation time (years) estimated from a linear LS regression of mammalian generation lengths (Eisenberg 1981) onto body mass (Eisenberg 1981; Silva and Downing 1995): $y = 0.2415x + 4.2276$.
* Samples with molar shape data considered in this study.
** Samples lying at tree nodes.

Appendix 2 Differences in first lower molar shape between samples used in this study. Numbers at the top left are those same distances divided by the pooled population standard deviation (0.025). Distances that are consistent with a Brownian-motion random walk are flagged at both the Bonferroni-adjusted 0.0007 level (*) and the 0.05 level (**). All distances are consistent with Brownian motion at the 0.0007 level, but only six can be distinguished from stasis at the 0.05 level

	V. acutus Wa-1	V. acutus Wa-2	D. proteus Wa-0	D. proteus Cf-3	D. proteus Cf-2	D. protenus Wa-6	D. protenus Wa-5	D. protenus Wa-4	D. protenus Wa-3	D. protenus Wa-2	D. protenus Wa-1	D. leptomylus Wa-2	D. leptomylus Wa-1
D. leptomylus Wa-1	7.32*	7.32*	3.90*	2.71*	3.62*	4.88*	4.54*	5.10*	5.29*	5.57**	4.19**	5.36**	
D. leptomylus Wa-2	7.09*	8.05*	2.81**	3.81*	3.20*	4.55*	3.30*	2.69*	2.18*	2.74*	4.52*		0.13**
D. protenus Wa-1	7.03*	7.86*	4.24**	3.38*	3.27*	3.86*	3.05*	3.28*	3.83**	3.73**		0.11*	0.10**
D. protenus Wa-2	6.78*	8.04*	3.70*	4.15*	3.17*	4.18*	3.22*	2.42*	2.34*		0.09**	0.07*	0.14**
D. protenus Wa-3	6.59*	7.43*	3.63*	3.90*	2.81*	3.55*	2.66*	2.03*		0.06*	0.10*	0.05*	0.13*
D. protenus Wa-4	6.13*	7.11*	3.50*	3.47*	3.05*	2.96*	1.87*		0.05*	0.06*	0.08*	0.07*	0.13*
D. protenus Wa-5	6.14*	6.85*	3.78*	3.45*	3.26*	2.84*		0.05*	0.07*	0.08*	0.08*	0.08*	0.11*
D. protenus Wa-6	6.84*	7.34*	4.28**	3.22*	3.01*		0.07*	0.07*	0.09*	0.10*	0.10*	0.11*	0.12*
D. proteus Cf-2	7.17*	7.68*	2.58*	1.83*		0.08*	0.08*	0.08*	0.07*	0.08*	0.08*	0.08*	0.09*
D. proteus Cf-3	7.17*	7.52*	2.26*		0.05*	0.08*	0.09*	0.09*	0.10*	0.10*	0.08*	0.10*	0.07*
D. proteus Wa-0	7.81*	8.40*		0.06*	0.06*	0.11*	0.09*	0.09*	0.09*	0.09*	0.11**	0.07*	0.10*
V. acutus Wa-2	2.85*		0.21*	0.19*	0.19*	0.18*	0.17*	0.18*	0.19*	0.20*	0.20*	0.20*	0.18*
V. acutus Wa-1		0.07*	0.20*	0.18*	0.18*	0.17*	0.15*	0.15*	0.16*	0.17*	0.18*	0.18*	0.18*
	V. acutus Wa-1	V. acutus Wa-2	D. proteus Wa-0	D. proteus Cf-3	D. proteus Cf-2	D. protenus Wa-6	D. protenus Wa-5	D. protenus Wa-4	D. protenus Wa-3	D. protenus Wa-2	D. protenus Wa-1	D. leptomylus Wa-2	D. leptomylus Wa-1

Appendix 3 Branch lengths between samples considered in this study. Numbers in the lower right half are in years, numbers in the upper right are in generations

	V. acutus Wa-1	V. acutus Wa-2	D. proteus Wa-0	D. proteus Cf-3	D. proteus Cf-2	D. proteus Wa-6	D. protenus Wa-5	D. protenus Wa-4	D. protenus Wa-3	D. protenus Wa-2	D. protenus Wa-1	D. leptomylus Wa-2	D. leptomylus Wa-1
D. leptomylus Wa-1	7,878,000	8,138,000	130,000	442,000	780,000	2,340,000	195,000	1,560,000	1,040,000	520,000	260,000	260,000	
D. leptomylus Wa-2	8,138,000	8,398,000	390,000	702,000	1,040,000	2,600,000	2,210,000	1,820,000	1,300,000	780,000	520,000		201,550
D. protenus Wa-1	7,878,000	8,138,000	130,000	442,000	780,000	2,080,000	1,690,000	1,300,000	780,000	260,000		392,347	190,797
D. protenus Wa-2	8,138,000	8,398,000	390,000	702,000	1,040,000	1,820,000	1,430,000	1,040,000	520,000		159,021	551,368	349,818
D. protenus Wa-3	8,658,000	8,918,000	910,000	1,222,000	1,560,000	1,300,000	910,000	520,000		343,234	502,255	894,602	693,052
D. protenus Wa-4	9,178,000	9,438,000	1,430,000	1,742,000	2,080,000	780,000	390,000		335,484	678,718	837,739	1,230,086	1,028,536
D. protenus Wa-5	9,568,000	9,828,000	1,820,000	2,132,000	2,470,000	390,000		239,264	574,748	917,982	1,077,003	1,469,350	1,267,800
D. protenus Wa-6	9,958,000	10,218,000	2,210,000	2,522,000	2,860,000		220,339	459,603	795,087	1,138,321	1,297,342	1,689,689	1,488,139
D. protenus Cf-2	7,098,000	7,358,000	650,000	338,000		1,867,206	1,646,867	1,407,603	1,072,119	728,885	569,864	583,791	785,341
D. protenus Cf-3	7,436,000	7,696,000	312,000		241,429	1,625,777	1,405,438	1,166,174	830,690	487,456	328,435	342,362	543,912
D. protenus Wa-0	7,748,000	8,008,000		240,000	481,429	1,385,777	1,165,438	926,174	590,690	247,456	88,435	102,362	303,912
V. acutus Wa-2	260,000		7,992,330	7,752,330	7,510,901	9,378,107	9,157,768	8,918,504	8,583,020	8,239,786	8,080,765	8,296,242	8,094,692
V. acutus Wa-1		406,250	7,586,080	7,346,080	7,104,651	8,971,857	8,751,518	8,512,254	8,176,770	7,833,536	7,674,515	7,889,992	7,688,442

Appendix 4 Critical values for pairwise shape comparisons. Values in the upper left are lower critical values, below which a random walk is rejected in favor of stabilizing selection; in the lower right are upper critical values, above which a random walk is rejected in favor of directional change. All observed distances (Appendix 2) fall below the directional critical values and all but six fall below the stasis values

	V. acutus Wa-1	V. acutus Wa-2	D. proteus Wa-0	D. proteus Cf-3	D. proteus Cf-2	D. proteus Wa-6	D. proteus Wa-5	D. proteus Wa-4	D. proteus Wa-3	D. proteus Wa-2	D. proteus Wa-1	D. leptomylus Wa-2	D. leptomylus Wa-1
D. leptomylus Wa-1	22.7	23.3	4.5	6.0	7.3	10.0	9.2	8.3	6.8	4.8	3.6	3.7	
D. leptomylus Wa-2	23.0	23.6	2.6	4.8	6.3	10.6	9.9	9.1	7.7	6.1	5.1		265.6
D. proteus Wa-1	22.7	23.3	2.4	4.7	6.2	9.3	8.5	7.5	5.8	3.3		370.6	258.4
D. proteus Wa-2	22.9	23.5	4.1	5.7	7.0	8.7	7.8	6.7	4.8		235.9	439.3	349.9
D. proteus Wa-3	23.4	24.0	6.3	7.5	8.5	7.3	6.2	4.7		346.6	419.3	559.6	492.5
D. proteus Wa-4	23.9	24.4	7.9	8.8	9.7	5.5	4.0		342.7	487.4	541.5	656.2	600.0
D. proteus Wa-5	24.2	24.8	8.8	9.7	10.5	3.8		289.4	448.5	566.8	614.0	717.1	666.1
D. proteus Wa-6	24.5	25.1	9.6	10.4	11.2		277.7	401.1	527.5	631.2	673.9	769.0	721.1
D. proteus Cf-2	21.8	22.4	5.7	4.0		808.4	759.2	701.9	612.6	505.1	446.6	452.0	524.3
D. proteus Cf-3	22.2	22.8	4.0		290.7	754.4	701.4	638.9	539.2	431.1	339.1	346.2	436.3
D. proteus Wa-0	22.5	23.1		289.8	410.5	696.5	638.7	569.4	454.7	294.3	175.9	189.3	326.2
V. acutus Wa-2	5.2		1672.6	1647.3	1621.4	1811.8	1790.4	1766.8	1733.3	1698.3	1681.8	1704.1	1683.2
V. acutus Wa-1		377.1	1629.5	1603.5	1576.9	1772.1	1750.2	1726.1	1691.8	1655.9	1639.0	1661.8	1640.5
	V. acutus Wa-1	V. acutus Wa-2	D. proteus Wa-0	D. proteus Cf-3	D. proteus Cf-2	D. proteus Wa-6	D. proteus Wa-5	D. proteus Wa-4	D. proteus Wa-3	D. proteus Wa-2	D. proteus Wa-1	D. leptomylus Wa-2	D. leptomylus Wa-1

Chapter 12

Ancestral states and evolutionary rates of continuous characters

Andrea J. Webster and Andy Purvis

ABSTRACT

In this chapter we review a number of methods recently developed to estimate ancestral character states for continuous characters. Though parsimony and maximum likelihood methods appear very different, we highlight deep similarities between them. We show how an assumption of many of the methods (consistency of rate of change) can be tested, and show that it does not hold for an example data set of carnivore body masses. The methods are then compared in terms of the accuracy of their predictions; we use primate body mass and conodont *Pa* element size to assess how well the ancestral estimates from each method agree with fossil evidence. No method was found to give precise and accurate estimates, and adding complexity to models does not necessarily improve the ancestral estimates obtained.

Introduction

Most chapters in this book focus on how morphometric data and morphometric methods can inform systematics and phylogenetics. In this chapter, we will assume that a phylogeny is available, and will look at how it can be used together with morphometric characters to make inferences about ancestral states and rates of evolution.

The most direct way of finding out character states of ancestors is to examine fossils of individuals from the ancestral species. However, this approach is often extremely problematic: only a tiny proportion of individuals leave fossils, accurate placement of these fossils within a phylogeny is difficult, and many interesting characters do not fossilise. Faced with these difficulties, many biologists have explored ways of using information about living species to make inferences about what their ancestors were like. These inferences are then used to choose between alternative scenarios. Were the ancient Galapagos finches granivores, insectivores or folivores? How large were the ancestral primates? Were ancestral foraminifera smaller than their descendants?

In this chapter, we review the range of methods currently used to estimate ancestral states for continuous characters. Many of these methods have been developed within the last three years, and there has been rapid growth in their use. Our intention here is not to provide an exhaustive review of the literature (two recent reviews are available: Cunningham *et al.* 1998; Pagel 1999). Rather, we look at the assumptions of the various methods, especially the assumption of rate constancy, and point out deep similarities between methods developed under the apparently very different philosophies of parsimony and maximum likelihood methods. We use three data sets – primate and carnivore body mass and palmatolepid conodont *Pa* element size – to illustrate how key assumptions of some methods can be tested and to assess whether the ancestral estimates from any method agree with what is inferred from the fossil record. This is the first comparison of a wide range of methods against measurements from fossils, though some previous studies have used multiple methods to estimate ancestral characters (e.g., Butler and Losos 1997; Martins 1999). In our data sets at least, assumptions are commonly violated, no method gives precise and accurate estimates, and extra complexity does not necessarily give better results. We discuss the implications of these findings for the inference of ancestral states in general.

The methods

There are several methods used in the literature to estimate the ancestral character states for a continuous variable, given a phylogeny and the descendent character states. These methods also often provide an estimate of the rate of change of the trait in question. We examine a number of these methods, listed in the first column of Table 12.1. It is possible to roughly divide the methods into two groups: those using a parsimony criterion to predict ancestral states (linear, squared change and weighted

Table 12.1 A summary of which methods calculate the ML rate and ancestral reconstruction under the model specified by the method [Computer programs used to implement each method are also indicated (not an exhaustive list)]

Method	ML rate under specified model	Global ML ancestral reconstruction	Can be implemented in
One-parameter	✓	✓ SE	ML – Ancml[1] GLM – Ancestor[*,5]
Independent contrasts WITH branch length information	✓[†]	✗	CAIC[2]
Independent contrasts WITHOUT branch length information	✓[†] BL = 1	✗	CAIC[2]
Two-parameter	✓	✓ SE	GLM – Ancestor[*,5]
Linear parsimony	✗	✗	MacClade[3]
Squared change parsimony	✓	✓	MacClade[3]
Weighted squared change parsimony	Not valid to calculate rate	✓	PDSQCHP[4]

SE indicates that standard errors are calculated. BL = 1: branch lengths must equal 1 for the calculated rate to be the ML rate.
* cannot be used to calculate rate estimates.
† signifies the result must be squared to be equivalent to the ML rate.
1 Schluter et al. (1997),
2 Purvis and Rambaut (1995),
3 Maddison and Maddison (1992),
4 Garland et al. (1997),
5 Martins and Hansen (1997).

squared change parsimony), with the remaining methods using a statistical model of evolution (e.g., Brownian motion). Parsimony methods make no explicit assumption about how change occurs over time: they merely follow the dictum of Ockham's razor, that the favoured explanation of the data is the one requiring the least evolutionary change. Their criterion is simply to minimise some measure of change over the tree. These methods can therefore be used to estimate ancestral states but cannot generally be used to say anything about the rate at which this evolution occurs. Methods involving statistical models, unlike parsimony methods, make some explicit assumptions about the way evolution occurs, so impose a framework onto which the data must be fitted. This allows not only ancestral estimates to be made, but also allows inferences involving the rate of evolution.

Although these methods are conceptually approaching the problem from two different angles, the differences are sometimes more apparent than real; any model which reconstructs ancestral traits from descendants must involve measures of the descendants' traits and, frequently, an important consideration is the time which has elapsed between the ancestor and its descendants. Although statistical and parsimony methods seem to use different reasoning to estimate ancestral reconstructions, algorithms are shared, so methods which appear very different blur together where models and criteria overlap.

As parsimony methods can often be viewed as simplification of statistical methods, we will consider the statistical methods first. The following available methods can be loosely grouped together under the heading of Brownian motion:

1. one-parameter maximum likelihood model (Schluter *et al.* 1997),
2. independent contrasts (Felsenstein 1985) (a) with and (b) without branch length information, and
3. two-parameter maximum likelihood models (Martins 1994).

These methods make use of a Brownian motion (random walk) model of evolution in some way (for details of this model see Box 12.1). Neither of the independent contrast methods was ever intended to produce meaningful estimates of ancestral characters. However they have been used in the past to infer ancestral states, and as such must be considered in this review.

The one-parameter maximum likelihood method

This method, outlined by (Schluter *et al.* 1997), uses the Brownian motion model in a maximum likelihood (ML) framework to calculate the ML rate parameter for the model as well as the ML set of ancestral reconstructions. These two estimates are calculated by two different processes. The ML rate parameter (β), is calculated by finding the difference in the particular trait between the descendants from each branching point in the phylogeny, weighting this change proportionally by the inverse of the square root of the branch length between them, and then finding the rate which maximises the likelihood of observing these weighted differences. Exact knowledge of the ancestral state is irrelevant at this point because of Felsenstein's 'pulley principle' (see Box 12.2). In this case β is the only parameter in the model that needs to be estimated.

The Brownian motion model has been used to describe the random evolution of traits; it corresponds to random genetic drift. Alternatively, it may apply to traits evolving in response to selection, where the direction of selection varies often. The model makes three assumptions:

1 that the probability of change is independent of both prior and current character states,
2 that transitions along any branches are independent of changes elsewhere in the tree, and
3 that rates of change are constant (aside from stochastic variation) throughout time and along all branches.

Total change along each branch of the phylogeny is equal to the rate of accumulation of variance (β) multiplied by the amount of time (t) that has elapsed.

So:

Total squared change $= \beta t + \varepsilon$, where ε is the error term and β is the rate parameter.

Box 12.1 The Brownian motion model of evolution.

The likelihood of a particular tree is determined by the sum of its branch lengths, so far example (a) this is $x + y$ or $4 + 7 = 11$.

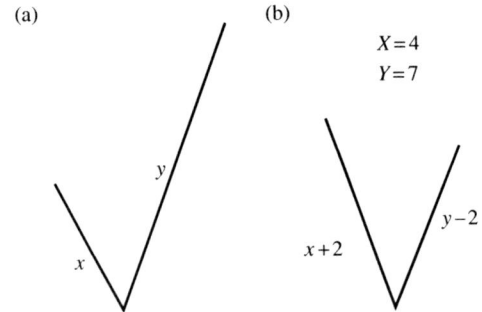

If you consider the fork to act as pulley, the likelihood of this fork occuring will be the same regardless of where the ancester is placed, as the total branch length subtracted from one branch is always exactly added back to the other branch. This is shown in (b); the total branch length is now $(x + 2) + (y - 2)$ or $6 + 5 = 11$.

This principle can be extended to a three way fork – the pulley can be rolled onto any branch. The likelihood is now effectively being estimated for an unrooted tree. Hence, the most likely tree, and consequently the most likely rate of evolution can be determined without an accurate estimate of the ancestral character state.

Box 12.2 Felsenstein's Pulley Principle (Felsenstein 1981).

Maximum likelihood ancestral character states are estimated in a similar way. Initially, ancestral states are computed as weighted averages of the values of descendants, with the weights being inversely proportional to branch length between ancestor and descendant. An iterative algorithm then moves through the phylogeny from the root upwards adjusting the ancestral values so the set of estimated values is statistically the most likely set of ancestral states when the whole phylogeny is considered. This is shown to be the case when the rate of accumulation of variance is minimised (Maddison 1991); that is, when the sum of the square of the weighted difference between ancestors and descendants over the whole tree is minimised (weights are again inversely proportional to branch length). The result is the global ML reconstruction of ancestral states for this particular phylogeny.

It is also possible to calculate the standard error associated with an ancestral reconstruction by calculating the standard deviation of the marginal distribution of the ancestral state. This is equivalent to the standard error of the ancestral ML estimate.

Independent contrasts

Independent contrasts (Felsenstein 1985) is a method which allows comparative data to be analysed as statistically independent points; it removes pseudo-replication in the data that is due to phylogeny. When branch length information is being used, an independent contrast is the difference between two sister clades, divided by a multiple of the expected standard deviation of that difference. The standard deviation, under Brownian motion, is proportional to the square root of the branch length. An independent contrast is therefore a measure of absolute change over some function of time, so is in many ways like a rate (Garland 1992). Hence a mean rate of change throughout the phylogeny can be calculated. The mean rate of absolute change calculated over the phylogeny can then be directly equated to the rate parameter when squared, as the rate parameter is the accumulation of variance over time, or linear change squared:

$$\beta = \frac{\text{Accumulated variance}}{\text{Branch length}}$$

$$\text{Independent contrast} = \frac{\text{Absolute linear change}}{\sqrt{\text{Branch length}}} = \frac{\sqrt{\text{Accumulated variance}}}{\sqrt{\text{Branch length}}}$$

Independent contrasts are therefore another way of calculating the ML rate of change in a trait.

Although the method gives the ML estimate of rate, it does not calculate the global ML ancestral reconstructions. The nodal reconstructions are created as an intermediate stage in the calculation of phylogenetically independent contrasts and are simply a weighted average value of the descendant character states, where the weights are inversely proportional to branch lengths. The algorithm uses the model of Brownian motion and its assumptions, but the reconstruction of the nodal values is only a *local* solution – only direct descendants are considered when nodal states are assigned. This is all that is required in order to calculate independent contrasts and because of the pulley principle, all that is required to calculate the ML rate, but as a result the nodal reconstructions cannot be defended as estimates of ancestral character states. These states are the states that are reconstructed by the one-parameter model in its first pass down the tree.

Where there is no branch length information, independent contrasts are calculated in the same way but with branch lengths all set to the same value; branch lengths must be made to equal one when rate is to be considered. If all branches are assumed to be equal and of length one (an implicit assumption that the amount of change along a branch is independent of its length), the rate parameter will be that calculated by squared change parsimony (see below) and by the one-parameter model where all branch lengths are set to one. As when branch length information is used, the ancestral reconstructions are not the global ML solutions, but will be those calculated by the first pass down the tree in the one-parameter model.

Two-parameter model

More complex models are available, with two parameters rather than one. These are extensions of the Brownian motion model of evolution. One example is the Ornstein–Uhlenbeck model (Martins 1994), in which the random walk is constrained, such as would occur when a trait is subjected to a stabilising selection pressure. The restraining force can be thought of in two ways; either as a force which restrains species from wandering too far from a central fixed point under random genetic drift, or as species remaining close to an optimum whose variation is restricted by this constraining force (termed α). This model is now a two-parameter model with the parameters α and β. As the constraining force (α) becomes close to zero, the model becomes equivalent to the one-parameter Brownian motion model. This model calculates the ML rate of evolution in the same way as in the one-parameter model, but the ancestral states and rate calculated are the reconstruction for the advanced two-parameter model; they are different to the those for the one-parameter model. Tests of significance are used to assess whether inclusion of α significantly improves the fit of the model to the data. If not, then a one-parameter model is preferred.

Generalised linear models (GLM) can also be used to calculate the ML rate and ancestral reconstructions for both the one and two parameter models (Martins and Hansen 1997). The methods estimate the ancestral trait values as a linear combination of the phenotypes of the species at the tips of the phylogenies. Ancestral states are assigned the value of the average of all the descendant states weighted by a measure of how closely the species are related to the ancestor in question. Although this also gives the ML reconstruction under the Brownian motion model, the standard errors it assigns to the ancestral estimates may differ slightly from those calculated by the iterative algorithms.

Parsimony

All parsimony methods (described by Maddison (1991) and Swofford and Maddison (1987)) use the criterion that the preferred ancestral reconstruction is that which minimises some measure of evolutionary change over the phylogeny as a whole. They all use iterative algorithms in their implementation. There is no explicit model of evolution specified in the dictum of parsimony but, as we shall see, some of these methods are equivalent to some ML approaches. The three methods examined here are:

1 weighted squared change parsimony (Maddison 1991),
2 squared change parsimony (Maddison 1991), and
3 linear change (Swofford and Maddison 1987).

Weighted squared change parsimony

In weighted squared change parsimony, the measure of change minimised is the sum of squared changes divided by branch length, that is,

$$\sum \frac{\text{change}^2}{\text{branch length}}$$

The ancestral reconstructions which minimise this measure are the most parsimonious under this criterion. But not only this: these ancestral reconstructions are also the ML reconstructions under Brownian motion model of evolution. This can be explained because branch length in the parsimony method is analogous with time in the Brownian motion model. The square of linear change with time is variance. So this method is in effect finding those ancestral reconstructions which minimise the accumulation of variance with time (β in the one-parameter model).

It is not strictly correct to calculate a rate parameter with weighted squared change parsimony as this would involve imposing some assumptions of how a character trait evolves. Unless one specifies the nature of the process, it is meaningless to estimate the rate at which it occurs. It is impossible to have a maximum likelihood rate without specifying the conditions for which it is the solution.

Squared change parsimony

In squared change parsimony, branch length information is not considered to provide any further contribution to the ancestral reconstruction and the sum of the squared change over all branches is the unit minimised. By assuming that branch lengths are unimportant, they are effectively set at 1 (any change along the branch is in units of per branch rather than per branch length). Squared change parsimony is a special case of weighted squared change parsimony, where the branch lengths equal 1. Therefore, without imposing anything further to the model (in fact by reducing the amount of information it uses, by saying branch lengths are equal), we are able to calculate ancestral reconstructions and rate of the evolution of the trait. The ancestral states are again the ML reconstructions under the one-parameter model with equal branch lengths as variance is still being minimised, while the rate parameter is that under a one-parameter model where branch lengths are all equal to one. As previously mentioned, this is also the rate calculated by independent contrasts when no branch length data are used.

Linear parsimony

Linear parsimony minimises absolute total change, rather than any function of squared change. This method will not estimate the ML rate or the ML ancestral states under any useful circumstances.

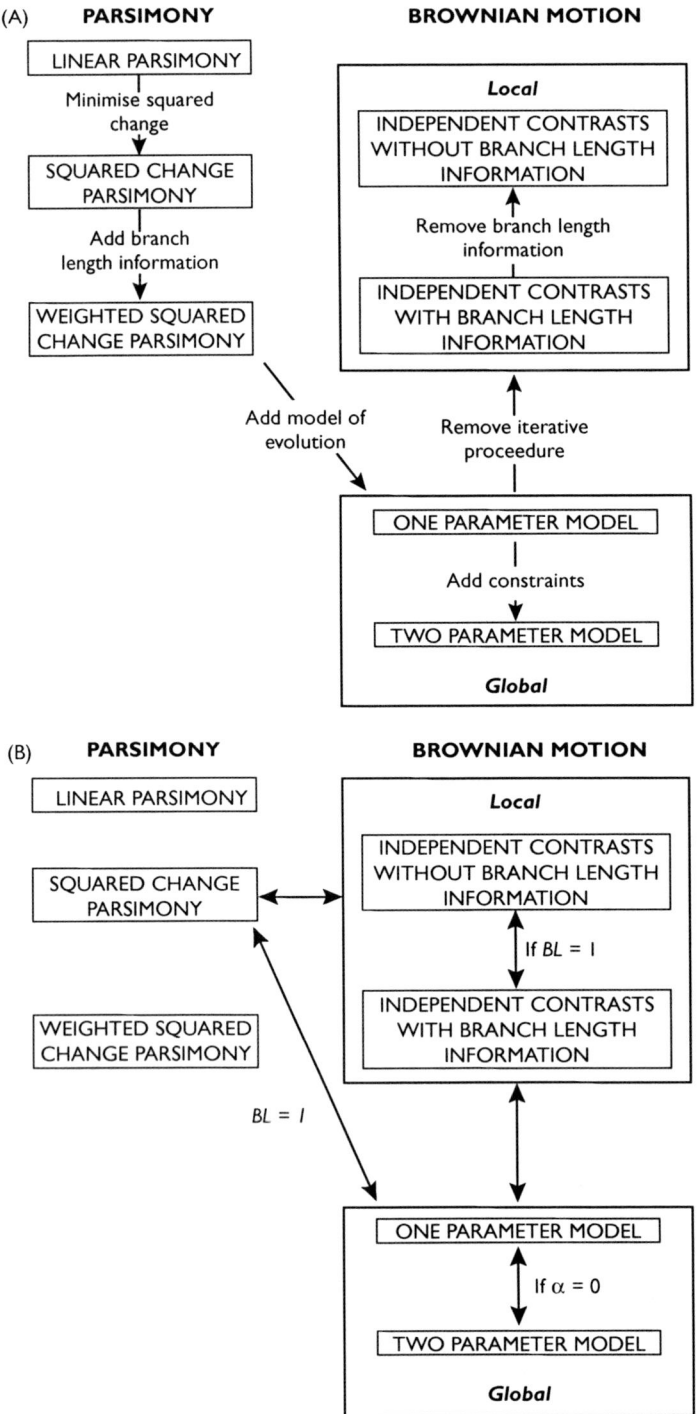

Figure 12.1

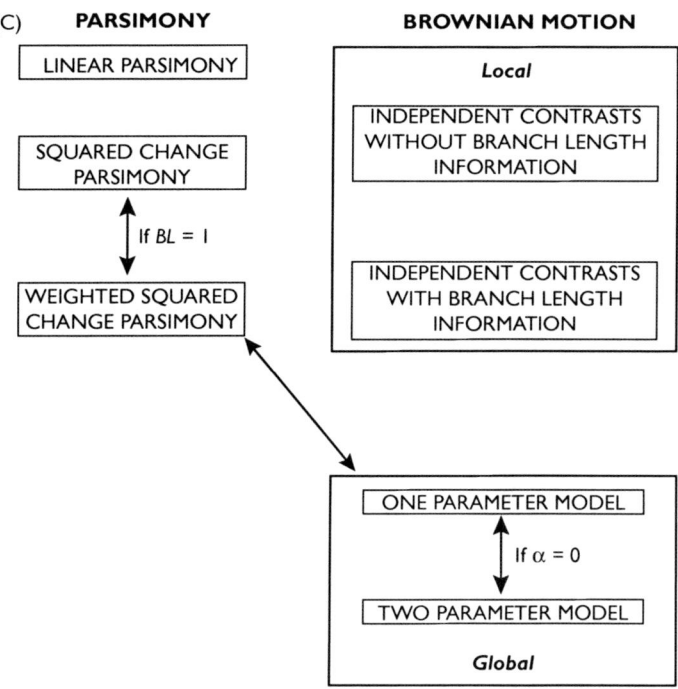

Figure 12.1 This diagram relates the methods in terms of (A) their methodology, (B) the similarity of their rate estimates and (C) the similarity of the ancestral state estimates. In (B) and (C) the arrows link identical estimates, and any necessary conditions are stated by the arrows (*BL* = branch length).

Table 12.1 and Figure 12.1 summarise how each method relates to the others and which program can be used to implement them.

Practical application of the methods

When faced with so many different methods, it is difficult to decide which gives the most accurate representation of historical characteristics. Another important consideration, frequently overlooked, is which method best suits the data available – if the data violate assumptions set down by the model, the conclusions will be flawed before the calculations are even started. We present some case studies to first test the assumptions of Brownian motion models with some typical data sets, and second, to examine the accuracy of each method when compared to actual fossil data.

Validity of using Brownian motion models

Before using the Brownian motion-based models to reconstruct ancestral character states it is important to check that their ability to make predictions is not compromised by the data that is being entered into them. If the data used break the assumptions of the model, the ancestral reconstructions calculated from them could be misinformative; little is known about how badly violation of assumptions affect performance.

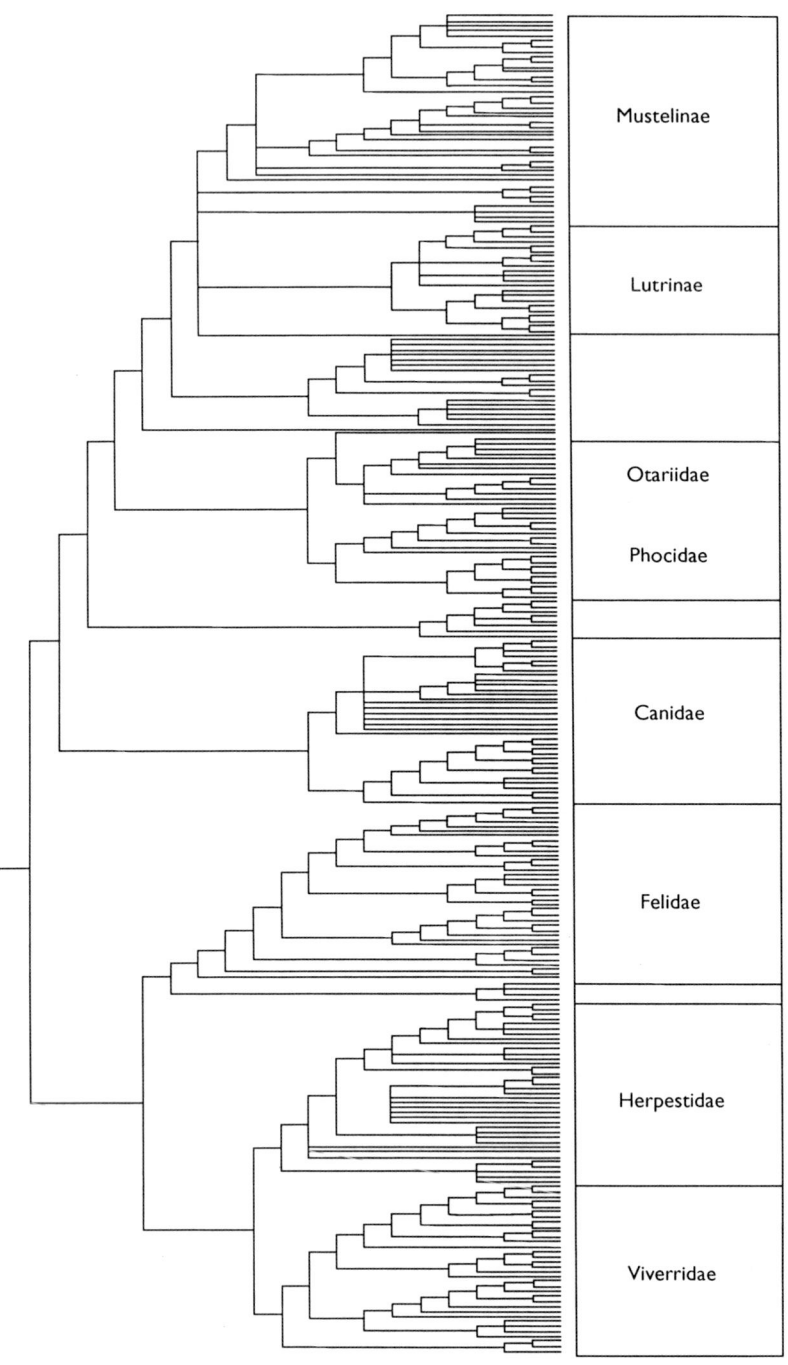

Figure 12.2 Carnivore phylogeny showing clade groupings (branch lengths not shown).

It is possible to test whether the rate of evolution of traits is consistent throughout the phylogeny, one of the assumptions of the one-parameter model. We initially tested this assumption using body mass as the continuous trait and the carnivore phylogeny (Bininda-Emonds *et al.* 1999), split into seven clades as shown in Figure 12.2. We used independent contrasts, generated by CAIC (Purvis and Rambaut 1995), to obtain our rate estimates: as expected, the one-parameter ML method produced exactly the same estimates when it could be used.

The treatment of polytomies in these analyses could affect the rate estimates. There may be some systematic bias in the variance associated with the multiple nodes when compared to bifurcating nodes, therefore making better resolved clades seem to have systematically different rates of evolution when compared to less resolved clades. We dealt with polytomies in four different ways as follows:

1 Soft polytomies (i.e., polytomies taken to represent ignorance of the true pattern of relationship), using all within-clade standardised linear contrasts to calculate mean rate.
2 Soft polytomies, using only those standardised linear contrasts not directly involved in a polytomy to calculate mean rate.
3 Hard polytomies (i.e., polytomies taken to represent multi-way speciation events), using all within-clade standardised linear contrasts to calculate mean rate.
4 All species causing polytomies removed, using all within-clade standardised linear contrasts to calculate mean rate.

The rates of phenotypic evolution are shown in Table 12.2. We then tested to see whether there was a significant difference between these rates of evolution of body mass. In order to normalise the distribution of the contrasts, they were log transformed, then we used ANOVA and Kruskal–Wallis tests to determine whether there was a difference in rate of evolution between clades. The results of these tests are shown in Table 12.2. Fisher's individual error rate test was used to determine exactly which clades were significantly different from each other.

Despite the different ways in which polytomies are treated, there is a significant difference in rate of evolution of body mass between some of the clades in the analysis. Assigning hard polytomies produces the highest number of significant differences among clades; using soft polytomies and excluding rates derived from a node with more than two descendants produces the fewest. All analyses agree that the pinnipeds (Phocidae and Otariidae) evolve at a much slower rate than the felids, and that viverrids evolve at a much slower rate than the felids and the mustelines.

There is the chance that the difference in rates is an artifact of data error causing independent contrasts to over estimate difference between closely related species; small measurement error between closely related species can be hugely exaggerated (Purvis and Webster 1999). If this explanation is correct, the age of a node should be a significant predictor of the rate estimate obtained from it. We therefore carried out an Ancova test with age of clade as the covariate. In none of the previously analysed datasets did age of clade fall out as a significant covariate, and the significant difference in rates did not disappear (Table 12.3). Similar analysis of the primate phylogeny ((Purvis 1995), shown in Figure 12.3) revealed the same result: clades differ significantly in their rate of body size evolution (cebids evolve at a slower rate than cercopithecines). These

Table 12.2 Rate estimates for each Carnivore clade as shown in Figure 12.2.

Phylogeny type	Rate estimates (arbitrary units)							Significance tests	
	Mustelina	Lutrinae	Phocidae/Otariidae	Canidae	Felidae	Herpestidae	Viverridae	ANOVA (p)	Kruskal–Wallis (p)
Soft polytomies – all contrasts	0.037	0.043	0.006	0.009	0.024	0.009	0.003	0.000	0.000
Soft polytomies – contrasts not involved in polytomies	0.023	0.009	0.006	0.006	0.024	0.011	0.002	0.001	0.000
Hard polytomies – all contrasts	0.032	0.041	0.005	0.007	0.023	0.006	0.003	0.000	0.000
No polytomies – species removed	0.024	0.007	0.006	0.008	0.024	0.009	0.003	0.001	0.001

Estimates are shown for the four different ways of treating polytomies. Results of tests looking for a significant difference in rate of evolution between clades are also shown.

Table 12.3 Results of an ANCOVA test looking for a significant difference in rate of evolution between Carnivore clades (as above) with age of clade as the covariate

Phylogeny type	ANCOVA tests	
	Age (covariate) (p)	Clade (p)
Soft polytomies – all contrasts	0.3	0.002
Soft polytomies – contrasts not involved in polytomies	0.3	0.004
Hard polytomies – all contrasts	0.2	0.001
No polytomies – speicies removed	0.5	0.004

results lead us to conclude that those methods requiring the data to uphold assumptions made by the Brownian motion model cannot be used with confidence across the whole of the carnivore or primate phylogenies. This is likely to extend across many other phylogenies in many other analyses.

However, this does not render these methods useless in the situations where rate is not constant across the whole phylogeny. We suggest these methods should be used repeatedly over separate subsets of the tree where rates are not significantly different. For example, although there is a significant difference in rate *between* the five large clades marked on the primate phylogeny (see Figure 12.3), there is no significant variation *within* them: t-tests between sister clades at all points marked in Figure 12.3 with a black dot showed no significant differences (results not shown). Therefore, although it would not be sound to apply one-parameter methods over the whole tree, they can be used to predict ancestral traits over smaller subsets where rate is not significantly different.

Biological inferences from rate variation

Rates calculated along pairs of sister branches may also be of interest in themselves. While not being shown to be significantly different within clades in the carnivore and primate phylogenies, rates do show some variation. The rate associated with the galagos, appears to be faster than the others in the clade. This is probably due to an error in the phylogeny; the ancestor of the galagos probably evolved long ago than the 1.8 million years ago suggested by the tree. However most of the other unusual rates are probably indicators of real biological phenomena. For instance, many of those species displaying unusual rates of body size change have been identified as phyletic dwarfs or giants (Webster 2001).

Accuracy of ancestral state estimates

Which of the methods gives the most accurate picture of past events? A comparison against fossil evidence is needed, but the need for these methods has arisen because of lack of detail in the fossil record. Even where fossil material is present, only a fraction

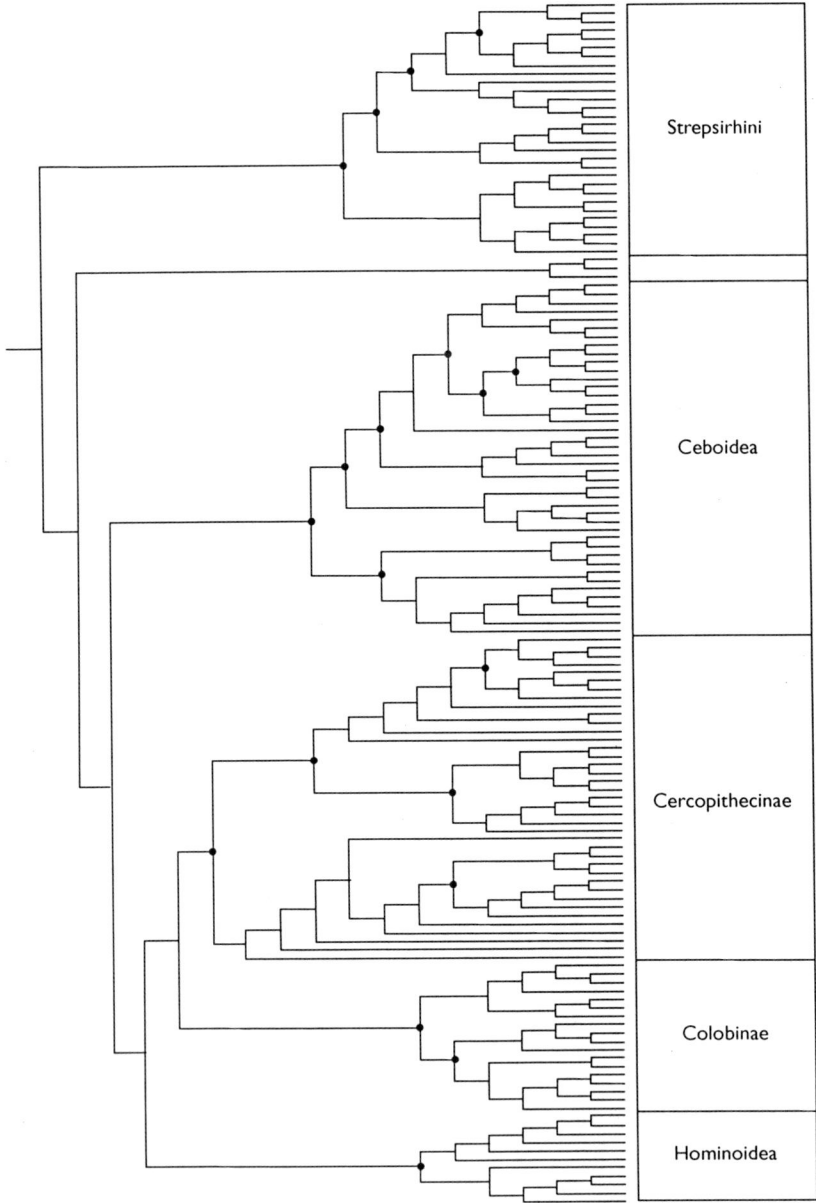

Figure 12.3 Primate phylogeny showing clade groupings (branch lengths not shown). • indicates where t-tests were performed.

of the characteristics of the organism have been preserved. For these reasons ancestral and descendent relationships are frequently sketchy and ancestral characteristics at best an educated guess with confidence limits. A few reasonably well-established dated phylogenies which include putative ancestor species do exist however. We use

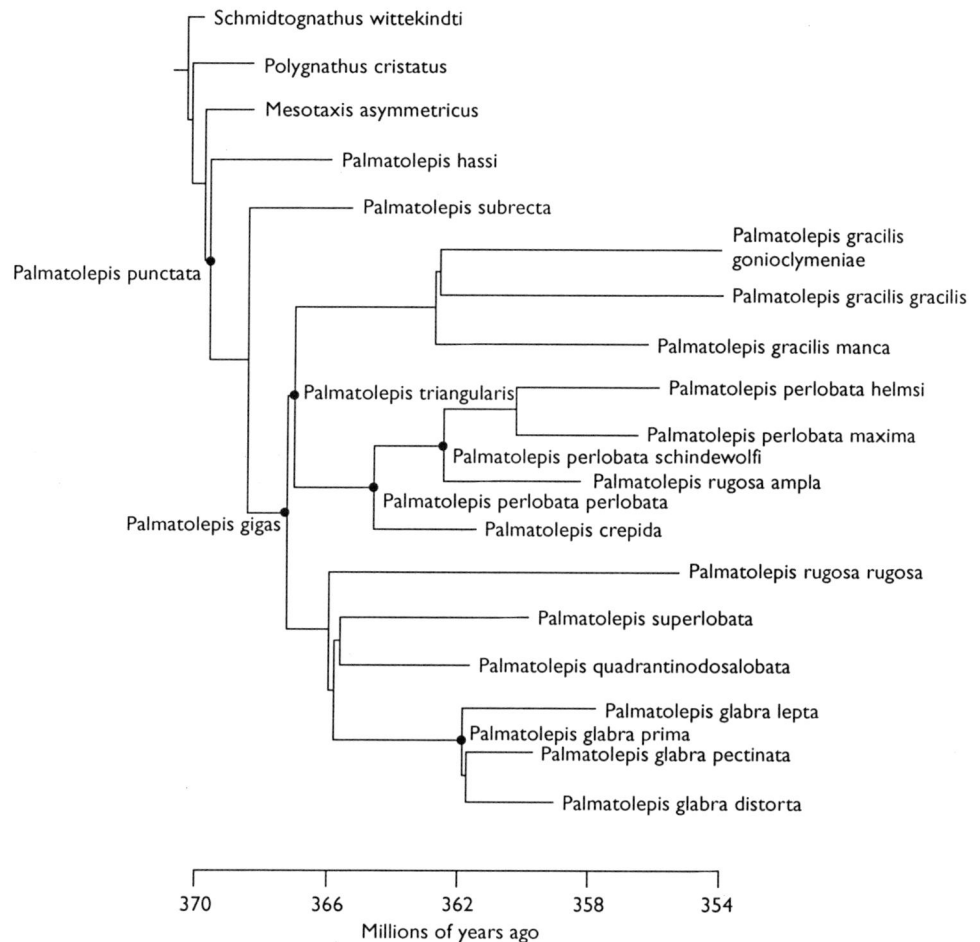

Figure 12.4 Palmatolepis phylogeny.

two: palmatolepid conodonts (Sweet 1988) and New World monkeys (Fleagle 1999; Purvis 1995).

The subset of the conodont phylogeny we used is shown in Figure 12.4. We chose to use length, width and area measurements of conodont *Pa* elements (Table 12.4). The *Pa* element is thought to have been equivalent to a molar in a complex feeding basket situated in the head of the conodont animal (Donoghue and Purnell 1999) and it is likely that element scaled with the animal's size. For each species, the length and width of a *Pa* element was taken, using photographs from the literature. Length was measured at its maximum point and width taken at the maximum point perpendicular to this. Area was calculated roughly as length multiplied by width. By contrast, the New World monkey example has an ancestral history which is less certain (and as such this dataset required more assumptions to be made). In particular, there is uncertainty about the placement of *Protopithecus* and *Dolichocebus*. The phylogeny we used is

Table 12.4 Size data used in the conodont analyses

Species	Length (µm)	Width (µm)	Area (mm²)
Schmidtognathus wittekindti	911.93	244.91	223.34
Polygnathus cristatus	1985.60	866.40	1720.32
Mesotaxis asymmetricus	1381.48	635.56	878.01
Palmatolepis hassi	823.43	552.29	454.77
Palmatolepis subrecta	1097.33	651.33	714.73
Palmatolepis gracilis gonioclymeniae	1430.00	680.00	349.87
Palmatolepis gracilis gracilis	3030.00	1110.00	432.35
Palmatolepis gracilis manca	636.29	245.14	292.82
Palmatolepis perlobata helmsi	1792.80	993.60	155.98
Palmatolepis perlobata maxima	1112.00	388.80	3363.30
Palmatolepis rugosa ampla	984.00	355.56	1781.33
Palmatolepis crepida	802.57	364.86	972.40
Palmatolepis rugosa rugosa	1393.33	784.44	1092.99
Palmatolepis superlobata	889.33	528.00	884.05
Palmatolepis quadrantinodosalobata	1043.33	847.33	469.57
Palmatolepis glabra lepta	1352.44	300.89	406.94
Palmatolepis glabra pectinata	848.94	288.09	244.57
Palmatolepis glabra distorta	1198.40	468.00	560.85
Palmatolepis punctata	2209.33	1341.33	2963.45
Palmatolepis gigas	2317.33	740.80	1716.68
Palmatolepis triangularis	2248.00	1082.67	2433.83
Palmatolepis perlobata perlobata	1552.67	750.67	1165.54
Palmatolepis perlobata schindewolfi	1392.89	637.33	887.73
Palmatolepis glabra prima	1573.33	645.33	1015.32

shown in Figure 12.5 and the body mass data on which the analysis is based is shown in Table 12.5.

After ensuring that rates of evolution were not significantly different within either phylogeny, we calculated ancestral character states using the methods listed and as implemented in the programs listed in Table 12.1 Although five different methods are used to calculate ancestral reconstructions, more are represented by these results. This is due to some methods such as weighted squared change parsimony (not used here) and the one-parameter ML method producing the same ancestral reconstruction.

Although results were simple to interpret for conodonts, as all ancestors were placed at nodes on the phylogeny, the primate phylogeny was more complicated. The fossil primates are not placed at nodes. Therefore once nodal estimates of body size were calculated, it was necessary to extrapolate back along the branch to the point at which the fossil ancestor was placed. Although there is no certainty that these fossil ancestors are direct ancestors of extant species, this extrapolation is still valid for most methods: even if the extinct species should be placed on a small side branch, the best estimate of their size under the Brownian motion model of evolution is the size at the point that the side branch leaves the main branch.

Although this interpolation is valid for one-parameter methods, because it is linear, it is not valid for the two parameter model involving the Ornstein–Uhlenbeck process (which is not linear); similarly it is not valid to interpolate standard errors for any method. These estimates were therefore eliminated from this analysis.

Ancestral character states 263

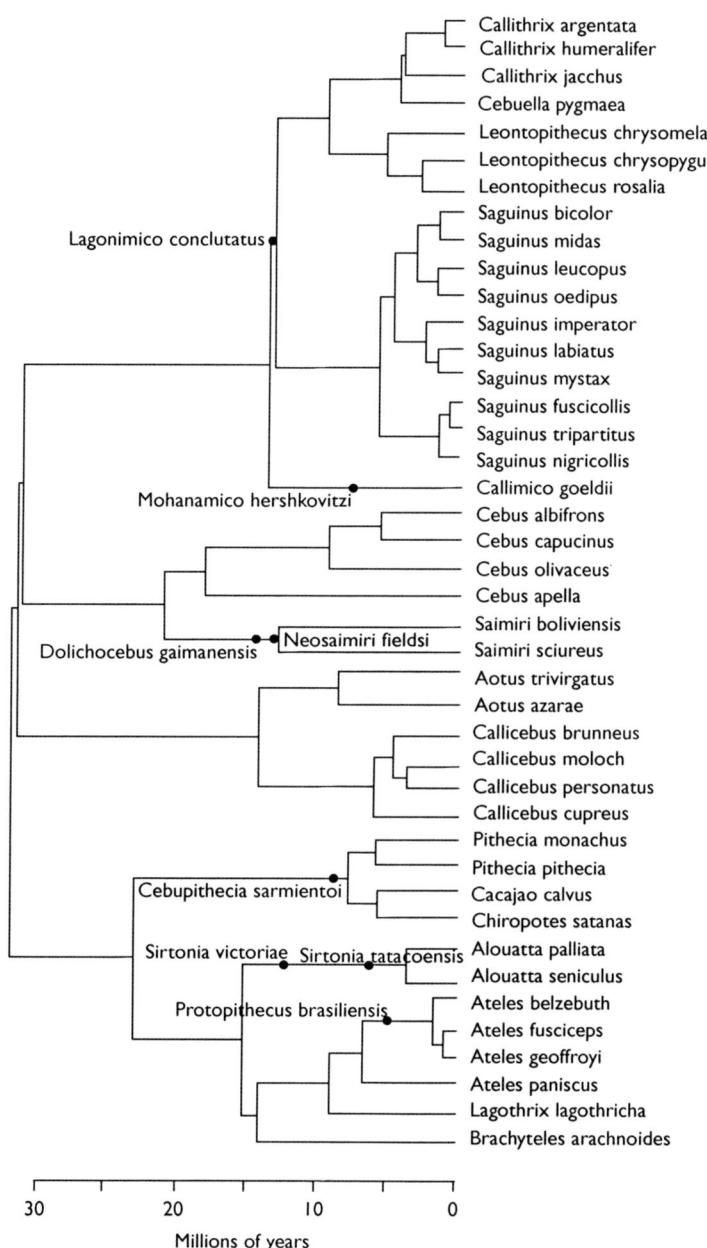

Figure 12.5 New world monkey phylogeny.

To compare the accuracy of the point estimates, we ranked the methods under two different criteria. We first rated the methods according to overall combined accuracy: the estimate of ancestral size at each node was compared to the estimate from the fossil record, and the residual sum of squares (RSS) calculated. The second measure

Table 12.5 Body size data used in the primate analysis. Extant species data taken from Purvis (2000) and fossil species data taken from Fleagle (1999)

	Body mass (g)
Extant species	
Callithrix argentata	364.31
Callithrix humeralifer	350.02
Callithrix jacchus	290.62
Cebuella pygmaea	129.28
Leontopithecus chrysomelas	550.04
Leontopithecus chrysopygus	615.23
Leontopithecus rosalia	592.29
Saguinus bicolor	430.09
Saguinus midas	424.96
Saguinus leucopus	456.69
Saguinus oedipus	492.26
Saguinus imperator	412.82
Saguinus labiatus	497.20
Saguinus mystax	564.53
Saguinus fuscicollis	403.83
Saguinus tripartitus	376.91
Saguinus nigricollis	418.64
Callimico goeldii	566.23
Cebus albifrons	2560.61
Cebus capucinus	2983.94
Cebus olivaceus	2818.61
Cebus apella	2757.28
Saimiri boliviensis	772.01
Saimiri sciureus	762.80
Aotus trivirgatus	956.23
Aotus azarae	896.95
Callicebus brunneus	806.74
Callicebus moloch	980.44
Callicebus personatus	1523.86
Callicebus cupreus	1031.74
Pithecia monachus	1943.02
Pithecia pithecia	1667.37
Cacajao calvus	3422.07
Chiropotes satanas	2966.09
Alouatta palliata	7273.55
Alouatta seniculus	6707.62
Alouatta belzebul	6210.52
Ateles fusciceps	9009.19
Ateles geoffroyi	6522.42
Ateles paniscus	8094.98
Lagothrix lagothricha	6568.23
Brachyteles arachnoides	10981.86
Fossil species	
Lagonimico conclutatus	1300
Mohanamico hershkovitzi	1000
Dolichocebus gaimanensis	2700
Neosaimiri fieldsi	840

Table 12.5 (Continued)

	Body mass (g)
Cebupithecia sarmientoi	2200
Sirtonia victoriae	10000
Sirtonia tatacoensis	5800
Protopithecus brasiliensis	23500

Table 12.6 Tests of accuracy of the ancestral estimates calculated by each method for the primate phylogeny (a) and conodont phylogeny, (b) for each row the best ranking method is highlighted in bold

	Linear parsimony	Squared change parsimony	One-parameter	Two-parameter	Independent contrasts – equal branch lengths	Independent contrasts – real branch lengths
(a)						
All estimates included						
Total RSS	5.63	**3.51**	3.85		3.70	3.71
Median rank	2	2.5	3.5		3	3
Two estimates excluded						
Total RSS	**1.35**	1.38	1.74		2.06	2.23
Median rank	2	2	3		4	4.5
(b)						
Width						
Total RSS	**1.22**	1.48	1.90	1.90	1.86	1.63
Median rank	**1.5**	3.5	4	4	5	3.5
Length						
Total RSS	1.50	1.58	1.59	1.59	**1.49**	1.70
Median rank	3	3	4	5	**2.5**	6
Area						
Total RSS	**5.05**	5.86	6.62	6.62	6.42	6.35
Median rank	**1**	3	4.5	4.5	3	3.5

of accuracy involved ranking methods at each node individually, basing judgement on the square of the residual: each method was then assigned its median ranking, thus reducing the influence of single outliers. The results of these comparisons are shown in Table 12.6. Those methods giving an indication of confidence in the ancestral estimates can be analysed further, by calculating what percentage of fossil sizes fell within the confidence limits of each prediction. The results of this analysis are in Table 12.7.

Some of the differences among methods merit discussion. The linear parsimony estimates are often ranges; there is more than one most parsimonious assignment. This method tends to concentrate change, so ancestral values are static over much of the tree. Squared change parsimony gives a single most parsimonious reconstruction, and conversely forces change to spread out over the tree. Unlike with linear parsimony, ancestral values can lie outside the descendent range when using squared change parsimony.

Table 12.7 Percentage of fossil measurements which fall within confidence intervals calculated for conodont size

	One-parameter (%)	One-parameter (GLM) (%)	Two-parameter (GLM) (%)
Width	33	100	33
Length	50	100	50
Area	33	100	33

The ML reconstruction under Brownian motion takes the branch length into account, and gives standard error estimates for the point estimates. Because change on short branch length is weighted to be more important, the model predicts ancestral characters to be closer to the ancestor at the end of smaller branches, so further from those on longer branches. This is something not seen in the parsimony methods. The two-parameter model produced results very similar to the one-parameter model as the restraining force was estimated to be close to zero.

In primates, squared change parsimony gives the best prediction according to the residual sum of squares with both parsimony methods scoring well in the median ranking. Linear parsimony scores badly in the RSS because of two extremely bad predictions. When the two fossils which have dubious placings are removed, linear parsimony appears most accurate (using both RSS and median rank), followed by squared change parsimony.

In our more comprehensive analysis on the Palmatolepidae, linear parsimony gave the best point estimates and the lowest RSS for *Pa* width and area. With *Pa* length, linear parsimony is second to estimates calculated during independent contrast calculations. The more complicated one and two parameter models consistently do less well in the ranking.

It is also possible to look at standard errors of the nodal estimates given by the one and two parameter models (both GLM and Ancml standard error estimates are included for the one-parameter model). Although the one-parameter model appears successful when looking at the GLM estimates of the standard error, it is unlikely to be wrong considering the confidence interval places the size of the *Pa* element somewhere between the size of a mitochondrion and that of a rugby ball. The other estimates are more precise, so they would be more useful, but they are not accurate. Furthermore, simulations carried out by Martins (1999) indicate that even these wide standard errors could be underestimates and that, in reality, confidence intervals could be significantly larger than are indicated by all methods tested in this study.

Discussion

The methods used in these analyses, although involving different philosophies, are deceptively similar – something which is not always appreciated. These methods cannot be used as support for the accuracy of each other's results, except as a consistency check. It is also important that these methods are not used with data which violate their implicit assumptions. Methods must not be considered a blanket solution for entire phylogenies, especially where phylogenies are large. Although ML methods are considered to give a more accurate picture with more information, clades cannot be added to a phylogeny at the expense of the integrity of the methods.

Reconstructing ancestral states will always be difficult. As many authors point out, a large number of traits may be used to correctly reconstruct a tree, but single traits cannot be reconstructed from a phylogeny even when that phylogeny is perfect as no single trait fits a tree exactly. At the present time, and drawing conclusions from the analyses we carried out, it appears, increasing method complexity has not improved the ability of the methods to estimate ancestral states. If anything, simple parsimony methods, and in particular linear parsimony, outperform the more complicated one and two-parameter ML methods. It is likely that in the future, with more accurate models of how traits evolve, ML methods will improve, but this study underlines the importance of testing evolutionary methods against available palaeontological data.

Acknowledgements

We would like to thank Kate Jones, Paul Agapow and Austin Burt for comments on the manuscript. The study was supported by NERC grants GT4/96/164/T and GR3/11526.

References

Bininda-Emonds, O. R. P., Gittleman, J. L. and Purvis, A. (1999) 'Building large trees by combining phylogenetic information: a complete phylogeny of the extant Carnivora (Mammalia)', *Biological Review*, 74, 143–175.

Butler, M. A. and Losos, J. B. (1997) 'Testing for unequal amounts of evolution in a continuous character on different branches of a phylogenetic tree using linear and squared-change parsimony: and example using Lesser Antillean *Anolis* lizards', *Evolution*, 51, 1623–1635.

Cunningham, C. W., Omland, K. E. and Oakley, T. H. (1998) 'Reconstructing ancestral character states: a critical reappraisal', *Trends in Ecology and Evolution*, 13, 361–366.

Donoghue, P. C. J. and Purnell, M. A. (1999) 'Mammal-like occlusion in conodonts', *Paleobiology*, 25, 58–74.

Felsenstein, J. (1981) 'Evolutionary trees from gene-frequencies and quantitative characters – finding maximum-likelihood estimates', *Evolution*, 35, 1229–1242.

Felsenstein, J. (1985) 'Phylogenies and the comparative method', *American Naturalist*, 125, 1–15.

Fleagle, J. G. (1999) *Primate adaptation and evolution*, London: Academic Press.

Garland, T. (1992) 'Rate tests for phenotypic evolution using phylogenetically independent contrasts', *American Naturalist*, 140, 509–519.

Garland, T., Martin, K. L. M. and Diaz-Uriarte, R. (1997) 'Reconstructing ancestral trait values using squared-change parsimony: plasma osmolarity at the origin of amniotes', in Sumida, S. S. and Martin, K. L. M. (eds) *Amniote origins: completing the transition to land*, San Diego: Academic Press, pp. 425–502.

Maddison, W. P. (1991) 'Squared-change parsimony reconstructions of ancestral states for continuous-valued characters on a phylogenetic tree', *Systematic Zoology*, 40, 304–314.

Maddison, W. P. and Maddison, D. R. (1992) *MacClade*, Sunderland, MA: Sineaur Associates.

Martins, E. P. (1994) 'Estimating the rate of phenotypic evolution from comparative data', *American Naturalist*, 144, 193–209.

Martins, E. P. (1999) 'Estimation of ancestral states of continuous characters: a computer simulation study', *Systematic Biology*, 48, 642–650.

Martins, E. P. and Hansen, T. F. (1997) 'Phylogenies and the comparative method: a general approach to incorporating phylogenetic information into the analysis of interspecific data', *American Naturalist*, 149, 646–667.

Pagel, M. (1999) 'Inferring the historical patterns of biological evolution', *Nature*, 401, 877–884.

Purvis, A. (1995) 'A composite estimate of primate phylogeny', *Philosophical Transactions of the Royal Society of London Series B – Biological Sciences*, 348, 405–421.

Purvis, A., Gittleman, J. L., Cowlishaw, G. and Mace, G. M. (2000) 'Predicting extinction risk in declining species', *Proceedings of the Royal Society of London Series B – Biological Sciences*, 267, 1947–1952.

Purvis, A. and Rambaut, A. (1995) 'Comparative analysis by independent contrasts (CAIC): an Apple Macintosh application for analysing comparative data', *Computer Applications in Bioscience*, 11, 247–251.

Purvis, A. and Webster, A. J. (1999) 'Phylogenetically independent comparisons and primate phylogeny', in Lee, P. (ed.) *Comparative primate socioecology*, Cambridge: Cambridge University Press, pp. 44–70.

Schluter, D., Price, T., Mooers, A. O. and Ludwig, D. (1997) 'Likelihood of ancestor states in adaptive radiation', *Evolution*, 51, 1699–1711.

Sweet, W. C. (1988) *The conodonta*, Oxford Monographs on Geology and Geophysics, Oxford: Clarendon Press.

Swofford, D. L. and Maddison, W. P. (1987) 'Reconstructing ancestral character states under Wagner parsimony', *Mathematical Biosciences*, 87, 199–229.

Webster A. J. (2001) 'Ancestral body size and the evolutionary ecology of phyletic dwarfs', Ph.D. thesis, University of London.

Chapter 13

Modelling the evolution of continuously varying characters on phylogenetic trees
The case of Hominid cranial capacity

Mark Pagel

ABSTRACT

I describe a generalised least squares model for analysing trait evolution on phylogenetic trees, and apply the model to characterise brain-size evolution in the hominids. The model incorporates the conventional Brownian-motion or random-walk model of trait evolution, but can also estimate a directional component to trait evolution – such as would arise if the trait were getting bigger or smaller through evolutionary time. This also makes it possible to reconstruct ancestral states that fall outside the range of observed (extant) values, something that cannot occur with simple Brownian-motion models. The model can also estimate three scaling parameters relevant to testing hypotheses about the tempo and mode of trait evolution: is it punctuated or gradual, does it proceed at a constant rate or speed up (slow down), and are the similarities among species what we would expect? Applied to hominid trait evolution, the model detects the well-known increase in brain size in this group, and as a result, estimates ancestral states more accurately than the random-walk model. The model further suggests that brain-size evolution has been gradual, but that its rate of increase has increased over time (i.e., it is accelerating).

Introduction

Given a collection of species, information on their attributes, and a phylogeny that describes their shared hierarchy of descent, the prospect is raised of reconstructing the characteristics of the ancestors to these species. This is an intriguing idea, holding out as it does the possibility of glimpsing the past and of seeing how the present came about.

The attraction is more than just curiosity. Some ecological and evolutionary theories require a specific order and direction of evolution from ancestors to descendants. Cope's famous 'law' proposes that as species give way to their descendants body size tends to increase, yielding an evolutionary trend toward increasing size. Where no theory exists to make a prediction, reconstructed ancestral states provide, if accurate, ideas about how and why creatures evolved as they did. Omland (1994, 1997), for example, investigates empirical patterns of directional evolution in morphological traits of ducks. Schluter, Mooers and colleagues (Schluter 1995; Schluter *et al.* 1997; Mooers *et al.* 1999; Mooers and Schluter 1999) have been instrumental in calling attention to these and other possibilities inherent in reconstructing ancestral states on phylogenies. Golding and Dean (1998) emphasise the potential for reconstructing ancient genes and proteins.

The attributes of species come in all shapes and sizes, but they can be broadly categorised into two classes. The so-called 'discrete' traits adopt a finite and typically small number of states and may or may not be ordered. The presence or absence of some feature is a binary discrete trait. Living solitarily, in a relationship with one other, or in a group could be an ordered discrete trait. More often, traits are defined in such a way as to constitute a continuously varying feature of the organism or its environment. Wing length, geographic range size, body size, brain volume, age at maturation, running speed, and body temperature are all examples of continuously varying traits.

Here, I shall confine myself to discussing the statistical reconstruction of the probable ancestral character states of continuously varying traits, discrete traits having recently been discussed elsewhere (Schluter 1995; Yang et al. 1995; Koshi and Goldstein 1996; Schluter et al. 1997; Mooers and Schluter 1999; Pagel 1999a,b). Continuously varying traits also have received attention in recent work (Schluter et al. 1997; Garland et al. 1999; Mooers et al. 1999), but I discuss a relatively new maximum-likelihood model that can detect and characterise directional trends of evolution (see Pagel 1997, 1999b). This method can detect features of trait evolution not available to other methods, such as gradual versus punctuational change, accelerating versus decelerating trait evolution, and whether phylogenetic effects are present. By virtue of being capable of detecting directional trends the method raises the attractive possibility of reconstructing ancestral states to fall outside of the range of values observed amongst species. That is, it may be possible to infer the historical existence of traits never directly observed. I use a recent phylogeny proposed for the Hominid species (Foley 1998) and information on cranial capacity to illustrate the method in the context of brain-size evolution.

General theory for statistical models of continuously varying traits

The conventional approach to reconstructing ancestral states of continuous traits proceeds by choosing those values for the ancestral states that minimise some criterion of the total amount of evolution on the tree (e.g., W. P. Maddison 1991; D. R. Maddison 1994). This is the method of maximum parsimony, and the criterion that is minimised is usually the square of the amount of inferred change along the branches of the phylogeny – hence 'squared-change' parsimony. Other criteria are possible, such as minimising the absolute value of change.

The chief weaknesses of this approach are that it assumes that character change is rare, and it fails to incorporate any stochastic element into the process of evolution. Character change may not be rare (e.g., Schluter 1995; Pagel 1999b), and it is important to document the expected uncertainty in our estimates of ancestral states to know what alternative values can be safely ruled out. Maximum parsimony methods do not provide estimates of this uncertainty, although an error-rate of sorts may be possible in principle to calculate (Maddison 1995).

The principle alternative to parsimony methods is to adopt a statistical approach to reconstructing ancestral states (e.g., Pagel 1997, 1999a,b; Schluter *et al.* 1997; Garland *et al.* 1999; Mooers *et al.* 1999; Mooers and Schluter 1999). For traits that naturally vary along a continuous scale constant-variance random-walk models (sometimes called Brownian motion) provide a useful framework within which to model character evolution. In the conventional random-walk model, traits evolve each instant of 'time' dt with a mean change of zero and unknown and constant variance σ^2. Time may be chronological or some other unit of divergence such as genetic distance. The evolutionary process is presumed to unfold independently at each instant of time and along each of the branches of the phylogeny.

This framework makes it possible to calculate the uncertainty associated with estimates allotted to different parts of a phylogenetic tree. It is an important capability when attempting to estimate ancestral states: trait values of species or lineages that have diverged more from the root are expected to have larger variances and thus are less reliable observations for reconstructing the past, other things equal. To see why, consider that the expected variance of a given species' trait value is $t\sigma^2$, where t records the total path length (time or distance) from the root to that species. This is the variance in the trait that would be expected were the process of evolution to be re-run many times from the same starting point. The starting point is the value of the trait at the root of the tree, and is estimated from the data.

Schluter *et al.* (1997) apply this basic model to reconstructing ancestral wing lengths in scrubwrens and to ecological diversification in lizards. Garland *et al.* (1999) investigate the basic constant-variance model and variations on it for a range of characters and taxa. Ancestral states obtained from the popular independent-contrasts approaches for comparative studies (Felsenstein 1985; Harvey and Pagel 1991; Garland *et al.* 1992; Pagel 1992) are equivalent to those obtained from the constant-variance random-walk model. Similarly, ancestral states estimated from 'local' squared-change parsimony, in which only the species immediately descendant from a node are used to estimate the state of the node, are also equivalent to those obtained from the constant-variance model.

A limitation of the standard constant-variance model is that, by presuming that traits evolve according to an unbiased random-walk (neutral-drift), it cannot detect any directional trends of trait evolution along the branches of the tree (Pagel 1997, 1999b). Historical trends such as a phyletic increase in size, or more generally, greater amounts of change in the traits of lineages that have diverged more, will be masked. This model along with independent-contrast and squared-change parsimony approaches always estimates the ancestral state at the root of the phylogeny as falling somewhere within the range of observed values in the species data, as it has no route by which to place them outisde of this range.

By comparison, here I shall show how a *directional* constant-variance model for continuous traits can detect historical trends of trait evolution, and use them to develop more plausible and accurate estimates of ancestral states. Unlike the neutral random-walk model, the directional model presumes that there has been a bias in evolution such that traits evolve at each instant of time dt with a mean of β and unknown and constant variance σ^2 (Pagel 1997, 1999b). Note that here β is used to signify the bias in the random walk, and not the variance of trait evolution as in Schluter *et al.* (1997). The directional model, in effect, examines the correlation between the species' trait values and the total phylogenetic distance or path length from the root of the tree. If a directional trend exists, species that have diverged more from the root, will tend also to have changed more in a given direction, that is, be larger, or mature earlier, and so on.

The parameter of directional evolution is directly interpretable as the direction and amount of change in a character per unit time or other unit of divergence. When a significant trend exists, the estimate of the instantaneous variance of trait evolution will be small, and consequently the directional model will reconstruct ancestral states with narrower confidence intervals than the neutral-drift model. The directional model can also, in such circumstances, reconstruct the character state at the root of the tree to lie outside of the range of observed values in the data. When no directional trend exists, the estimate of β will not differ statistically from zero, and the directional model collapses to the random-walk model.

A framework for modelling continuously varying traits evolving on phylogenies

The random-walk and directional models both estimate the instantaneous component of variance σ^2, as defined above. It turns out that σ^2 is determined by choosing in the random walk model a value of α, where α is the trait value assigned to the root of the tree. In the directional model, σ^2 is found by choosing values of α and β. For both models, the values of the parameters are chosen to minimise the variance of the observed species values, taking into account their patterns of relatedness. In the random-walk model the variance is calculated around α, whereas in the directional model the variance is calculated around $\alpha + \beta t_i$ where t_i is the total path length leading to the ith species.

The parameters are found within a statistical framework known as generalised least squares or GLS (e.g., Pagel 1997; Schluter et al. 1997) that allows one to take into account the patterns of relatedness amongst species. Pagel (1997) further describes how to estimate in the same GLS framework, the parameter of directional evolution, β (a computer program is available from the author to perform all of the calculations reported here).

Closely related species will tend to have similar trait values, even under a random walk, owing to sharing most of their evolutionary history. This makes the phylogeny a fundamental component of any exercise in reconstructing ancestral states, or more broadly of any comparative statistical method. It specifies via the shared and unique evolutionary trajectories the expected patterns of similarity amongst all pairs of species. These patterns can be represented in a variance–covariance matrix, V. The matrix V is a square matrix with the n main diagonal elements representing the expected variances of the n species, and the $n(n-1)/2$ off-diagonal elements specifying expected covariances amongst the possible pairs of species. The variance of a given species trait value is presumed to be directly proportional to the total path length from the root to a given species. The off-diagonal elements are proportional to the length of the shared branches between any pair of species. Figure 13.1 shows a hypothetical phylogeny of three species and the variance–covariance matrix that phylogeny implies.

Statistical models conventionally presume that observed patterns of similarity in trait values are directly proportional to the topology of the phylogenetic tree. This

Modelling continuously varying characters 273

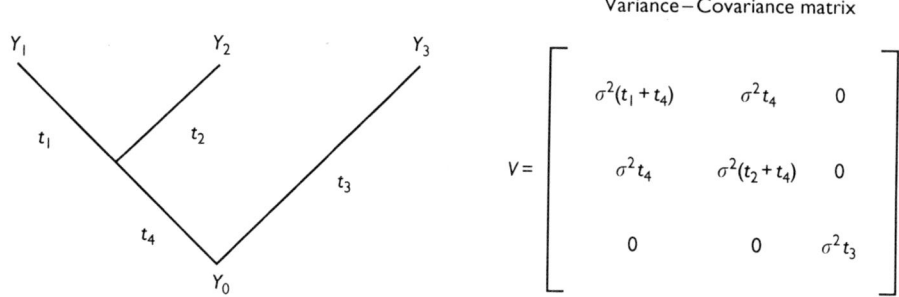

Figure 13.1 A hypothetical phylogeny of three species with branch lengths shown in units of time (t). A trait Y is measured on each of the three species. The matrix V is the variance–covariance matrix presumed to describe the trait values, under a constant-variance random-walk model of evolution. The main diagonal elements describe the variance of a species trait. The variance is assumed to be directly proportional to the sum of the branch lengths from the root to the tips. The expected covariance between any two species is assumed to be proportional to the sum of their shared branch lengths.

means that were evolution to be re-run many times, the pattern of variance of traits and correlation between the realised trait values of pairs of species in the phylogeny is expected to be proportional to the relevant path lengths as represented in V. I have suggested elsewhere (Pagel 1999b) that the occasional poor performance or suggestion thereof, of some comparative methods (e.g., Westoby et al. 1995; Ricklefs and Starck 1996; Price 1997; Harvey and Rambaut 1998) may arise when the relationship presumed to hold between the topology and patterns of trait similarity is violated.

Usefully, though, this assumption of isomorphy between V and the realised patterns of similarity can be tested. Three phylogenetic scaling parameters that bear directly on it are estimable from the combination of the species' data and the phylogeny (Pagel 1997, 1999b). They can then be used to scale the phylogeny in response to patterns in the data, and to do so in such a way as to make the phylogeny optimally conform to the presumed constant-variance model of evolution. That is, where comparative methods presume an underlying constant-variance model directly proportional to branch lengths, the scaling parameters allow one to find the branch-length basis in which the traits most closely conform to the constant-variance model. This means that the assumption of constant variance can be tested directly and the topology suitably modifed where it is rejected.

Phylogeny scaling parameter

Define λ as a parameter that estimates the extent to which the phylogeny correctly predicts patterns of similarity among species. It is a scalar that multiplies the off-diagonal elements of V, and can range from 0 to 1. The default value of 1.0 leaves the off-diagonals unchanged, and corresponds to the patterns of trait similarity being directly proportional to the elements in V. This is the value of λ implicit in all conventional statistical models of trait evolution (e.g., Schluter et al. 1997) and in comparative methods such as independent contrasts (e.g., Felsenstein 1985; Harvey and Pagel 1991; Pagel 1992).

When λ is 0.0, all of the off-diagonal elements are zero but the main diagonal elements are unchanged. This corresponds to the belief that the patterns of trait similarity amongst species are independent of phylogeny. Topologically, it is equivalent to treating the data as if they arose from a 'star' or 'big-bang' phylogeny in which all species emerge simultaneously from a common ancestor. Intermediate values of λ correspond to patterns of trait similarity following, to a greater or lesser degree, the outlines of the topology. Thus, λ measures the extent to which it is necessary when investigating trait evolution to take the phylogeny into account (see Pagel 1999b).

Path-length scaling parameter

The parameter δ differentially scales the unique and shared path lengths in the phylogeny (Pagel 1999b) in response to patterns of trait evolution. It acts as a power to which all of the elements of V are raised. The longest paths in a phylogeny, are, on average, those along the main diagonals of V, that is, the paths leading to species. The off-diagonals record the shared path lengths, corresponding on average to the earlier (older) portions of the phylogeny. If the lengths of the main diagonals are directly proportional to the magnitude of the trait, then δ is 1.0, and trait evolution has been gradual. This is the default value implicitly presumed in models of trait evolution applied to phylogenies.

If the traits at the end of the largest main diagonals have diverged disproportionately more from the root, then δ is greater than 1.0. Values of δ greater than 1.0 lengthen the longer paths in V disproportionately to the shorter paths. Because the longer paths are typically those leading to species, this signifies that later evolution in the tree has had a greater effect. Equivalently, it implies that the rate of trait evolution per unit branch length has accelerated. If the main diagonal elements are found to be too long given the trait values, then δ is less than 1.0. This acts to compress the longest elements of V more than the shorter elements. It corresponds to earlier evolutionary paths having contributed more to trait evolution, as might be expected of an adaptive radiation. Trait evolution has been gradual but decelerating. Values of δ different from 1.0 correspond to scaled-gradualism (Pagel 1999b): the trait has evolved gradually but according to a different scale from that implied by the untransformed path lengths.

Branch-length scaling parameter

A third parameter, κ, scales the relationship between the individual branch lengths (as opposed to the total path lengths as with δ) and trait evolution (Pagel 1994). It is found as the power to which individual branch lengths should be raised to maximise the fit of the model of evolution to the data. When κ is 1.0, trait evolution is directly proportional to branch lengths and evolution is gradual. Values of κ less than 1.0 signify proportionally more evolution in shorter branches. In the extreme, when κ is 0.0, all branches have the same length of 1.0 and by definition the amount of evolution per branch is independent of branch length. This is the pattern that might be expected from some models of punctuational evolution: rapid evolution at or near speciation followed by longer epochs of stasis. If κ is greater than 1.0, longer branches contribute proportionally more to trait evolution. This parameter, then, captures elements of the 'mode' of trait evolution (Pagel 1994, 1999b).

All existing models of trait evolution implicitly adopt values of λ, δ, and κ equal to 1.0. I will show below how these assumptions can be tested. Where they prove to be at odds with the data, I will show how the parameters can be used to improve the fit of the model to the data. The broader point is that important signatures of trait evolution reside in the combination of the trait values and the phylogeny, and these can be used to test hypotheses about underlying evolutionary processes (Pagel 1997, 1999b).

Hypothesis testing

All parameters are estimated in a maximum-likelihood framework (Edwards 1972), such that the values of the parameters are chosen to make the observed data most likely given the model of evolution. The likelihoods of different models provide a direct measure of their relative goodness of fit to the data. Conventionally, one compares the log-likelihoods of two models. If they are special cases of one another, the difference in their log-likelihoods is distributed approximately as a χ^2 variate with degrees of freedom equal to the difference in the number of parameters in the two models. For example, if none of the scaling parameters is invoked, the neutral drift and directional models differ by one parameter (the neutral drift model implicitly assumes that $\beta = 0.0$). Similarly, a model in which λ (or δ or κ) is presumed to be 1.0 is a special case of a model in which λ (or δ or κ) is allowed to take its maximum-likelihood value. Pagel (1994, 1997) discusses how to test the likelihoods when the models are not special cases of each other.

An application to the evolution of Hominid cranial capacity

The Hominids are those species that branched perhaps five million years ago from the common ancestor to modern chimpanzees and humans. Their phylogenetic relationships are uncertain, many phylogenetic hypotheses have been advanced, and the debate about their phylogenetic relationships is sometimes intemperate (Wood and Collard 2000). Here, I use one recent phylogeny, derived from Foley's (1998 and person. comm.) proposals (Figure 13.2). The phylogeny is based upon cranial and post-cranial characters, indices of tool use, and geography. The lengths of the branches are in units of millions of years. I have placed *Australopithecus afarensis* (commonly known as 'Lucy') arbitrarily close (0.001 million years) to the root of the tree, dated to about 3.1 million years ago (mya). Some authors (e.g., Wood and Collard 2000) now prefer to place *H. habilis* and *H. rudolphensis* as either emerging along with *A. africanus* from the root node, or even classify them with the *Australopithecines*. Wood and Collard (2000) review these and many of the other phylogenetic hypotheses advanced for the Hominids.

The tips of the tree refer, except in the case of modern humans, to fossil 'species'. In several cases a fossil species is found over a considerable proportion of the total tree and so I do not mean to imply, for example, that *Homo ergaster* arose only about 0.8 mya, going extinct 0.6 mya. Rather, *H. ergaster* probably arose at or near the common ancestor to the later *H. ergaster* and *H. erectus*, and may have existed until around 0.6 mya (Foley 1998). At some point it branched into an independent lineage

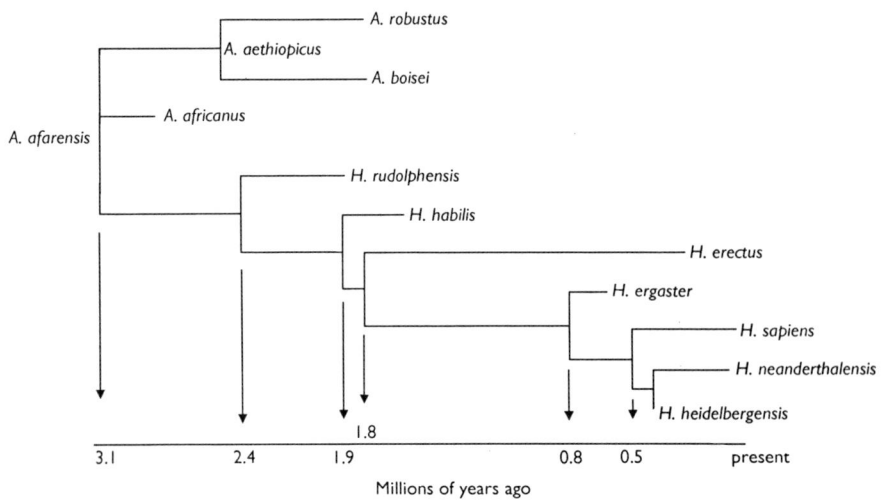

Figure 13.2 A phylogeny of the Hominid species, based upon Foley (1998 and pers. comm.). Branch lengths are in units of millons of years of evolution. See the text for further description.

that paralleled one that would eventually lead to the common ancestor of modern humans and Neanderthals. Fossil evidence for *H. rudolphensis* appears about 2.4 mya even though the tip is shown at about 1.8 mya.

As all but one of the Hominid species are extinct, the total path lengths from the root are different, even though measured as units of time. This makes it possible to seek relationships between cranial capacity and time. When analysing only extant species, the total amount of time from the root will generally be the same for all of them. In these circumstances, branch lengths in units of genetic divergence can be used in place of time. Indeed, such units may in general be more relevant than time to underlying trait evolution, representing something like the 'opportunity for selection' (Pagel 1994).

Table 13.1 shows the species, the Hominid cranial capacities in cubic centimetres, and the date attributed to the cranial elements from which the cranial capacity was measured. I shall use these data, in combination with the phylogeny to investigate the evolution of cranial capacity in the Hominids. I have used only one data point per species, the one conforming to the last observation of that species. For some species, such as *H. erectus*, cranial capacity measures exist for nearly its entire temporal range. Including those points in the analysis could lead to different conclusions, a point to which I shall return below. In particular, some authors (e.g., Aiello and Wheeler 1995; Ruff *et al.* 1997) suggest that Hominid cranial capacity is characterised by distinct evolutionary 'grades' rather than being a single trajectory, and that *H. erectus* cranial capacity was relatively stable for up to 1 million years. However, I have not included those points here because my interest is to explore how using only a single set of 'species' values can be informative (or not) about the past.

Table 13.2 shows the likelihood ratios of three different statistical models of the evolution of cranial capacity and estimates of parameters relevant to these models. The cranial capacity data were logarithmically transformed before analysis. This

Table 13.1 The Hominid data set (data from Aiello and Dunbar 1993)

Species	Date, mya	Endocranial volume, cc
Homo sapiens	0	1450
H. neanderthalensis	0.03	1512
H. heidlbergensis	0.4	1198
H. erectus (Java)	0.25	1100
H. ergaster (Africa)	0.90	908
H. habilis	1.5	673
H. rudolphensis	1.89	752
A. robustus	1.8	530 ($n = 4$)
A. boisei	1.78	504 ($n = 2$)
A. aethiopicus	2.5	410
A. africanus	2.74	452 ($n = 6$)
A. afarensis	3.1	433 ($n = 3$)

makes the function relating unlogged cranial capacity to time exponential in time. The conventional constant-variance random-walk model fits the data least well. The phylogeny-scaling parameter λ is estimated to be 1.0 indicating that there is a strong link between phylogenetic relatedness and cranial capacity. The values of κ and δ are both greater than 1.0. This indicates that trait evolution has been gradual but faster in longer branches (κ) or paths (δ).

The constant-variance directional model fits the data substantially better than the random-walk model. This model differs from the random-walk model only by the parameter β, implying that β is significantly different from zero for these data, and confirming a significant statistical trend towards increasing cranial capacity over time. The directional model finds that the equation relating predicted cranial capacity to time is exp[6.07 + 0.33 * time]. This also implies a gradualist interpretation of brain-size evolution. The interpretation can be checked by fitting the branch-length scaling parameter κ. The maximum-likelihood estimate of κ for these data is now 1.18 and its 95 per cent confidence intervals include 1.0 but exclude zero. The value of κ has declined because variance that was previously attributed to the random walk has now been accounted for by β.

A value for κ near 1.0 supports a gradualist interpretation for these data. By excluding 0.0 from its 95 per cent confidence intervals the κ scaling parameter detects no evidence for a punctuational interpretation of Hominid brain-size evolution. Strictly speaking one cannot rule out a punctuational view, however. Each branch of the tree may hide many undetected speciation events, such as would arise if extinction rates are high. If each of these were associated with a burst of evolution, then evolution could appear gradual when observed only from the perspective of the longer branches. With branch lengths measured in units of time, this scenario is probably unlikely. Speciation events are thought to influence genetic or phenotypic divergence under some punctuational models and thus longer branches, where branch length records divergence, may be confounded with more undetected speciation. However, the number of speciation events will not influence branch length measured in units of time.

The third result in Table 13.2 includes the path-length scaling parameter δ within the directional model. This model significantly improves upon the unscaled-directional model, fitting a value of δ significantly greater than 1.0. This implies that the

Table 13.2 Likelihoods and values of parameters for models of evolution as estimated from Hominid cranial capacity data

Model	Likelihood	λ	κ	δ	Likelihood ratio[a]	p-value	σ^2 = variance[b]	α = root[c]	β = slope/my
Random-walk	−36.56	1.0	4.12 (2.00–8.15)	3.01 (1.72–4.08)			0.094	433 ± 0.01	
Directional	−28.71	1.0	1.185 (0.43–2.6)		7.86	0.00007 versus Drift	0.027	433 ± 0.007	0.33 ± 0.06
Scaled-directional	−25.43	1.0		1.55 (1.12–2.27)	3.28	< 0.010 versus Directional	0.014	433 ± 0.0009	0.20 ± 0.03

[a] Calculated as the difference between the two log-likelihoods. Twice these values is distributed approximately as a χ^2 variate with 1 degree of freedom.
[b] The variance (σ^2) is the estimated instantaneous rate of evolution, modelled as the variance of the distribution of step-lengths in the random walk (see text).
[c] The small errors of prediction arise because the species A. afarensis was placed arbitrarily close to the root of the tree. See Figure 13.5 for errors of prediction that arise as successive species are removed from the tree.

rate of brain-size evolution has increased over time, with the most rapid rates of evolution being those leading to Neanderthals and modern Humans (the longest total paths). The equation relating predicted cranial capacity to time in this model is $\exp[6.07 + 0.20 * \text{time}^{1.55}]$: longer total path lengths make a greater contribution than shorter ones.

The estimates of the instantaneous variance of evolution – the variance parameter – differ as expected among the three models, with the random-walk model returning the largest value. Superficially this would seem to imply that the random-walk model finds that evolution has progressed by larger steps than in the other two models. In fact, the correct interpretation is that the random-walk model mistakenly attributes all of the variance in the trait (cranial capacity) to the random-walk. By comparison, the two-directional models attribute large proportions of the variance to a systematic trend of trait evolution. As a consequence, the estimate of error variance is reduced. This will translate into more accurate estimates of reconstructed ancestral states (below).

The statistical approach has confirmed that there is an effect of phylogenetic relatedness on similarity in cranial capacity ($\lambda = 1.0$), that cranial capacity has increased steadily rather than following a random-walk ($\beta > 0.0$), that its evolution has been gradual and not punctuational ($\kappa > 0.0$), and that it has increased its rate of evolution over time ($\delta > 1.0$). The curves estimated for the directional and scaled-directional models are plotted in Figure 13.3. Figure 13.3 shows that the unscaled-directional model yields an exponentially increasing curve, but even this fails to capture the rapid increase in brain size that becomes evident for later *Homo* species. The scaled-directional model better captures the accelerating trend, showing

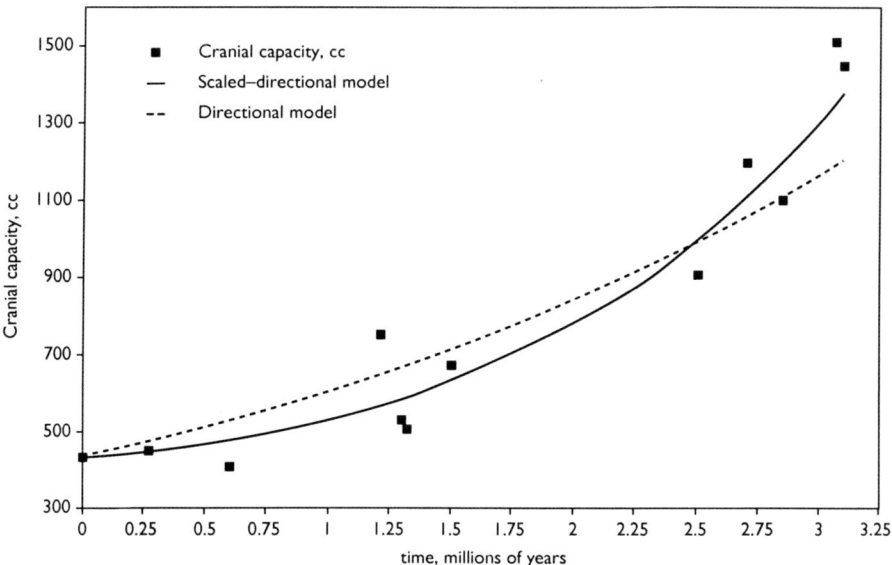

Figure 13.3 Plot of Hominid cranial capacity versus the total time since the root of the tree. The dashed curve represents the best fitting line from the unscaled-directional model: predicted cranial capacity = $\exp[6.07 + 0.33 * \text{time}]$; the solid curve represents the scaled-directional model: predicted cranial capacity = $\exp[6.07 + 0.20 * \text{time}^{1.55}]$. See text for detail of the models.

that Hominid brain size has increased at a faster than exponential rate over the past three million years.

The scaled-directional model accounts for over 98 per cent of the variance in cranial capacity. Still, there are some large deviations from the scaled-directional curve, most conspicuously that of *H. rudolphensis*. There are also *H. ergaster* crania of around 800–900 cc not long after *H. rudolphensis* (e.g., Aiello and Dunbar 1993; Ruff *et al.* 1997). Whether these deviations represent adaptive divergence from an underlying modal trajectory of evolution, or are indicative of entirely different trajectories is unknown.

It is straightforward to calculate the period-doubling time for cranial capacity from the two-directional model curves. These are plotted in Figure 13.4. The unscaled-directional model is a simple exponential curve, and therefore it has a constant period-doubling time, estimated here at about once every two million years. The scaled-directional model indicates that the rate of evolution has increased through time. Its period-doubling time begins at approximately the same point as the unscaled model, but decreases to approximately once every 1 million years by the time of the common ancestor to humans and Neanderthals.

All three models predict the root to be 433 cc (Table 13.2). This result is uninteresting, arising as it does because *A. afarensis* with a cranial capacity of 433 cc was arbitrarily placed so close to the root. It is a characteristic of any statistical approach that shorter branch lengths are given greater weight, denoting as they do that the value observed at the end of the branch is unlikely to change substantially were the process re-run. Nevertheless, the directional and scaled-directional models, as expected, return smaller errors of prediction than the random-walk model (Table 13.2).

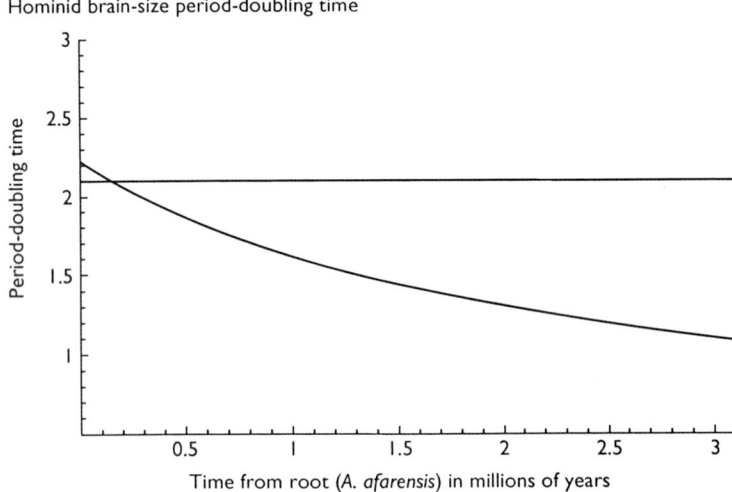

Figure 13.4 The predicted period-doubling time for Hominid cranial capacity as derived from the unscaled- and scaled-directional models. The horizontal line corresponds to the unscaled-directional model. It is a simple exponential and thus has a constant period-doubling time. The curve corresponds to the period doubling time for the scaled-directional model. It declines as time from the root increases, reflecting the accelerating pace of change in Hominid cranial capacity in later lineages.

Modelling continuously varying characters 281

Of greater interest is to ask how well these models reconstruct the value at the root of the tree when it is not known. Here, that question can be investigated by successively removing species from the phylogeny of Figure 13.2 beginning at the root, and asking the models to estimate the root only from the subset of species remaining. As the root is known in this example, it is possible to draw some conclusions about the relative performance of the various models. The interesting comparison is that between the random-walk model and the directional model. The random-walk model estimates of the ancestral state, as mentioned, are equivalent to those obtained from 'local' squared-change parsimony and from independent-contrast comparative methods.

Figure 13.5 shows the estimated cranial capacity for the species at the root of the Hominid phylogeny derived from re-estimating the random-walk and directional models based upon successively smaller subsets of the data. The horizontal line in the Figure is placed at 433 cc, corresponding to *A. afarensis*. The estimates derived from the random-walk model steadily increase as species are deleted, and they always fall within the range of cranial capacities observed in the data. Further, its confidence intervals exclude the value of 433 cc. By comparison the directional-model estimates

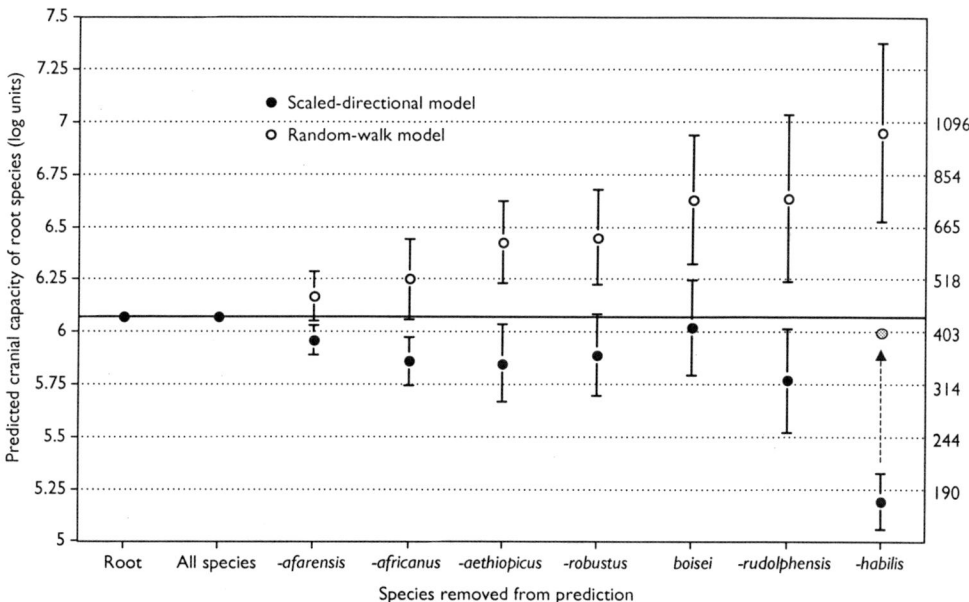

Figure 13.5 The predicted ancestral cranial capacity at the root of the Hominid phylogeny based upon samples in which increasing numbers of species have been deleted from the data set. The root value is 433 cc. Both models estimate the root value at 433 cc when all species are included (only directional model symbol is shown; see text for explanation of prediction). Beginning with 'afarensis' and moving from left to right, the data set excludes all those species before and including the one listed below the estimate. Thus, the estimates associated with 'africanus' exclude *A. afarensis* and *A. africanus* from the data. Estimates from the random-walk model systematically diverge from 433 cc while those for the directional model do not show a systematic bias, and have narrower confidence intervals. The dashed line points to the estimate of the root derived from the directional model when the path-length scaling parameter, δ, is used for this subset of the data (see text).

are always closer to 433 cc and its confidence intervals are narrower but still overlap, in two instances, the 433 line. The directional model also correctly estimates the ancestral state to fall outside of the range of values observed in the subset. These results are encouraging in light of the relatively small number of data points in the subsets.

Both the random-walk and directional models have quite bad performances when all species up to and including *H. habilis* are removed. The directional-model estimates would be improved were the scaling parameter δ included. In the subsets of data used for Figure 13.5 the 95 per cent confidence intervals for δ overlapped 1.0, and therefore I derived estimates of the root solely from the unscaled-directional model. Including the scaling parameter in the directional model, despite the fact that it is not significant, increases its estimate of the root, based on this subset, to 403 cc.

The models estimated from the complete data set can also be used to reconstruct the probable ancestral states of the remaining interior nodes of the tree. Figure 13.6 shows the Hominid phylogeny with ancestral states reconstructed from the random-walk model and from the scaled-directional model. This comparison is drawn to highlight the differences between the conventional 'Brownian motion' approach to reconstructing ancestral states and one that seeks to make use of additional information in the data.

The most obvious difference between the two models is that the random-walk model reconstructs nodes from a 'top-down' perspective: it begins at the tips and works

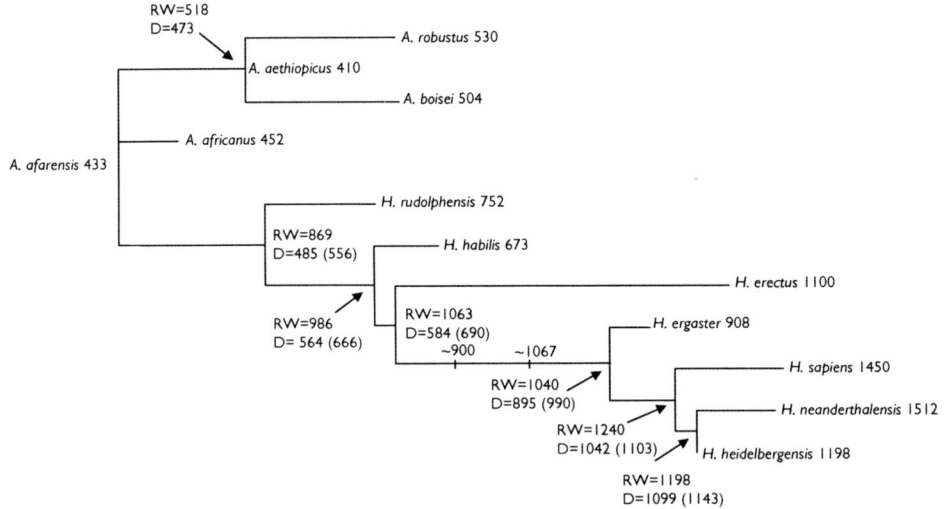

Figure 13.6   The predicted ancestral states at internal nodes of the Hominid phylogeny, as derived from two models: RW = the random-walk model; D = the scaled-directional model (Table 13.2). The values in parentheses for the directional model correspond to the reconstructed values obtained when the *Australopithecine* species are removed from the calculation of the equation linking cranial capacity to time. The random-walk model returns estimates of the ancestral states that are equivalent to those obtained from squared-change parsimony or from independent contrast methods. The actual (fossil) cranial capacities of the species are shown next to the species names. The fossil value for *H. ergaster* corresponds to a point sampled 0.90 mya and is probably too low by at least 100 cc for a putative time period of 0.60 mya. The two points along the branch leading to *H. ergaster* and other *Homo* species correspond to fossil *H. ergaster* crania measured around 1.2 and 1.5 mya.

backwards, using only the two immediate descendants of each node to estimate the value at the node. This means that the random-walk model consistently reconstructs ancestral cranial capacities to be larger than one of the descendants. By comparison, the directional model works from a 'bottom-up' perspective: it identifies a temporal trend (if one exists) and then reconstructs ancestral values as a function of the starting point (the root) and the distance from the root to the node being reconstructed.

The predictions of the random-walk model seem particularly implausible, implying as they do at least six, often substantial, reversals of the trend towards increasing brain size. The directional model predictions avoid this, but seem too low for the nodes connecting to *H. rudolphensis*, *H. habilis*, and possibly *H. erectus*. These may arise in part because the directional model is fitted to the entire data set, and it appears that the two *Australopithecine* species are characterised by a shallower slope relating cranial capacity to time. Deleting these species and re-estimating the directional-model increases yields the model *cranial capacity* $\sim \exp[6.07 + 0.36 * \text{time}]$. This equation produces the reconstructed values for the *Homo* clade as shown in parentheses in Figure 13.6. These seem much more in line with the trend of the data, and the one reversal this model implies is probably within the measurement error of the data.

Discussion

The Hominid cranial capacity data in combination with a phylogeny provide an unusual data set with which to compare the performance of various approaches for reconstructing ancestral character states of continuously evolving traits. What emerges from this data set is that conventional models that do not allow for the possibility of directional trends can be badly misled when reconstructing ancestral values, if such trends exist. The random-walk model investigated here is widely used, being equivalent to the model that underpins independent-contrast and squared-change parsimony methods. Its performance in predicting the root of the tree deteriorated rapidly as smaller and smaller subsets of the data were used. Further, it gave what would appear to be implausible reconstructions for the interior nodes of the tree.

By comparison, the directional model detected the temporal trend toward increasing cranial capacity and used it to make what would appear to be more accurate estimates of the ancestral character-state at the root of the Hominid phylogeny. Its performance did not deteriorate as rapidly in subsets of the data, and its reconstructions of interior nodes seemed more plausible than those derived from the random-walk model.

Directional trends such as those observed in the cranial capacity data should always be tested for as a first step in reconstructing the past. They are interesting in their own right, and as shown here, may be important for developing accurate estimates of ancestral states. Simple random-walk models are doomed always to reconstruct the past in the image of the present, a feature of this approach that elsewhere I have referred to as 'recapitulating the present' (Pagel 1991). Directional approaches can escape this trap and reconstruct previously unobserved values at ancestral nodes.

The scaling parameters λ, κ, and δ make it possible to ask whether the data and phylogeny fit the implicit assumptions of the model of evolutionary change. This is important, as several authors have now questioned the general utility of independent contrast techniques for analysing comparative data (Westoby *et al.* 1995; Ricklefs and Starck 1996; Price 1997; Harvey and Rambaut 1998). Under some models of evolution trait values can be partially or completely independent of phylogeny, and most evolutionary change may take place in earlier branches of the phylogeny. Under these circumstances, independent-contrast techniques can be shown to perform worse than techniques that do not use a phylogenetic correction. I have argued elsewhere (Pagel 1999b) that these situations can be detected by estimating the phylogeney scaling parameter λ and the path-length scaling parameter δ.

Estimating these parameters for the Hominid cranial capacity data, the branch-length scaling parameter κ allows one to reject a punctuational model of evolution as a description of these data on the phylogeny. Then, by estimating δ, a non-linear transformation of the path lengths is seen significantly to improve the fit of the model, implying that rates of evolution have increased over time. The importance of these results lies in demonstrating that signatures of past evolutionary trends are present in the data and can be used to assess the appropriateness of a given evolutionary model as a representation of those data. Where a given model is shown to be inadequate, for example the random-walk and unscaled-directional models here, I have shown that it may be possible to re-scale the phylogeny to find a better description of the data.

To my knowledge this is the first phylogenetic analysis of Hominid cranial capacity data, despite the fact that the issue of the temporal trend of Hominid cranial capacity has been much discussed and debated (see Pilbeam and Gould 1974; Aiello and Wheeler 1995; Ruff *et al.* 1997, and references therein). Apart from being the correct approach statistically (e.g., Table 13.2), the phylogeny focuses attention on the evolutionary pathways relevant to particular hypotheses. If one is interested in cranial capacity increase in *Homo* per se, then, for example, the Australopithecines in Figure 13.2 that arise after *A. afarensis* are not pertinent. In fact, as Figure 13.6 shows, removing them improves the reconstructed ancestral states for *Homo*.

More radical pruning of the phylogeny could apply were one interested in characterising the evolutionary trajectory leading to, say, *H. sapiens*. The phylogeny shows that numerous side branches of the *Homo* clade are not relevant to this trajectory. One can, in fact, trace the unique path leading from the root to *H. sapiens*, and ignore all other information. This might be dubbed the 'backbone' hypothesis to signify that one has analysed only the direct line of descent leading to *H. sapiens*. Doing this suggests (see Figure 13.3 and additional fossil data on Figure 13.6) that cranial capacity has increased linearly with time, at least in the path leading to *H. sapiens*. This is in contrast to the strongly curvelinear result reported in Table 13.2 and Figure 13.3, although the number of data points is small.

I have assumed throughout that a single variance of evolutionary change applies equally throughout the tree. Mooers *et al.* (1999) discuss a model in which this variance component is allowed to vary in every branch. Although not a practical model of evolution, requiring at least as many parameters as branches in the tree, it provides a useful general model against which to test more restricted models. If this general model does not fit the data significantly better than a restricted one, it gives evidence in favour of the simpler model. I have treated the phylogeny as being known without error. This, of course, will seldom be true and so it may be desirable to incorporate uncertainty about the phylogeny into estimates of ancestral states and other parameters of evolution. A new set of techniques based upon Markov-Chain-Monte-Carlo (MCMC) methods may make this possible. MCMC methods afford the possibility in principle of integrating parameter estimates (or estimates of ancestral states) over a random sample of phylogenies from the space of all possible phylogenies for a given set of species (Wilson and Balding 1998; Larget and Simon 1999; Lutzoni et al. 2001).

The general prospect of reconstructing novel ancestral states receives encouraging support from the Hominid data. Statistical models similar to those described here have already been employed to question widely-held beliefs about events ranging from the origin of life, to the Cambrian explosion and gene-culture evolution in human societies (Pagel 1999b). Whether the directional trends that make it possible to infer novel historical states are common or are confined only to some kinds of traits is unknown. Nevertheless, where such reconstructions are derived from statistical models that provide ways of assessing their appropriateness to the data, they can increasingly stand alongside other methods as independent branches of enquiry into the nature of past events of biological evolution.

Acknowledgements

Leslie Aiello, Mark Collard, Robert Foley and Marta Lahr have all helped me sort through the Hominid data and the myriad phylogenies proposed to describe their evolution. They may not, however, agree with everything I have said here. I am supported by the Leverhulme Trust, the NERC and the BBSRC.

References

Aiello, L. C. and Dunbar, R. I. M. (1993) 'Neocortex size, group size, and the evolution of language', *Current Anthropology*, 34, 184–193.
Aiello, L. C. and Wheeler, P. (1995) 'The expensive tissue hypothesis', *Current Anthropology*, 36, 199–221.
Edwards, A. W. F. (1972) 'Likelihood', Cambridge: Cambridge University Press.
Felsenstein, J. (1985) 'Phylogenies and the comparative method', *American Naturalist*, 125, 1–15.
Foley, R. (1998) 'The context of human genetic evolution', *Genome Research*, 8, 339–347.
Garland, T., Jr., Harvey, P. H. and Ives, A. R. (1992) 'Procedures for the analysis of comparative data using phylogenetically independent contrasts', *Systematic Biology*, 41, 18–32.
Garland, T., Jr., Midford, P. E. and Ives, A. R. (1999) 'An introduction to phylogenetically-based statistical methods, with a new method for confidence intervals on ancestral values', *American Zoologist*, 39, 374–388.
Golding, G. B. and Dean, A. M. (1998) 'The structural basis of molecular adaptation', *Molecular Biology and Evolution*, 15, 355–369.
Harvey, P. H. and Pagel, M. (1991) *The comparative method in evolutionary biology*, Oxford: Oxford University Press.
Harvey, P. H. and Rambaut, A. (1998) 'Phylogenetic extinction rates and comparative methodology', *Proceedings of the Royal Society of London (B)*, 265, 1691–1696.
Koshi, J. M. and Goldstein, R. A. (1996) 'Probabilistic reconstruction of ancestral protein sequences', *Journal of Molecular Evolution*, 42, 313–320.
Larget, B. and Simon, D. L. (1999) 'Markov chain Monte Carlo algorithms for the Bayesian analysis of phylogenetic trees', *Molecular Biology and Evolution*, 16, 750–759.
Lutzoni, F., Pagel, M. and Reeb, V. (2001) 'Major fungal lineages derived from lichen-symbiotic ancestors', *Nature*, 411, 937–940.

Maddison, W. P. (1991) 'Squared-change parsimony reconstructions of ancestral states for continuous-valued characters on a phylogenetic tree', *Systematic Zoology*, **40**, 304–314.

Maddison, D. R. (1994) 'Phylogenetic methods for inferring the evolutonary history and process of change in discreetly valued characters', *Annual Review of Entomology*, **39**, 267–292.

Maddison, W. P. (1995) 'Calculating the probability distribution of ancestral states reconstructed by parsimony on phylogenetic trees', *Systematic Biology*, **44**, 474–481.

Mooers, A. Ø. and Schluter, D. (1999) 'Support for one and two rate models of discrete trait evolution', *Systematic Biology*, **48**, 623–633.

Mooers, A. Ø., Vamosi, S. M. and Schluter, D. (1999) 'Using phylogenies to test macroevolutionary hypotheses of trait evolution,' *American Naturalist*, **154**, 249–259.

Omland, K. (1994) 'Character congruence between a molecular and a morphological phylogeny for dabbling ducks (anas)', *Systematic Biology*, **43**, 369–386.

Omland, K. (1997) 'Correlated rates of molecular and morphological evolution', *Evolution*, **51**, 1381–1393.

Pagel, M. (1991) 'Constructing every animal', Review of Paleobiology (eds J. Damuth and Brian MacFadden) *Nature* (Lond.), **351**, 532–533.

Pagel, M. (1992) 'A method for the analysis of comparative data', *Journal of Theoretical Biology*, **156**, 431–442.

Pagel, M. (1994) 'Detecting correlated evolution on phylogenies: a general method for the comparative analysis of discrete characters', *Proceedings of the Royal Society (B)*, **255**, 37–45.

Pagel, M. (1997) 'Inferring evolutionary processes from phylogenies,' *Zoologica Scripta*, **26**, 331–348.

Pagel, M. (1999a) 'The maximum likelihood approach to reconstructing ancestral character states of discrete characters on phylogenies', *Systematic Biology*, **48**, 612–622.

Pagel, M. (1999b) 'Inferring the historical patterns of biological evolution', *Nature*, **401**, 877–884.

Pilbeam, D. and Gould, S. J. (1974) 'Size and scaling in human evolution,' *Science*, **186**, 892–901.

Price, T. (1997) 'Correlated evolution and independent contrasts', *Philosophical Transactions of the Royal Society of London (B)*, **352**, 519–529.

Ricklefs, R. E. and Starck J. M. (1996) 'Applications of phylogenetically independent contrasts: A mixed progress report,' *Oikos*, **77**, 167–172.

Ruff, C., Trinkaus, E. and Holliday, T. W. (1997) 'Body mass and encephalization in Pleistocene *Homo*', *Nature*, **387**, 173–176.

Schluter, D. (1995) 'Uncertainty in ancient phylogenies', *Nature*, **377**, 108–109.

Schluter, D., T. Price, A. Ø. Mooers and Ludwig, D. (1997) 'Likelihood of ancestor states in adaptive radiation,' *Evolution*, **51**, 1699–1711.

Westoby, M., Leishman, M. R. and Lord, J. M. (1995) 'On misinterpreting the phylogenetic correction', *Journal of Ecology*, **83**, 531–534.

Wilson, I. and Balding, D. (1998) 'Genealogical inference from microsatellite data', *Genetics*, **150**, 499–510.

Wood, B. B. and Collard, M. (2000) 'The changing face of the genus *Homo*', *Evolutionary Anthropology*, **8**, 195–207.

Yang, Z., Kumar, S. and Nei, M. (1995) 'A new method of inference of ancestral nucleotide and amino acid sequences', *Genetics*, **141**, 1641–1650.

Chapter 14

Summary

Peter L. Forey

In the introduction to this book we asked four questions of the relationship between morphometrics and phylogeny reconstruction. We now need to review how fully those questions have been answered – if at all – and to enquire if the attempt prompted new enquiries.

The first two questions asked whether continuously distributed variables can be used in our attempts at phylogeny reconstruction. The consensus among our authors appears to be yes – under certain conditions. And it is in specifying those conditions that authors have explored the concepts of homology in systematics and morphometrics as well as the nature of a character as used in phylogenetic systematics.

Humphries (Chapter 2) points out that in phylogenetic reconstruction – which today almost universally means phylogenetic systematics or cladistics – homology is a theory that specifies relationships between taxa. The discovery of individual homologues is an empirical procedure that consists of character recognition as well as subsequent cladistic analysis. It is at the level of character recognition and character coding that morphometrics and phylogeny reconstruction interact. In phylogeny reconstruction character recognition is usually described as primary homology (Humphries, Chapter 2) and consists of three stages: (1) some estimation of topographic identity and/or compositional identity; (2) measurement of these 'identical' features and grouping the measurements; (3) comparing the measurements to construct a data matrix to be analysed (Brower and Schawaroch 1996; Stevens 2000).

The first stage is clearly a key stage, yet it is one which causes some confusion between morphometrics and phylogenetic systematics. For instance, in criticising an earlier paper of Zelditch *et al.* (1995) which proposed that landmarks (points) necessarily embody the concept of homology Macleod (1999) pointed out that the dorsal fin placed along the back of a swordfish, an ichthyosaur and a whale could be described as 'triangular' in all. In morphometric terms and the mathematical descriptions (e.g., landmark-referenced triangles) that we may wish to apply to describe the shape, the triangle would be considered homologous. In terms of systematics such comparisons would not even satisfy the first stage of homology recognition since these structures are not compositionally the same in all three taxa. Therefore, to use morphometrics to describe those fins as equal or non-equal in their triangularity is meaningless in terms of phylogeny reconstruction. If we were to compare the triangular dorsal fins of a goldfish, a minnow and a carp, then such comparisons would be meaningful since all are compositionally the same (they have bony fin rays and a scaled web) and have the same relationships to surrounding structures (they are supported by endoskeletal radial bones which from articular surfaces with the fin rays). Clearly, this information does require some prior knowledge (see below).

Additionally, any key positions that we may take as landmarks in order to describe differences in shape or aspect ratio between the taxa, such as the positions of articulation of the first and last fin rays, could be justified as biological primary homologues. In contrast, it needs to be stressed that the 'points' used as landmarks for morphometric analysis cannot be regarded as theoretical equivalents to primary homologues. Geometric landmarks are not themselves inherently biological homologues. Instead they are points, chosen because of their geometric suitability and used to describe spatial arrangements from other landmarks. In morphometrics it is the space between the landmarks that is homologous. As Bookstein identified landmarks may be of three kinds. So-called Type 1 landmarks are points located at junctions between different tissues or structures. Type 2 landmarks are points at the maxima of curvature and Type 3 landmarks are points interpolated between the principal landmarks, sometimes used to more accurately describe an outline.

Nevertheless, although the axioms of morphometric analysis do not recognise biological homologues, it is perfectly possible that Type 1 landmarks may correspond to primary homologues in phylogenetic systematics. Therefore, the Swiderski *et al.* claim (Chapter 6) that it is acceptable to use as phylogenetic characters morphometric figures that are anchored to biologically homologous structures seems reasonable. But this does imply three constraints on using landmark-based morphometric variables as phylogenetic characters. The first is that a geometric landmark is, theoretically, an infinitesimal point and it is unlikely to be truly homologous in spatial terms between the taxa sampled. The second is that such variables may only be chosen from taxonomic samples taken from within relatively low taxonomic ranks where there can be little ambiguity about the biological primary homology of the landmarks (since homology is a theory always subject to test – Chapter 2 – there will always be some ambiguity). The third flows from the second because it means that we must have some idea of the phylogeny to start with. Thus, we may be justified in using morphometric variables of dorsal fin shape as data for phylogeny reconstructions of different species of ichthyosaurs, or different species of whales but not for phylogeny reconstruction of different species of tetrapods. This is because our prior phylogenetic analysis has already demonstrated non-homology of the dorsal fins of ichthyosaurs and whales. It should be pointed out, however, that even in cladistic analysis using qualitative characters some idea of a phylogeny is usually accepted in order to choose the outgroup taxon used to root the tree although this is not a theoretical requirement.

Morphometrics, like initial recognition of characters in phylogenetic systematics, is atemporal – it is agnostic to history (phylogeny). However, accepting the constraints imposed by phylogeny, morphometrics can repay us in our attempts at phylogeny reconstruction and it can do this at the second stage of character recognition – the measurement of identical features. 'Identical' here means features that we have already judged to be identical based on topographic similarity or composition. The various ways in which shape can be measured (e.g., Bookstein Shape Coordinates, opencurve eigenshape analysis, partial warps, creases etc. – see chapters by Macleod, Swiderski *et al.*, Bookstein, Rohlf, Polly) can help us to differentiate one shape group from another, or to determine whether there really is a further shape to be considered (Macleod, Chapter 7), or indeed if the shape variation among the taxa we are considering is truly continuous or discontinuous. For instance, we may differentiate by eye a leaf margin in some taxa as being round in some and arch-shaped in others. Morphometric description of the margin may further reveal details about the symmetry of this curvature as a further variable which may or may not co-vary with round or arch-shaped.

And with morphometric variables there are other factors to be taken into account. Rae (Chapter 4) mentions the problem of scaling caused by absolute size of the structure being assessed. He gives an example elaborated in Rae and Koppe (2000) where the volume of the maxillary sinus, which formerly was thought to be variable among anthropoids and hence of potential systematic value, is similar in all if scaled to cranial size. Such correlation between the size of one structure and another raises another issue that we must be aware of in using morphometric data – this is character dependence. In a cladistic analysis each column of data – the character with characters states – is assumed to be independent of each other and to imply relationships between taxa. These relationships may be consistent or conflict with relationships revealed by another character. Clearly, if one character is dependent on another this assumption is violated. Thus, to use the volume of the maxillary sinus as one character and the cranial capacity as another would effectively be entering the same data twice and weighting that data. Morphometric variables are not unique in this respect but they may be particularly prone to correlation and covariance since a shape difference between taxa in one parameter may automatically imply a difference in another.

Felsenstein (Chapter 3), in typical lucid style, gives an excellent account of how qualitative (morphometric) characters can be analysed with qualitative data (e.g., molecular data) by using the trees produced by qualitative data to estimate covariance among the quantitative data and then apply the results to select maximum likelihood trees for all taxa including fossils. Thus, Felsenstein takes an iterative view by using qualitative data to infer something about the evolution of the quantitative data which can then be fed back into a total evidence approach.

Of course, morphometric variables are used freely at the level of diagnosis of one species from another. In fish taxonomy, for instance, it is common practice to distinguish species A from species B on parameters such as relative head length, numbers of vertebrae, numbers of fin rays in the different fins and the positions of these fins relative to one another. What authors have explored in this book is the potential use of similarity between one species and another in any or all of such variables as phylogenetically important characters. This exploration is perfectly justifiable since the variation we exploit to distinguish taxa must have a phylogenetic basis (Cole et al., Chapter 10; Polly, Chapter 11). Arguments suggesting they have no theoretical place in phylogenetic reconstruction (Pimental and Riggins 1987; David and Laurin 1996) are, in our view, misplaced.

Having decided that it is both possible, in theory, and desirable in practice to include morphometric data into phylogenetic analysis it becomes necessary to discuss ways in which descriptions of shape may be captured most faithfully. This is included especially in chapters by Bookstein (Chapter 8), MacLeod (Chapter 7) and Swiderski et al. (Chapter 6), but it is a recurring theme through others. As MacLeod and Swiderski et al. point out, it is landmark-based morphometrics that are the most often used and it is key to Swiderski et al.'s method of coding characters for phylogenetic analysis in which they equate the landmark points with homologues. Swiderski et al. translate the relative positions of the landmarks into abstract characters useful for phylogenetic analysis by comparing the aspects of landmark variation patterns, which is the character, with a standard. The character states are the deviations from that standard which, in the Swiderski *et al.* case, is taken as an outgroup taxon (in traditional morphometric analysis the standard is taken as the mean of all shapes to be compared). The deviations used are mathematical descriptions of the distortion or partial warp of a thin plate spline, which is a method for expressing shape difference in different regions of the form. What perhaps needs more careful attention is the danger of equating the mathematical distortion of the landmarks with a theory about what actually happened in the evolution of one shape from another. We run the risk of equating the use of landmark morphometrics with transformational homology in phylogenetic systematics. It is possible that at some level they do coincide. The difficulty is in knowing at what level they fail to do so.

Furthermore, as pointed out by MacLeod (1999) the use of landmarks alone does not necessarily accurately describe the shape between them and can, in some cases, be positively misleading (MacLeod 1999: figure 20). Therefore, the recommendations made by MacLeod that the outline between landmarks is systematically important and that morphometric strategies be developed to represent such data (e.g., Swiderski *et al.*, McLeod) is an important extension of the landmark method. One such extension is that the outline between landmarks be divided into many equally-spaced semi-landmarks (Bookstein 1997; McLeod 1999), each of which can be compared from individual to individual or taxon to taxon as part of an open-curve. There is no implication that each corresponding semi-landmark from taxon to taxon is homologous and perhaps this is one of the drawbacks of such methods. Nevertheless, there is no doubt that it does provide a more accurate description of the shape between accepted end-point positions and it is considerably more detailed and subject to repeatability tests than the *ad hoc* estimations of shape that systematists regularly make. MacLeod also challenges the idea of using partial warps to describe the variation in landmark positions between taxa, by describing a new relative warp method. This relative warp approach, offered here for the first time, appears to be more successful at recovering a phylogenetic signal than partial warp analysis.

One of the issues surrounding landmark-based morphometrics is raised by Bookstein (Chapter 8) who is concerned about how many landmarks may be necessary to fully describe a shape difference between individuals or between species. Bookstein recognises that our morphometric descriptions of shape may be constrained by the variables we choose to measure at the outset of our analysis and hence may not capture the full or even the greatest shape difference. In its place Bookstein has developed a method of creases which is a method which attempts to compare shapes in multidimensions simultaneously and to localise the greatest shape difference along the crease.

Most authors, who question the use of morphometric variables, do so at the level of the difficulty of translating such variables into the standard codings required for cladistic analysis (Bookstein 1994). Historically, part of the difficulty with using morphometric data is they are inherently variable along some kind of scale. The issue of whether the variation of a particular feature of an organism is continuous or discontinuous is clearly of more than passing interest because this has been seen as one of the primary reasons for rejecting morphometric variables for phylogenetic analysis. Several of the authors in this book raise this issue.

It is true that much of the variation that is used for phylogenetic analysis is inherently continuous even though our character state matrices suggest that only part of that continuum is expressed in the taxa studied. For example, in six species of the fish genus *Alosa* (shads) the number of gill rakers (fine comb-like bony filaments on the gill arches used for filter feeding) upon the lower part of the first gill arch as recorded by Whitehead (1985) ranges in total from 18 to 73. However, two species have 18–24, a further two have 38–52 and the additional two have 59–73 (the variation within each group arises because the number of gill rakers increases to a maximum throughout growth). Thus although there is no apparent reason why species should not display intervening numbers they do not and the actual distributions are non-overlapping. Of course, it is always possible that were more species known in this particular feature the separation of these six species into three distinct groups would break down. But it is worth remembering that this can also happen with non-continuously distributed variables. For instance, the presence or absence of a certain bone in the skull roof of fishes can be subject to considerable individual variation (Hilton and Bemis 1999).

There is another issue raised in connection with continuous variables encountered in morphometric analysis. For morphometric data, whether they be meristic measurements or interlandmark ratios it is rare for two or more of the sampled individuals for any one taxon to be identical. Therefore, some consideration needs to be given to exactly which of the measurements represents the conditions that will be coded for in any one taxon. Of course, if the terminal units of our analysis are individuals there is no problem but phylogenetic analysis (as opposed to human genealogy) is rarely, if ever, done at this level. We may take a mean or modal value, or perhaps a range within one standard deviation or, when using landmarks, even employ some averaging criterion such as Procrustes fit. But the fact remains that such morphometric variables are abstractions which are not repeatable in any one individual and they are seen to be subject to sampling problems and as such they have been criticised. To some extent this criticism of morphometric data is unreasonable since sampling can be equally problematic when using qualitative data for phylogenetic analysis. Thus, the presence or absence of a particular feature may be variable within the population and there may be doubt about which of the variants to use for analysis. Additionally our sampling may only detect one or other condition, and that condition may be the extreme minority variant. Wiens (2000) gives a good account of the problems and some solutions to using frequency data of different character states in phylogenetic analysis.

One new technique offered in this book is the method devised by Cole *et al.* (Chapter 10). These authors suggest that we may use a parametric bootstrapping technique to sample the total variation expressed by the studied taxa. Bootstrapping has long been applied to the problems of sampling discrete characters following the recommendations and techniques outlined by Felsenstein (1985) and a similar parametric approach is regularly used in the phylogenetic analysis of molecular data. In this application of the bootstrap Cole *et al.* use the morphometric information of each of the taxa as represented by scatters of landmark constellations. The bootstrap is now applied by sampling the taxa with replacement to replicate the original total number of taxa and some statistical value applied to each of the nodes on the phenogram recording what percentage of times that particular node is recovered in repeated phenetic analyses.

A further aspect of morphometric variables often discussed is how samples of values ranged along a continuum are divided up into discrete codes. There is a wealth of literature on this subject and this is referred to in many of the chapters but especially by Reid and Sidwell (Chapter 5; see also Wiens 2000). All of the methods have strengths and weaknesses but one of the interesting conclusions arrived at by Reid and Sidwell is that different methods result in different ways of describing the variation and these may have different consequences for the results of phylogenetic reconstruction. Rae (Chapter 4) advocates the use of homogeneous subset coding. Since the methods are conventions it is not surprising that there is no agreement on one technique or another. At the same time it would be wrong to reject these attempts simply because of the lack of an agreed method. A simple perusal at data matrices compiled for qualitative data also shows lack of agreement among different authors in translating observations into codes for phylogenetic analysis (e.g., see Hawkins 2000). But perhaps one of the most pertinent messages to come out of these discussions is the difficulty of using coding to express overlapping variables in continually distributed data (MacLeod, Chapter 7). Such codings may be used but only after the individual taxa have been recognised. Even here they imply that the taxa will have polymorphic codings unless it be accepted that individuals of two or more taxa which have the extreme (overlapping) variants are ignored during the succeeding phylogenetic analysis.

The results of the exploration of using morphometric variables as characters to construct a phylogeny show that there is no inherent reason why they should not be used but there remain problems and agreed criteria by which such variables are to be coded. Of course, this is no different from qualitative data.

A further question asked in the Introduction relates to ways in which morphometric information can supplement our understanding of the paths of evolution as deduced from phylogenetic information obtained from non-morphometric data. As Felsenstein (Chapter 3) suggests quantitative characters (and this includes morphometric variables) will be of interest for themselves rather as characters to be used in phylogenetic analysis. In its simplest form we can simply map morphometric data on to an already existing phylogenetic tree. An example of this is given by Forey (1991) where he mapped various metric parameters of the vertebral column (numbers of vertebrae, spacing of vertebrae) on to a phylogeny of coelacanths. This study showed that during coelacanth evolution there was both an increase in the number of vertebrae as well as a progressive crowding of the vertebrae within the abdominal region of the vertebral column.

A more sophisticated use of morphometric data is exemplified by Polly (Chapter 11) who plots rates of evolution of shape changes – in this case the changes in cusp patterns on the first lower molars in species of a primitive group of early Cenozoic carnivores. Rates of morphological evolution are questions increasingly asked, particularly because of the potential to make comparisons with rates of molecular evolution. Polly's study is important because it is rare to be able to incorporate tight stratigraphic control that is beneficial to date divergence events. More often, in the absence of a good stratigraphic record, trees are simply divided up into equal length branches with corresponding potential error for miscalculating rates of change. But here, it is possible to record actual time intervals between ancestor and descendant or between common ancestors and daughter taxa which can be used to calculate the rate of change per generation time. Such knowledge is clearly important, especially when applying parsimony methods to reconstruct ancestral states or rates of evolution (Webster and Purvis, Chapter 12).

Unfortunately, such precise dating of divergence times is not generally available and it therefore becomes difficult to calculate rates of change independent of some prior model of rate of change. A slightly different approach may be taken by trying to estimate the ancestral conditions of morphometric variables rather than the rate *per se* and this is done by Webster and Purvis and also by Felsenstein (Chapter 3). Of course, when trying to estimate ancestral conditions the rate of evolution along different branches theoretically bears on this attempt. In consequence maximum likelihood methods may at first seem most reasonable since models of rates can be built in. However, Webster and Purvis conclude that the parsimony methods, which take the paths of least change, generally perform better at recovering probable ancestral states. Rohlf (Chapter 9) develops one particular parsimony method to estimate the shape characteristics at the internal nodes (ancestral positions) of a phylogeny. He applies this to wing venation pattern described by landmarks positions over a phylogeny of mosquitoes. Thus, the shape change in venation can be followed across the phylogeny.

This book has taken up a particularly difficult challenge in bringing together the two disciplines of morphometric analysis and phylogenetic systematics. Historically, these two have had little dialogue, yet they have much to offer each other as this book has demonstrated. We cannot claim unanimity for including all morphometric data into phylogenetic analysis or the precise way in which any of that data is to be incorporated. Nor can we claim that a phylogenetic perspective is necessary for morphometrics. What we do claim is that where the two overlap it results in benefits to both, that the dialogue has been productive and should continue.

References

Bookstein, F. L. (1994) 'Can biometrical shape be a homologous character?', in Hall, B. K. (ed.) *Homology: the hierarchical basis of comparative biology*, San Diego, CA: Academic Press, pp. 197–227.

Bookstein, F. L. (1997) 'Landmark methods for forms without landmarks: localizing group differences in outline shape', *Medical Image Analysis*, 1, 225–243.

Brower, A. V. Z. and Schawaroch, V. (1996) 'Three steps of homology assessment', *Cladistics*, 12, 265–272.

David, B. and Laurin, B. (1996) 'Morphometrics and cladistics: measuring phylogeny in the sea urchin *Echinocardium*', *Evolution*, 50, 348–359.

Felsenstein, J. (1985) 'Confidence limits on phylogenies: an approach using the bootstrap', *Evolution*, 39, 783–791.

Forey, P. L. (1991) '*Latimeria chalumnae* and its pedigree', *Environmental Biology of Fishes*, 32, 75–97.

Hawkins, J. A. (2000) 'A survey of primary homology assessment: different botanists perceive and define characters in different ways', in Scotland, R. W. and Pennington, R. T. (eds) *Homology and systematics: coding characters for phylogenetic analysis*, London: Taylor & Francis, pp. 22–53.

Hilton, E. J. and Bemis, W. E. (1999) 'Skeletal variation in shortnosed sturgeon (*Acipenser brevirostrum*) from the Connecticut River: implications for comparative osteological study of fossil and living fishes', in Arratia, G. and Schultze, H.-P. (eds) *Mesozoic fishes 2 – systematics and the fossil record*, München: Dr Friedrich Pfeil, pp. 69–94.

MacLeod, N. (1999) 'Generalizing and extending the eigenshape method of shape space visualization and analysis', *Paleobiology*, **25**, 107–138.

Pimental, R. A. and Riggins, R. (1987) 'The nature of cladistic data', *Cladistics*, **3**, 201–209.

Rae, T. and Koppe, T. (2000) 'Isometric scaling of maxillary sinus volume in hominoids', *Journal of Human Evolution*, **38**, 411–423.

Stevens, P. F. (2000) 'On characters and character states: do overlapping and non-overlapping variation, morphology and molecules all yield data of the same value?', in Scotland, R. W. and Pennington, R. T. (eds) *Homology and systematics: coding characters for phylogenetic analysis*, London: Taylor & Francis, pp. 81–105.

Whitehead, P. J. P. (1985) 'FAO species catalogue. Vol. 7. Clupeoid fishes of the world (suborder Clupeoidei). Part 1 – Chirocentridae, Clupeidae and Pristigasteridae', *FAO Fisheries Synopsis No. 125*, **7**, 1–303.

Wiens, J. J. (2000) 'Coding morphological variation within species and higher taxa for phylogenetic analysis', in Wiens, J. J. (ed.) *Phylogenetic analysis of morphological data*, Washington: Smithsonian Institution Press, pp. 115–145.

Zelditch, M. L., Fink, W. L. and Swiderski, D. L. (1995) 'Morphometrics, homology, and phylogenetics: quantified characters as synapomorphies', *Systematic Biology*, **44**, 179–189.

Index

"Struszia" mccartneyi 105–6, 108
4-simplex 142
adaptations 197–8, 200
adaptive zones 197
Aedes 183–5
agreement subtree 123, 125
allometry 16
Alouatta 211–13
Alouatta palliata 263–4
Alouatta seniculus 212, 263–4
alveolus 212
amniotic membranes 2
amplitude 84
anagenetic evolution 111
anatomical feature 73, 85, 87, 90
anatomy 211
ancestor (common) 168, 196, 220–1, 234–5, 237
ancestors 10, 31, 69, 185
ancestors (hypothetical) 11
ancestral character state estimation 6
ancestral character states *see* states (ancestral character)
ancestral nodes 233
ancestry 101
ANCOVA 259
angles 175
Angophora 21
angular process 82, 87
angularity 2
Anopheles 183–4
ANOVA 56, 257–8
anthropoid primates 47
Aotus trivirgatus 263–4
apes 171
Aramigus weevils 68
articular process 89–90
asymmetry 93
Ateles 212–14
Ateles belzebuth 263–4
Ateles fusciceps 263–4

Ateles geoffreyi 212, 263–4
Ateles paniscus 263–4
Atelinae 211
autapomorphic states *see* states (autapomorphic)
autapomorphies 198–9, 213
autapomorphs 16
autocorrelation analysis 221
Avalanchurus 105
Avalanchurus simoni 104–5, 108–9, 121–2, 125
average image 183
average shape 177, 188
axial furrow 122
Ayala, F. J. 201

backbone 14
Banksia 21
baseline 74, 92
basicranial 68
bat 2, 9
beetles 39
behavior 196, 211
bending energy 168, 172
bending-energy matrix 77, 119, 127, 179
Bighorn Basin, Wyoming 220, 223–4, 227–8, 243
Billevittia adraini 105, 108–9, 125
binary characters *see* characters (binary)
biometry 145
birds 2, 9
black-tailed prairie dog 75
body covering 102
body mass 220, 228–9, 243, 257; *see also* size (body)
bone 2
Bookstein shape coordinates *see* coordinates (Bookstein shape)
Bookstein, F. L. 6, 125
bootstrap trees *see* tree (bootstrap)
bootstrapping 35, 194, 201

bootstrapping (nonparametric) 202, 214
bootstrapping (parametric) 202, 204–5, 209, 214–15
botanists 53
boundary (Paleocene–Eocene) 227
boundary curves 5
Brachyteles 211–14
Brachyteles arachnoides 212, 263–4
Brady, R. H. 9
branch length(s) 220, 229, 232, 234–5, 245, 249, 251–3
branch-and-bound analysis 125
breadths 3
bregma 167
Brownian motion 28–9, 31, 34, 37–8, 244, 249, 250–2, 255, 259, 262, 266

Cacajao calvus 263–4
Callicebus brunneus 263–4
Callicebus cupreus 263–4
Callicebus moloch 263–4
Callicebus personatus 263–4
Callimico goeldii 263–4
Callithrix argentata 263–4
Callithrix humeralifer 263–4
Callithrix jacchus 263–4
calva 169
canalis opticus 167
Canidae 256, 258
canids 221
canine tooth 117
canonical correlation analysis 3
canonical variates analysis (CVA) 181, 187
Cantius 229
Carnivora 223, 228
carnivorans 6, 220–2
carnivores 198, 230, 247–8, 256–7, 259
Cartesian grid 3
catastrophe theory 139, 148
cebids 257
Ceboidea 260
Cebuella pygmaea 263–4
Cebupitheca sarmientoi 263–5
Cebus albifrons 263–4
Cebus apella 263–4
Cebus capucinus 263–4
Cebus olivaceus 263–4
centroid size *see* size (centroid)
centroids 45, 48
Cercopithecinae 260
cercopithecines 257
character analysis 11
character coding 18, 40, 53–4
character coding problem 39
character correspondence 17

character evolution 22
character formulation 12
character state identity 12
character state(s) 1, 8, 10, 13, 18, 46–7, 49, 81, 100, 103, 105, 113
character transformation 9
character uncoding problem 27, 40–1
characters 5, 8, 10, 12–13, 15, 17–18, 22, 27, 47, 49, 53, 56, 64, 76, 87, 95, 104, 129, 130, 134–135, 169, 187, 215
characters (ancestral) 248
characters (binary) 19
characters (cladistic) 14
characters (congruent) 12
characters (continuous) 15, 18, 55–6, 65, 247
characters (discrete) 1, 27, 139
characters (incongruent) 12
characters (metric) 49
characters (morphological) 9, 12, 18, 33, 55–6, 100
characters (multistate) 13, 18–20
characters (non-continuous) 15
characters (overlapping morphological) 63
characters (overlapping) 53
characters (qualitative) 15
characters (quantitative) 5, 27
characters (systematic) 102–3, 119
character-state assignments 130
character-state coding 112
character-state definition 113
character-state transformations 120
cheek teeth 72
Chiropotes satanas 263–4
clade(s) 197, 212, 256, 259, 260
cladistic analysis 12–13, 45
cladistic relationships 46
cladists 10
cladogenesis 111, 134
cladogram 10, 11, 13, 119–20, 123, 194–6, 201, 205, 208, 212, 214, 223–4, 234
cladogram (consensus) 224
cladogram (parsimonious) 123, 125
cladogram root 213
Clarkforkian 224
classification(s) 10, 22
clustering (neighbor-joining) 233
clustering (UPGMA) 212, 233
coding (cladistic) 46
coding (contingent) 18
coding (divergence) 19, 47
coding (gap) 18–20, 47, 111, 113, 142, 167
coding (generalized gap) 19–20, 56, 60, 62–3
coding (homogeneous subset) 19–20, 47
coding (homologue) 18
coding (inapplicable data coding) 18
coding (logically related) 18

coding (metric) 45
coding (range) 20
coding (ratio) 18
coding (segment) 19–20, 47, 56, 63, 111, 113
coding (simple gap) 56, 61, 63
coding (unifying) 18
coefficients (cophenetic correlation) 207, 212
coefficients (regression) 201
Colobinae 260
common ancestor *see* ancestor (common)
commutativity 119
comparative biology 13–14, 141
comparative method 4, 189, 231
component/factor analysis 3
condyle 91
congruence 2, 8, 71, 200
conjunction 2, 8, 71
conodont(s) 247–8, 261–2, 266
consistency index (CI) 125
constraints (developmental) 200
continuous variable *see* variables (continuous)
convergence 11, 69, 198
coordinates (Bookstein shape) 74, 106
  (*see* shape coordinates)
coordinates (Kendall tangent space) 178, 182
coordinates (tangent space) 187
coronoid 90
coronoid process 74, 87, 89–90, 92, 96
corpus callosum 156–7, 160
correlation (character) 30
correlation 3, 114, 201
correlation matrix 88
correlations (developmental) 42
correlations (genetic) 42
correlations (selective) 42
correspondence (point-to-point) 2
correspondence (topological) 9, 17; *see also* homology (topographical)
Cosomys primus 229
counts 8
covariance 3, 88, 101, 114, 125, 145, 178, 194, 236
covariances (additive genetic) 30–1
covariances (phylogenetic) 172
covariance (selective) 31
Coyote 224
crabs 10
cranial size 48
cranidium 121
Cranston, P. S. 120
crease(s) 6, 146–7, 151, 154, 161, 163, 166, 168–70
creases (ontogenetic) 171
Crisp, M. 103
crista galli 167
Culex 183–4

Culiseata 183–4
Curculionidae 68
cursorial species *see* species (cursorial)
curvature 2
cusps 147, 151, 236
Cuvier, G. 9
Cynomys ludovicianus 75–6, 79–83

Darwin, C. 9, 10–11, 15
deformation 92
deformation grids 115
Deinocerites 183–4
descendants 10, 69
diastema 89
diatom(s) 53, 56, 61–2
Didymictis 222, 232, 237
Didymictis leptomylus 223–5, 233, 243–6
Didymictis protenus 224–5, 227, 233, 243–6
Didymictis proteus 224–5, 232–3, 243, 245–6
difference between means tests 111
directional selection *see* selection (directional)
discontinuities (morphological) 15, 18, 20, 113, 125–6, 131, 134, 139
discrete characters *see* characters (discrete)
discrete states *see* states (discrete)
distance (Euclidean) 206, 227–8, 232, 244
distance (Euclidean tangent) 227
distance (Manhattan) 180
distance (Procrustes tangent) 227, 230
distance (Procrustes) 141, 154, 165, 167–8, 177, 182, 221, 228, 236
distance measures 175
distances 46
distribution (character-state) 223
distribution (Dryden–Mardia) 141
distribution (log-rate–log-interval) 220, 230, 235–7
distribution (multivariate normal) 215
divergence (phylogenetic) 222
DNA 214
Dolichocebus 261
Dolichocebus gaimanensis 263–4
dorsal fin 2, 129
doublural notch 104
drift (random genetic) 250
Dryden–Mardia distribution *see* distribution (Dryden–Mardia)
Dürer, A. 3

ecological covariances 101
Efron, B. 201
egg type 102
eigenanalysis 77, 114, 125–6
eigenshape analyses 67, 69, 87–8, 95
eigenshape analysis (extended) 67, 69, 90, 92

eigenshape analysis (open-curve) 108
eigenshapes 88
eigenvalues 204
eigenvectors 204
Eldredge, N. 234–5
elliptical condyle 73
elliptical Fourier analysis 67, 69, 85
enamel knots 236
Eocene 22, 221, 223
Equus germanicus 229
Euclidean distance 177–8
Euclidean matrix distance analysis (EDMA) 189, 203, 205–6, 215
Euclidean space *see* space (Euclidean)
eutherian mammals *see* mammals (eutherian)
evolution 6
evolution (molecular) 221
evolution (parallel) 11, 69, 198
evolution (rates of) 247
evolution (theory of) 10
extended eigenshape analysis *see* eigenshape analysis (extended)
extinctions 197
eye 14
eye (color) 102
eye (shape) 128
eye (socket) 122, 124

Famennia bachae 121–2
Farris, J. S. 20, 47, 102
feathers 70
Felidae 256, 258
felids 221
Felsenstein, J. 5, 103, 201
fin rays 271
finches (Galapagos) 247
Fink, William 5
fish taxonomy *see* taxonomy (fish)
Fisher's individual error rate test 257
fishes (bony) 215
fixed cheek 122, 124
flower petal 103
flowering plants 12, 68
flowers 12
folds 147, 151
foraminifera 247
forelimbs 9
form matrix *see* matrix (form)
form-difference matrix *see* matrix (form-difference)
fossil record 198
fossils 10, 201, 221, 247–8, 259, 264
fossorial species *see* species (fossorial)
Fourier analysis 84, 94–5
Fourier harmonic series 94

Fourier harmonics 84–5, 87, 188
frequency (gene) 28–9
frequency (topological) 210
frogs 9
frontal bone 165
frugivore *see* species (frugivorous)
function 5
functional covariances 101
fur 2

G. exoticum 60
G. gibbii 58–60
G. pensacole 58–61
G. perthense 58, 60–1
G. turgidum 60–1
G. wansbeckii 58, 60–1
galagos 259
Gallon, F. 3
gap coding *see* coding (gap)
gap weighting 113
gap-coding methods 22
gaps (morphological) 20, 134, 139
genal spine 117
gene frequency *see* frequency (gene)
genealogical relationships 194
genealogy 11
generalized gap coding *see* coding (generalized gap)
generalized least squares (GLS) 130, 172, 177, 184, 189, 227
generalized linear model 3, 252
generalized Procrustes analysis (GPA) 177
generation time 220
generation(s) 222, 231
genetic covariation 38
genetic drift 28–30, 37
genetic isolation 46
genetics 5, 211
genomics 37
geometric mean 49
geometric morphometric synthesis 4
geometry 2, 4, 139
geometry (non-Euclidean) 76, 178; *see also* geometry (Procrustes)
geometry (Procrustes) 143, 160
glabella 105, 121–2, 166–7, 169
gradualist model 39
graticule 56
ground squirrels 74–5, 79

hair color 102
haldanes 230
Hennig program 53, 56, 61, 63
Hennig, W. 10, 11, 14
heritability 29

Herpestidae 256
heterochrony 194
hierarchy 12
higher plants 53, 56
higher taxa 1
holotypes 170
hominins 169
Hominoidea 48, 260
hominoids 163
Homo 165–9
Homo heidelbergensis 163
Homo neanderthalensis 163
Homo sapiens 165
homologies 17, 13, 70, 82, 104, 115, 118; see also coding, homologue
homologous correspondence 108
homologues 53, 65, 100, 114, 117, 127, 135, 176, 200, 214
homology 2, 5, 8–11, 16, 22, 54, 61, 65, 67, 69–71, 73, 76, 84, 87, 92, 95–6, 116, 118–19, 135, 139, 170, 215
homology (biological) 115, 126, 129, 135
homology (geometric) 115
homology (operational) 17
homology (primary) 2, 12, 14, 46; see also primary homologues
homology (secondary) 12, 15, 17
homology (taxic) 118
homology (topographical) 14; see also correspondence (topographical)
homology (transcendental) 116
homology (transformational) 12–13, 116
homoplasy 12, 68–9, 119, 194, 198–9, 214–15, 223
hoofed mammals see mammals (hoofed)
horse 9
HSO 50
Huelsenbeck, J. P. 201
humerus 2
Humphries, C. J. 5, 120
hyolaryngeal apparatus 213
Hyopsodus 229
hypertetrahedron 143
Hyracotherium grangeri 229

ichthyosaurs 2
image unwarping 183
incisor 72, 81
independent contrasts method 221, 248–9, 251, 254–5, 257, 265
independent variables see variables (independent)
individuals 1, 46, 69, 88
ingroup 71, 114
integers 5, 19–20, 22, 53

internal nodes 11
interval-scale variables see variables (interval-scale)
Ionoxalis 53, 56
isometry 48

jaw joint 73
jaws 10, 16, 81, 83
Johnson, G. D. 15

Kanisamys 229
Killer whale see whale (Killer)
Kluge, A. 47
Kruskal–Wallis test 257–8

Lagonimico conclutatus 263–4
Lagothrix 211–14
Lagothrix lagothricha 212, 263–4
landmark coordinates 176, 181
landmarks 3–6, 17–18, 67–68, 73–6, 78–81, 92, 95–6, 108, 115–17, 119, 129, 135, 141, 143–4, 151, 153, 155, 157, 163, 170, 172, 176, 177, 179, 182, 184, 189, 194–5, 202–3, 205, 212, 215, 221–2, 227, 236–7
landmarks (type 1) 116
landmarks (type 2) 116–17
landmarks (type 3) 116–17
Lanyon, S. 201
leaf lengths 22, 68
Least Weasel 224
leaves see lengths (leaf)
lengths (branch) 233
lengths (leaf) 22, 68
lengths (limb) 30, 71–2
Leonardo da Vinci 3
Leontopithecus chrysomelas 263–4
Leontopithecus rosalia 263–4
life history 196, 211
limb length see lengths (limb)
linear change 253
linearized Procrustes method 179
Littorina obtusata 229
Lutrinae 256, 258
Lynch, M. 119

Mackenziurus 105, 109, 121
Mackenziurus ceejayi 105, 108–9, 122, 124–5
Mackenziurus deedeei 104, 106, 108–9, 122
Mackenziurus joeyi 122
Mackenziurus johnnyi 122, 125
magnetostratigraphy 225
mammals 31, 221
mammals (eutherian) 47

mammals (hoofed) 9
mammary glands 2
man 9
mandibles 69, 95
mandibular symphysis 74
Manhattan distance *see* distance (Manhattan)
Mansonia 183–5
Markov chain 41
Markov Chain Monte Carlo analysis 27, 41, 214
Marmota flaviventris 75–6, 79–82, 86, 89–92
marmotine jaws 85
marmotines 73–4, 95
Marmotini 69
marsupials 198
mastoid processes 71
matrix (cardinality difference) 209
matrix (cardinality) 208
matrix (form) 206
matrix (form-difference) 206
maxilla 163
maxillary sinus 47
maxillary sinus size *see* size (maxillary sinus)
maximum likelihood 181, 187, 221, 233, 237, 248–9, 251–2, 262, 266–7
means 201
measurements 8, 69
measurements (meristic) 1
meristic data 215
Mesotaxis asymmetricus 261–2
metamorphosis 13
metric characteristics *see* characters (metric)
metric data 45–6, 51
miacids 223
Michelangelo 3
Mickevich, M. F. 68
microscope (Reflex®) 227
Microtus pennsylvanicus 229
midbrain 153
Mohanamico hershkovitzi 263–4
molar shape *see* shape (molar)
molars 72, 213
molecular clock 233
molecular data 12, 27, 31–3, 35, 48, 100
molecular phylogeny 1
molecular sequence 28, 45
molecular sequence data 50, 237
molecular systematists 100
molecules 1, 100
monkeys 211, 263
monkeys (howler) 211
monkeys (spider) 211
monkeys (woolly spider) 211
monkeys (woolly) 211
monophyletic group(s) 11–12, 70, 112
monophyletic taxa 69

monophyly 12, 69, 84
Monte Carlo simulation 56
morphological characters *see* characters (morphological)
morphological discontinuity *see* discontinuities (morphological)
morphological novelty 4
morphometric data 8
morphometric relationships *see* relationships (morphometric)
morphometrics (behavioral) 42
morphometrics (developmental) 43
morphometrics (functional) 42–3
morphospace *see* space (morphospace)
mosaic evolution 215
mosquitoes 175, 183–4
Mueller, L. D. 201
multiple regression, multivariate 187
multivariate analyses 16
multivariate morphometrics 46
Mus musculus 229
Mustelina 258
Mustelinae 256
Myrmecophagidae 68

nasal cavity 47
natural selection *see* selection (natural)
Naylor, G. 6, 119–20, 127
Neanderthals 169; *see also Homo neanderthalensis*
Neosaimiri fieldsi 263–4
niche (phylogenetic) 197
non-Euclidean geometry *see* geometry (non-Euclidean)
non-metric multidimensional scaling (NMMDSA) 183, 185
non-uniform component 179, 188
North American Land Mammal Ages 223, 243

occipital furrow 122
Ockham's razor 249
ontogenetic character 11, 17
ontogenetic sequences 71
ontogenetic series 171
ontogenetic studies 120
ontogeny 16, 188, 236
orientation 115, 176, 177
Origin of Species 10
Ornstein–Uhlenbeck model 252, 262
Ornstein–Uhlenbeck process 38
Orthopodomyia 183–4
Otariidae 256, 257–8
outgroup 71, 81, 114, 163, 188
outline segments 18

outlines 18, 68–9, 84–5, 87–8, 90–3, 95–6, 117, 157, 176
overlap analysis 111–12
ovules 12
Owen, Richard 8–9, 13, 115, 135
Oxalis 53, 56, 61–2, 64

Pa element 247–8, 261, 266
Paleocene 222–3
Paleogene 220, 224
paleontological data 31
Palmatolepidae 266
Palmatolepis 261
Palmatolepis crepida 261–2
Palmatolepis gigas 261–2
Palmatolepis glabra distorta 261–2
Palmatolepis glabra lepta 261–2
Palmatolepis glabra pectinata 261–2
Palmatolepis glabra prima 261–2
Palmatolepis gracilia gracilis 261–2
Palmatolepis gracilia manca 261–2
Palmatolepis gracilis gonioclymeniae 261–2
Palmatolepis hassi 261–2
Palmatolepis perlobata helmsi 261–2
Palmatolepis perlobata maxima 261–2
Palmatolepis perlobata perlobata 261–2
Palmatolepis perlobata schindewolfi 261–2
Palmatolepis punctata 261–2
Palmatolepis quadrantinodosalobata 261–2
Palmatolepis rugosa ampla 261–2
Palmatolepis rugosa rugosa 261–2
Palmatolepis subrecta 261–2
Palmatolepis superlobata 261–2
Palmatolepis triangularis 261–2
Pan 165, 166–9
Pan troglodytes 163
parallelism *see* evolution (parallel)
parameters (nuisance) 204
paraphyletic species 11–12, 50
parsimony 1, 40, 46, 49, 50, 53, 120–1, 126, 168, 180, 223, 233, 237–8, 247–9, 252, 267
parsimony (linear) 181, 248, 253–5, 265–7
parsimony (squared change) 6, 175, 180, 184, 187, 221, 248–9, 252–5, 265–6
parsimony (Wagner) 187
parsimony (weighted squared change) 248, 253–5
partial warp analysis 126
partial warp axes 127
partial warps 69, 77–8, 81–4, 96, 100, 119, 121, 125, 129, 134–5, 176, 179, 180, 187–8
path analysis 3
Patterson, C. 3, 15, 71, 135, 200

PAUP program 49, 53, 61
PAUP v.3.1. program 56
peaks (adaptive) 197
pectoral fin 117
Pee-Wee program 53, 56, 61, 63
pelvic fin 129
periodic function 84
perturbations (multivariate Gaussian) 205
petals 12
phena 112
Phenacolemur praecox 229
pheneticists 19
phenetics 101
phenogram 108–9, 195, 201, 204, 213
phenogram (UPGMA) 212
phenograms (bootstrap) 207, 213
Phocidae 256–8
PHYLIP program 35
phylogenetic analysis 53, 135
phylogenetic covariances 101
phylogenetic reconstruction 14, 53
phylogenetic signal 199, 210
phylogenetic systematics 8, 11, 45–6, 48, 135
phylogenetic tree *see* tree (phylogenetic)
phylogeny 135, 175
Pimentel, R. A. 14, 54, 68–9, 72, 102–3, 114–15, 118
pinnepeds 257
pistils 10
Pitheca monachus 263–4
Pitheca pithecia 263–4
plant systematists 53
Pleurosigmataceae 53, 56
polarity 16
polygenic suites 111
Polygnathus cristatus 261–2
polymorphic taxa 45
polymorphism 41, 45, 49–51
polyphyletic groups 11–12
polytotomies 257
polytotomies (hard) 257–9
polytotomies (soft) 257–9
populations 1, 221–2
porpoise 9
Portopithecus brasiliensis 263–5
posterior-border furrow 122
post-occipital suture 122
premolar 72
primary homologues 8, 13, 17
primary homology *see* homology (primary)
primates 2, 247–8, 257, 259–60, 262, 265
principal component axes 69
principal components 145, 171
principal components analysis (PCA) 77, 95–6, 114, 123, 126, 181, 183, 185–7, 189; *see also* eigenanalysis

principal strains 149, 154
principal warps 78, 82, 121
principle of connections 9
processes, stochastic 196
Procrustes distance *see* distance (Procrustes)
Procrustes grand mean form 172
Procrustes registration 155
Procrustes shape coordinates *see* shape coordinates (Procrustes)
Procrustes shape space *see* shape space (Procrustes)
Procrustes superimposition *see* superimposition (Procrustes)
Progonomys 229
proportions (bootstrap) 208, 211–12
prosthion 212
Protictis agastor 224–5, 233, 243–6
Protictis haydenianus 224–5, 233, 243–6
Protictis paralus 224–5, 233, 243–6
Protopithecus 261
pseudosample 202, 204–5, 212, 215
Psorophora 183–4, 249
pulley principle 249–51
punctuated equilibrium model 39

quantitative trait loci (QTL) 37

radius bone 117
random walk 221–2, 231–3, 237–8, 244
rates (evolutionary) 233, 252
rates (maximum likelihood) 253
ratios 2, 8, 45, 49, 175
rats 171
rescaled consistency index (RC) 125
reconstructions (ancestral) 252–3
reference form 78
regression (linear) 227, 230
regression analysis 3
regression line 3, 48
relationships (ancestral) 260, 262
relationships (descendent) 260
relationships (morphometric) 46
relationships (topological) 9
relative warp analysis 100, 108, 129
relative warps 124, 126–7, 130–5, 145
Remane, A. 10
Renaissance 3
reptiles 215
residuals 45, 48
retention index (RI) 125
Reyment, Richard 170
rhinion 212
Rieppel, O. C. 11
Riggins, R. 14, 54, 68–9, 72, 102–3, 114–15, 118

Rocky Mountains 223
Rohlf, F. J. 6, 81, 102, 125, 172
Roskies, R. Z. 87–90, 92, 94, 108

Sabethini 186
Saguinus bicolor 263–4
Saguinus fusicicollis 263–4
Saguinus imperator 263–4
Saguinus labiatus 263–4
Saguinus leucopus 263–4
Saguinus midas 263–4
Saguinus mystax 263–4
Saguinus nigricollis 263–4
Saguinus oedipus 263–4
Saguinus tripartitus 263–4
Saimiri boliviensis 263–4
Saimiri sciureus 263–4
salmon 2
sample means 20
scalar values 3
scaling 45
scapulae 221
schizophrenia 153, 157
schizophrenics 170
Schmidtognathus wittekindti 261–2
scores (partial warp) 178–9, 184, 187–8
selection 29, 198–9
selection (directional) 111
selection (natural) 28, 30
selection (stabilizing) 246
semaphoronts 16
semi-landmarks 108, 117, 157, 163, 165, 170
shape 3, 104
shape (mean) 188
shape (molar) 220, 230, 235, 242, 244
shape (reference) 187
shape (standard reference) 177
shape coordinates 67, 69, 76, 84, 92, 139
shape matrix 206
Shape Nonmonotonicity Theorem 119
shape space (Euclidean) 227
shape space (Kendall's) 76, 176–8
shape space (Procrustes) 144, 153, 166, 170, 173
shape spaces 175, 186, 188
shape statistics 175
shoulder girdle 2
shrews 9
similarity 8, 15, 17, 69, 71
similarity (homoplasious) 200
similarity (morphometric) 215
similarity (topological) 116, 135, 205, 209
singular value decomposition 108
sinuosity 93
Sirtonia tatacoensis 263–5

Sirtonia victoriae 263–5
size 3, 45, 49, 104, 115, 176, 177
size (body) 30, 211, 257, 259, 262, 264;
    see also body mass
size (centroid) 79, 155, 163, 166, 177
size (maxillary sinus) 47
size variables 104
skeleton 212
skulls 161, 165, 170–1, 211
smaller coronoid 89
Smith, A. B. 13, 17
Sokal, R. R. 102
space (Euclidean) 141
space (linear tangent) 178
space (morphometric) 69
space (morphospace) 46, 141
space (multidimensional) 4, 46
space (multidimensional morphospace) 46
space (Nature) 203–4, 214
space (shape) 228
space (tangent) 77, 188, 228
sparrow (American house) 229
special structures 10
specializations (dietary) 213
speciation 197
species 1, 46
species (cursorial) 221
species (evolutionary) 223
species (fossorial) 221
species (frugivorous) 213
Spermophilus 75
Spermophilus columbianus 80–1, 83, 95
Spermophilus franklini 76, 79–81, 83, 89,
    91, 95
Spermophilus spilosoma 76, 79–82, 91–2, 95
Spermophilus tridcemlineatus 76, 79–81, 83,
    86–7, 95
Spermophilus variegatus 72, 74–5, 77, 79–82,
    86–7, 95
spine (nasal) 212
splenium 159–61
squirrels 69
St. Hilaire, G. 9, 10
stamens 10, 12
stasis 220
state 104
states (ancestral) 175, 180–1
states (ancestral character) 247, 249–50,
    252–3, 255
states (apomorphic) 27
states (autapomorphic) 62
states (character) 250
states (derived) 27
states (discrete) 22, 188
stickleback (Norwegian) 229
stratigraphical data 6

stratigraphy 223
stratocladistics 223
stratogladogram 224
Strepsirhini 260
Struszia 105, 121
Struszia dimitrovi 104, 122, 125
Struszia epsteini 104, 122
Struszia harrisoni 121–2, 124–5
Struszia martini 106, 122
Struszia onoae 104, 122
Struszia petebesti 122, 125
subset coding 113
superimposition (Procrustes) 180
superposition (GLS) 222
superposition (Procrustes) 222
support (bootstrap) 208
suture (premaxilla-maxilla) 212
Swiderski, Donald 5
symplesiomorphy 11–12, 200
synapomorphy 2, 11–13, 15, 200, 214, 234
systematics 9, 11, 13
systematists 3, 20

tails (prehensile) 211
tangent form 149
taxa 10
taxic diagnoses 134
taxic homology 8, 12–13, 17, 22
taxon (ancestral) 197
taxonomic covariances 101
taxonomic polymorphism 49
taxonomy 170
teeth 73
Teleopea 55
temporalis muscle 72, 92, 94
tensor field 150
terminal taxa 20
tetrahedrons 139–41, 143, 163, 165, 167–8
tetropods 12
Thamnochurtus 56
Thiele, K. 54
thin-plate spline 76, 118, 139, 150, 154, 159,
    161, 165, 171, 182, 185, 188
thin-plate spline analysis 78, 92
Thompson, D'Arcy Wentwoth 3, 4, 16, 79,
    118
three-ear ossicles 2
threshold model 40–1
time (generation) 228
time-series (evolutionary) 220
tokogenetic relationships 46, 50
tomography 48
tooth row 79, 83
topographic identity 12, 15–16, 17
topographical position 12

topographical relations 9
topological correspondence 13
topological relations *see* relationships (topological)
topology 115, 133, 135, 194, 208, 220
topology (cladistic) 200, 207
topology (phenetic) 207
total evidence approach 32
Toxorhynchites 183–4, 186
tpsSmall program 227
trait (heritable) 220
trajectories (developmental) 189
trajectories (evolutionary) 189
transformation 8, 22
transformation-grid 3
transformation series 11, 103
transformational approach 13
transformations (evolutionary) 194
tree 21
tree (bootstrap) 205
tree (molecular) 32–3
tree (morphological) 32
tree (phylogenetic) 6, 50, 54, 183–4, 186, 195, 223–4, 249
tree length 61
tree shrews 2
tree topology 63
trees (gene) 50
trees (rooted) 12
trilobite 121, 135
trilobite pygydia 105, 109
trilobites (encrinurine) 121

uniform component 179, 188
Uranotaenia 183–4, 186

variable (nominal) 102–3, 134
variable (ordinal-scale) 102–3
variables 17
variables (Bookstein shape) 236
variables (continuous) 4, 8–9, 13, 16, 19–21, 111
variables (discontinuous) 111
variables (independent) 16

variables (interval-scale) 102, 104, 111
variables (meristic) 102
variables (morphometric) 3–4
variables (overlapping) 18, 22, 53–4, 65
variables (qualitative) 15–16
variables (ratio-scale) 102–4, 111, 113, 134
variables (shape) 175, 179, 187–8
variance 20
variance (phylogenetic) 233
vertebrae 14
vertebrate skeletons 8
viverravids 223, 229–30
Viverravus 222, 232, 235, 237
Viverravus acutus 223–5, 232, 233, 243–6
Viverravus laytoni 224–5, 233, 243–6
Viverravus politus 224–5, 233, 243–6
Viverravus rosei 224–5, 233, 243–6
Viverridae 220, 256, 258
viverrids 257

Wagner, G. P. 118
Wasatchian 223
weasels 223
Weller, S. J. 68
Weston, P. 103
whale 269–70
whale (Killer) 2
widths 3, 22
Wiley, E. O. 14
Williams, David 9
Williams, S. 55
wing shape 186
wings 2, 9
Woodger, J. H. 10
Wyeomyia 183–4, 186

yellow-bellied marmot 75

Zahn and Roskies function 91–2, 94–5
Zahn, C. T. 87–90, 92, 94, 108
Zelditch, M. 5
zygomatic arch 72

Systematics Association Publications

Systematics Association Publications
1. Bibliography of key works for the identification of the British fauna and flora, 3rd edition (1967)†
Edited by G. J. Kerrich, R. D. Meikie and N. Tebble
2. Function and taxonomic importance (1959)†
Edited by A. J. Cain
3. The species concept in palaeontology (1956)†
Edited by P. C. Sylvester-Bradley
4. Taxonomy and geography (1962)†
Edited by D. Nichols
5. Speciation in the sea (1963)†
Edited by J. P. Harding and N. Tebble
6. Phenetic and Phylogenetic classification (1964)†
Edited by V. H. Heywood and J. McNeill
7. Aspects of Tethyan biogeography (1967)†
Edited by C. G. Adams and D. V. Ager
8. The soil ecosystem (1969)†
Edited by H. Sheals
9. Organisms and continents through time (1973)†
Edited by N. F. Hughes
10. Cladistics: a pratical course in systematics (1992)*
P. L. Forey, C. J. Humphries, I. J. Kitching, R. W. Scotland, D. J. Siebert and D. M. Williams
11. Cladistics: the theory and practice of parsimony analysis (2nd edition) (1998)*
I. J. Kitching, P. L. Forey, C. J. Humphries and D. M. Williams

* Published by Oxford University Press for the Systematics Association
† Published by the Association (out of print)

Systematics Association Special Volumes

1. The new systematics (1940)
Edited by J. S. Huxley (reprinted 1971)
2. Chemotaxonomy and serotaxonomy (1968)*
Edited by J. C. Hawkes

3. Data processing in biology and geology (1971)*
Edited by J. L. Cutbill
4. Scanning electron microscopy (1971)*
Edited by V. H. Heywood
5. Taxonomy and ecology (1973)*
Edited by V. H. Heywood
6. The changing flora and fauna of Britain (1974)*
Edited by D. L. Hawksworth
7. Biological identification with computers (1975)*
Edited by R. J. Pankhurst
8. Lichenology: progress and problems (1976)*
Edited by D. H. Brown, D. L. Hawksworth and R. H. Bailey
9. Key works to the fauna and flora of the British Isles and northwestern Europe, 4th edition (1978)*
Edited by G. J. Kerrich, D. L. Hawksworth and R. W. Sims
10. Modern approaches to the taxonomy of red and brown algae (1978)
Edited by D. E. G. Irvine and J. H. Price
11. Biology and systematics of colonial organisms (1979)*
Edited by C. Larwood and B. R. Rosen
12. The origin of major invertebrate groups (1979)*
Edited by M. R. House
13. Advances in bryozoology (1979)*
Edited by G. P. Larwood and M. B. Abbott
14. Bryophyte systematics (1979)*
Edited by G. C. S. Clarke and J. G. Duckett
15. The terrestrial environment and the origin of land vertebrates (1980)
Edited by A. L. Pachen
16. Chemosystematics: principles and practice (1980)*
Edited by F. A. Bisby, J. F. Vaughan and C. A. Wright
17. The shore environment: methods and ecosystems (2 volumes) (1980)*
Edited by J. H. Price, D. E. C. Irvine and W. F. Farnham
18. The Ammonoidea (1981)*
Edited by M. R. House and J. R. Senior
19. Biosystematics of social insects (1981)*
Edited by P. E. House and J.-L. Clement
20. Genome evolution (1982)*
Edited by G. A. Dover and R. B. Flavell
21. Problems of phylogenetic reconstruction (1982)
Edited by K. A. Joysey and A. E. Friday
22. Concepts in nematode systematics (1983)*
Edited by A. R. Stone, H. M. Platt and L. F. Khalil
23. Evolution, time and space: the emergence of the biosphere (1983)*
Edited by R. W. Sims, J. H. Price and P. E. S. Whalley
24. Protein polymorphism: adaptive and taxonomic significance (1983)*
Edited by G. S. Oxford and D. Rollinson
25. Current concepts in plant taxonomy (1983)*
Edited by V. H. Heywood and D. M. Moore

26. Databases in systematics (1984)*
Edited by R. Allkin and F. A. Bisby
27. Systematics of the green algae (1984)*
Edited by D. E. G. Irvine and D. M. John
28. The orgins and relationships of lower invertebrates (1985)‡
Edited by S. Conway Morris, J. D. George, R. Gibson and H. M. Platt
29. Infraspecific classification of wild and cultivated plants (1986)‡
Edited by B. T. Styles
30. Biomineralization in lower plants and animals (1986)‡
Edited by B. S. C. Leadbeater and R. Riding
31. Systematic and taxonomic approaches in palaeobotany (1986)‡
Edited by R. A. Spicer and B. A. Thomas
32. Coevolution and systematics (1986)‡
Edited by A. R. Stone and D. L. Hawksworth
33. Key works to the fauna and flora of the British Isles and northwestern Europe, 5th edition (1988)‡
Edited by R. W. Sims, P. Freeman and D. L. Hawksworth
34. Extinction and survival in the fossil record (1988)‡
Edited by G. P. Larwood
35. The phylogeny and classification of the tetrapods (2 volumes) (1988)‡
Edited by M. J. Benton
36. Prospects in systematics (1988)‡
Edited by J. L. Hawksworth
37. Biosystematics of haematophagous insects (1988)‡
Edited by M. W. Service
38. The chromophyte algae: problems and perspective (1989)‡
Edited by J. C. Green, B. S. C. Leadbeater and W. L. Diver
39. Electrophoretic studies on agricultural pests (1989)‡
Edited by H. D. Loxdale and J. den Hollander
40. Evolution, systematics, and fossil history of the Hamamelidae (2 volumes) (1989)‡
Edited by P. R. Crane and S. Blackmore
41. Scanning electron microscopy in taxonomy and functional morphology (1990)‡
Edited by D. Claugher
42. Major evolutionary radiations (1990)‡
Edited by P. D. Taylor and G. P. Larwood
43. Tropical lichens: their systematics, conservation and ecology (1991)‡
Edited by G. J. Galloway
44. Pollen and spores: patterns of diversification (1991)‡
Edited by S. Blackmore and S. H. Barnes
45. The biology of free-living heterotrophic flagellates (1991)‡
Edited by D. J. Patterson and J. Larsen
46. Plant-animal interactions in the marine benthos (1992)‡
Edited by D. M. John, S. J. Hawkins and J. H. Price
47. The Ammonoidea: environment, ecology and evolutionary change (1993)‡
Edited by M. R. House
48. Designs for a global plant species information system (1993)‡
Edited by F. A. Bisby, G. F. Russell and R. J. Pankhurst

49. Plant galls: organisms, interactions, populations (1994)‡
Edited by M. A. J. Williams
50. Systematics and conservation evaluation (1994)‡
Edited by P. L. Forey, C. J. Humphries and R. I. Vane-Wright
51. The Haptophyte algae (1994)‡
Edited by J. C. Green and B. S. C. Leadbeater
52. Models in phylogeny reconstruction (1994)‡
Edited by R. Scotland, D. I. Siebert and D. M. Williams
53. The ecology of agricultural pests: biochemical approaches (1996)**
Edited by W. O. C. Symondson and J. E. Liddell
54. Species: the units of diversity (1997)**
Edited by M. F. Claridge, H. A. Dawah and M. R. Wilson
55. Arthropod relationships (1998)**
Edited by R. A. Fortey and R. H. Thomas
56. Evolutionary relationships among Protozoa (1998)**
Edited by G. H. Coombs, K. Vickerman, M. A. Sleigh and A. Warren
57. Molecular systematics and plant evolution (1999)
Edited by P. M. Hollingsworth, R. M. Bateman and R. J. Gornall
58. Homology and systematics (2000)
Edited by R. Scotland and R. T. Pennington
59. The Flagellates: unity, diversity and evolution (2000)
Edited by B. S. C. Leadbeater and J. C. Green
60. Interrelationships of the Platyhelminthes (2001)
Edited by D. T. J. Littlewood and R. A. Bray
61. Major events in early vertebrate evolution (2001)
Edited by P. E. Ahlberg
62. The changing wildlife of Great Britain and Ireland (2001)
Edited by D. L. Hawksworth
63. Brachipods past and present (2001)
Edited by H. Brunton, L. R. M. Cocks and S. L. Long
64. Morphology, shape and phylogeny (2002)
Edited by N. MacLeod and P. L. Forey

* Published by Academic Press for the Systematics Association
† Published by the Palaeontological Association in conjunction with Systematics Association
‡ Published by the Oxford University Press for the Systematics Association
** Published by Chapman & Hall for the Systematics Association